Edited by C. Reinhardt

Chemical Sciences in the 20th Century

Bridging Boundaries

Edited by C. Reinhardt

Chemical Sciences in the 20th Century

Bridging Boundaries

WILEY-VCH

Weinheim – New York – Chichester – Brisbane – Singapore – Toronto

Dr. Carsten Reinhardt
Lehrstuhl für Wissenschaftsgeschichte
Universität Regensburg
93040 Regensburg
Germany

Library of Congress Card No.:
applied for

British Library Cataloguing-in-Publication Data
A catalogue for this book is available from the British Library.

Die Deutsche Bibliothek – CIP Cataloguing-in-Publication Data
A catalogue record for this publication is available from Die Deutsche Bibliothek.

Composition Typomedia, Ostfildern

Cover Design Gunther Schulz, Fußgönheim

ISBN 978-3-527-30271-0

Foreword

Why do active chemists need the history of chemistry? And why some of us are resistant to Clio's art in the laboratory? These are the questions I want to think about.

First, because everything has a history. Things happened, in a chronology and influenced by a personal past (a chemist's advisors, his or her students), in communication with others, and in the setting of a society. The system of science uses the addiction to curiosity of moderately smart, fallible, and underpaid individuals in the labor of a micro-society whose aim is to generate reliable knowledge of the beautiful world within and around us. This gloriously successful European invention demands open publication and communication and mandates frequent dipping back and forth between theory and reality. The tying of such a structure to normal human aspirations (suppressing some along the way, with consequences easily anticipated) nearly guarantees that any interesting new finding will be tested by someone out to prove it wrong. Science, being process, has a history – of individuals, their tools, their communications. It is natural that we should want to know how Diels and Alder got to the reaction named after them. That we discover how discovery took place.

The second reason I see for doing history of chemistry is simply that it is interesting to see how ideas evolved. Even if one was in the middle of the fray. Or, maybe, just because one was there: My mother and I were busy surviving World War II in Galicia; the news we had of the war was fragmentary and propagandistic. What a joy it was to read years later Winston Churchill's history of the Second World War! Chemistry isn't war, but there is a lot of action in those 500 000 articles published each year.

One interesting aspect of doing the history of 20th century chemistry is that the events are likely to be close to the personal experience of chemist-readers. Or they may see their *Doktorvaters* in them. Since we are human, and prone to self-justification, that proximity in time is likely to lead the responder to focus on the critical – the fact missed, the factor misjudged.

Third, the human in us is absolutely insatiable in its interest for the personal. When the medium is inherently expressive, as in a novel, we just take it in (though we may wonder who that character is based on). But if the mode of expression in which we ply our trade excludes (as the scientific article regretfully does) writing of people, motivation, emotion, anger, then we simply love it when it is allowed to come back. In *Nachtisch* gossip, for instance. Or, to be serious about it, many of us will recall the tremendous impact of the threadbare two line biographies of organic chemists in "Fieser and Fieser". Students are starving for history, and good teachers know this.

Fourth, history humanizes. The social construction of science program antagonizes scientists, almost reflexively. I think the SCS approach deserves what it gets, in part because it sometimes clothes an antipathy to the organism in the cloak of trying to bring us to see the practice of science as being no different from any other human enterprise. Good history of chemistry (and analysis of real life chemical practice) is not aggressive. It comes out of love for the subject, and shows by example how science is embedded in culture, that scientists are people with foibles and mind sets (call them themata, call them prejudices) that influence what they create. Good history of chemistry relaxes scientists, makes them more tolerant (by a hair) of what the humanities have to say about science.

This last point leads me to think about why chemists are suspicious of history of chemistry (until such time as they try to do some history).

1. There is an arrogance bred by the macho practice of modern chemistry – becoming an administrator, poet, historian is like dropping out of the race. Those who can, do; those who can't . . .
2. Science has bought into the cult of the new, with a vengeance. References to old papers are in there only to stake out claims to novelty ("no one since 1912 has . . .") or to establish lines of authority. The Oedipal urge is heightened in the enterprise ("the only prior calculations on this molecule are by the unsatisfactory extended Hückel method"). Nothing new here, except perhaps the pace. Given this valorization of the new, it is especially difficult to enter the mind set of chemistry done two hundred years ago. Many chemists don't have the patience to partake of the world past. They don't see the value. Incidentally, one of the oft-cited uses of history is that we may learn from the past. The behavior of individuals and nations leads one to be skeptical of the idea. Sometimes we learn, sometimes (as in falling in love) it's good that we don't. It's the same for chemists – it may be good not to know that someone else had tried an experiment and it failed.
3. There is a skepticism among chemists that historians can acquire the cognitive structure of chemists and so "understand" them. This is sometimes a silly conceit, because many of the historians of chemistry have the "passport" of a Ph. D. or have practiced chemistry. Shall we listen to the chemists themselves? Autobiographies are often poor history (to be exceeded in their unreliability only by biographies – pathography or hagiography – by children of their parents). But I think historians should ask practitioners in the field not only for fact checking, but also for expressions of that intangible feeling that an analysis is off.

We are occasionally inebriated by the beauty of what we have created in science. It is as if in that moment of understanding we were speaking to the gods. And when one touches the sublime, moral considerations don't matter. Werner Heisenberg, in his wartime visits to the Netherlands and Denmark wanted to talk science; he had no idea what it meant to be a human being (one who by chance is a scientist) in an occupied country.

History helps here, as I was reminded in a conversation with Hunter R. Rawlings, Cornell's President and a classical scholar, expert on the Greek historian Thucydides. Rawlings (and Thucydides) would stress the moral utility of history. History tells us how human beings acted, and asks us to think about the motives and consequences of their actions. In reconstructing history, we move outside of ourselves, and – not abdicating the capacity to feel strongly about what transpired – we are pushed gently toward alternative perspectives, towards tolerance, towards empathy. This has real spiritual value.

Roald Hoffmann

Preface

Mature fields of historical scholarship – not unlike classical music in that regard – boast of a repertoire of standard pieces for which the sources are easily accessible, the main lines of interpretation firmly established, and for which the interpretation has reached a considerable level of refinement. This is equally true for the historiography of chemistry. Lavoisier's chemical revolution, the evolution and diversification of chemistry as a discipline, or the social history of the professional chemist constitute such standard pieces in its repertoire. And, as in classical music, the eighteenth and nineteenth centuries attract the largest audiences.

The history of twentieth-century chemistry is a comparatively recent field of research and cannot be regarded as a mere extension of traditional approaches, for modern chemistry differs from its earlier forms in at least three regards: First, entirely new patterns of interaction between science, industry, and economy have been built up; second, instrumentation played an increasing role and has eventually led to a profound transformation of the laboratory as a result of the electric and electronic revolutions; third, disciplinary identity and public image of chemistry are deeply affected by the breaking-down of traditional boundaries. Scholars working on the chemical sciences during this period need a keen sense of historical complexity and a considerable amount of scientific knowledge. Still, historians of modern chemistry are but a tiny minority among the professional historians of modern science. But things are beginning to change.

After all, the twentieth century is behind us. It has become history and therefore a proper object for historians to deal with. Only recently have historians of chemistry begun to meet this challenge. They convinced university administrators, grants committees and the scientific community alike that the history of modern chemistry is not only a fascinating topic to study, but indeed a prerequisite for understanding the modern world in all its complexity. Historians are clearly no prophets, but to deal with the future of science and technology in a sensible and responsible way requires a certain amount of historical literacy.

Local and institutional history, economic and social history, the history of industries and of individual corporations, the history of special branches of chemistry, or of chemical theories and discoveries, are traditional foci of interest for historians of modern chemistry. But soon the disciplinary approach was supplemented by cross-disciplinary studies and an interest in those patterns of interaction that arise when the traditionally separated spheres of the political, the economical, and the scientific merge to form new hybrid structures such as the military industrial complex or large technological systems. As a consequence, the historian of modern science needs the expertise of the economic and business historian, the historian of technology, and the political and cultural historian. In the beginning, however, there was little coherence among these different groups, and there was no common platform for discussion. Though several international journals exist for the history of twentieth-century science, there is no journal for the history of twentieth-century chemistry. The two or three history of chemistry journals that exist in the world include almost anything from alchemy to DNA, and the same heterogeneous collection prevails at most history of chemistry meetings organized by the national chemical societies.

In order to link these scattered activities and to create an international network of historians of chemistry, some time ago the European Science Foundation in Strasbourg initiated a five year research program on "The Evolution of Chemistry, 1789–1939" , which has yielded a series of conferences and a fair number of books. When this program came to an end in 1997, it was felt that this network should be used – in a sort of follow-up project – to focus interest in, and to stimulate research on, the history of twentieth-century chemistry.

In July 1997 the idea materialized during the XX. International Congress of History of Science in Liège, Belgium. A Commission on the History of Modern Chemistry (CHMC) was established by the Division of History of Science of the International Union of the History and Philosophy of Science, a body related to UNESCO through the International Council of Scientific Unions. By the end of the year and as a result of intensive e-mail discussions, CHMC's Executive Council and agenda were agreed upon, and two Nobel laureates, Manfred Eigen and Roald Hoffmann, offered to act as honorary patrons for the new endeavor.

Since then CHMC has organized two major international symposia and was involved in a number of more local ones. A conference "Between Physics and Biology: Chemical Sciences in the Twentieth Century", held at the Deutsches Museum in Munich in May 1999, marked the official opening. Guided by the idea that the disciplinary structure of science is part of our nineteenth-century heritage and clearly no longer apt to describe present-day science, chemistry offered a particularly fine example of how modern research is organized in an impressive array of sub-discipline and hybrid-discipline formation, inter-disciplinary cooperation, and new experimental systems based on specific methodologies, techniques, or substrates. Most of the papers presented in this volume originated from this conference. Yet, without Carsten Reinhardt's unique combination of patient encouragement, stimulating criticism and successful lobbying the present volume would not have reached the press. And it would clearly not have been completed as efficiently and handsomely without the support the project has received from Wiley-VCH from the very beginning. The interest one of the major international science publishers has taken in our project is a clear sign that the Commission on the History of Modern Chemistry is on the right track.

Instrumentation, another key feature of modern chemistry, came under scrutiny in the second international CHMC conference "From the Test-Tube to the Autoanalyzer: The Development of Chemical Instrumentation in the Twentieth Century", held at Imperial College in London in August 2000. The focus was on post-1945 developments exclusively. In July 2001 a third CHMC conference "Shifting Centers and Emerging Peripheries: Global Patterns in Twentieth-Century Chemistry" will follow in Mexico City, aimed at exploring how the great geo-political shifts such as post-colonialism, post-communism and globalisation have transformed chemistry and the chemical industry, particularly in non-European and developing countries, during the second half of the past century. The Chemical Heritage Foundation in Philadelphia, United States, plans to host the 2002 conference on "Industrial-Academic Relationships in the Chemical and Molecular Sciences". This meeting will include interactions with other non-academic research centers, such as national and military laboratories, experimental stations, and colonial institutions.

Thus, within less than a decade, these efforts have not only succeeded in building up an efficient network of scholars from various disciplines; they have also succeeded in linking hitherto scattered activities and in providing visibility for a fascinating new field of research. But as in music, it is unlikely that the twentieth century will ever completely take over the repertoire. Historical scholarship needs the distance as well as the close-ups. The otherness of science in times long gone-by may help us to define more clearly those features which make up the specific challenge and fascination of studying twentieth-century chemistry from a historical perspective.

Christoph Meinel
Commission on the History of Modern Chemistry

Table of Contents

List of Contributors

Prof. Dr. Bernadette Bensaude-Vincent
Département de Philosophie
Université de Paris X
200, Avenue de la Republique
92001 Nanterre, France
<bernadette.bensaude-vincent
@uparis10.fr>

Dr. Marika Blondel-Mégrelis
Centre National de la Recherche
Scientifique, V. M. R. 8590, Institut
d'Histoire et Philosophie des Sciences
et des Techniques
13, rue du Four
75006 Paris, France
<marika.blondel-megrelis@liberty
surf.fr>

Prof. Dr. Yasu Furukawa
Tokyo Denki University
College of Engineering
2–2 Kanda-Nishiki-cho
Chiyoda-ku
Tokyo 101, Japan
<yfuru@cck.dendai.ac.jp>

Prof. Dr. Kostas Gavroglu
Department of History and Philosophy
of Science
University of Athens
John Kennedy 37
Athens, Greece
<kgavro@cc.uoa.gr>

Dr. Andreas Karachalios
Arbeitsgruppe für Geschichte der
Naturwissenschaften
Fachbereich 17 – Mathematik
Johannes Gutenberg-Universität
55099 Mainz, Germany
<Kara@mathematik.uni-mainz-de>

Prof. Dr. Helge Kragh
History of Science Department
University of Aarhus
Ny Munkegade, Building 521
8000 Aarhus C, Denmark
<ivhhk@ifa.au.dk>

Dr. Peter J. T. Morris
Senior Curator, Experimental
Chemistry
Science Museum
London SW7 2DD, U. K.
<p.morris@ic.ac.uk>

Prof. Dr. Mary Jo Nye
Department of History
Milam Hall 306
Oregon State University
Corvallis, OR 97331, USA
<nyem@ucs.orst.edu>

Dr. Nikos Psarros
Institut für Philosophie
Universität Leipzig
Burgstraße 21
04109 Leipzig, Germany
<psarros@rz.uni-leipzig.de>

Dr. Nicolas Rasmussen
School of Science and Technology Studies
LG Morven Brown
University of New South Wales
Sydney, NSW 2052, Australia
<Nrasmussn@delphi.com>

Dr. Carsten Reinhardt
Lehrstuhl für Wissenschaftsgeschichte
Universität Regensburg
93040 Regensburg, Germany
<Carsten.Reinhardt@psk.uni-regensburg.de>

Dr. Xavier Roqué
Centre d'Estudis d'Història de les Ciències (CEHIC), Edifici Cc
Universitat Autònoma de Barcelona
Bellaterra
08193 Barcelona, Spain
<Xavier.Roque@uab.es>

Prof. Dr. Ruth Lewin Sime
Emeritus, Department of Chemistry
Sacramento City College
609 Shangri Lane
Sacramento, CA 95825 (USA)
<rodsime@csus.edu>

Prof. Dr. Ana Simões
Universidade de Lisboa
Departamento de Física
Campo Grande, C8, Piso 6
1749–016 Lisboa, Portugal
<Fasimoes@fc.ul.pt>

Dr. Anthony S. Travis
Sidney M. Edelstein Center for History and Philosophy of Science, Technology, and Medicine
Levy Building, Givat Ram,
Hebrew University
91904 Jerusalem, Israel
<travis@cc.huji.ac.il>

Dr. Brigitte Van Tiggelen
Centre Interfacultaire d'Etudes en Histoire des Sciences, Université Catholique de Louvain
SC/PHYS/FYMA
Chemin du Cyclotron 2
1348 Louvain-La-Neuve, Belgium
<vantiggelen@fyma.ucl.ac.be>

Disciplines, Research Fields, and their Boundaries

Carsten Reinhardt

The scientific icons of the twentieth century were the atomic bomb and the genetic code. Physics and biology were foremost in the course of events and the development of the modern world, at least in public perception. In contrast, chemistry was experienced mainly through its industrial uses, and made headlines because of the catastrophes connected with it: *Silent Spring*, Seveso, and Bhopal come to mind. As a result, the advances brought on by chemistry through its wide-ranging applications and uses were rarely celebrated as having anything to do with chemistry. But nuclear physics is not feasible without nuclear chemistry, and molecular biology – made visible by its very name – relies heavily on chemical concepts and methods. Looking back almost four decades, we see the efforts of leading chemists to secure a strong position for their science. Chemistry, according to a 1965 report on basic research in U.S. chemistry, is "one of the fundamental sciences, supplying key materials and principles that are interwoven throughout today's technology, natural sciences, and culture." [1] In the opinion of the scientists who prepared the report, led by the Harvard based chemist Frank H. Westheimer, chemistry stood midway between physics and the biological sciences.

This book addresses the bridging of boundaries between chemistry and the other "classical" disciplines of science, physics, and biology; and chemistry's connections to mathematics and technology. Boundaries presuppose the notion of an accumulation of separated, perhaps loosely connected, intellectual fields, and this notion is seemingly at odds with old and established views of the unity of science. According to this viewpoint, it could be argued that though chemistry might serve as the central science, this refers only to a middle position in the hierarchical chain of science. An influential philosophical representation of this view, put forward by Paul Oppenheim and Hilary Putnam in the late 1950s, was a pyramidal hierarchy of the sciences, arranged in reductive levels, which all trace back to the lowest level, physics of elementary particles. [2] This physical reductionism had as much to do with the scientific realism held by its proponents as with the political interests and powerful influence of elementary particle physicists during the Cold War. In defense, chemists expressed strong resentments against the centralization of the sciences, especially in relation to its impact on research policy in general and government funding in particular. Compared with the costs for research vessels in oceanography, radio-telescopes in astronomy, and accelerators in particle physics, small scale chemistry projects did not match the requirements for a funding strategy firmly aligned with Big Science. In response, the chemists raised the argument for science to be decentralized.[3] Consequently, the authors of the 1965 report on chemistry in the United States referred to scientific fields such as "biochemistry, geochemistry, chemical physics, atmospheric chemistry, molecular biology, and astrochemistry," when they discussed the interrelation of the sciences; and it is precisely to these fields that we turn our attention here.[4]

It was through the very existence of hybrid fields that many scientists hoped for a unification of the sciences. The statement of purpose of the *Journal of Chemical Physics*, first printed in its 1940 issue, proclaimed that "the artificial boundary between physics and chemistry has now been in actual fact completely eliminated."[5] Another advocate of this view, the molecular physicist John Clarke Slater, in his 1939 *Introduction to Chemical Physics*, considered the separation of physics and chemistry to be an unfortunate one. His plan was to forge a unified physics and chemistry, and this was to happen through quantum mechanics and the infiltration of physical instruments into chemical laboratories. The gap between physics and chemistry was – according to Slater – "a result of tradition and training, not of subject matter."[6] Four decades later, drawing on the same unifying trends that Slater had identified, members of the Physics Survey Committee of the National Science Council put forward their view of a conceptual unity of the physical sciences. Interestingly, this went hand in hand with the large scale movement of scientists from chemistry into the more fundamental and prestigious area of chemical physics, while physicists tended to avoid working in fields with the suffix chemistry. Moreover, the committee stated that physicists and chemists in academic institutions were "overly concerned with defining their own traditional subjects," thus making it difficult for scientists in the interface areas to establish themselves.[7] This artificial and unfortunate separation, still influential in 1973, was linked to the disciplinary roots of chemistry and physics, which stretch back to the late eighteenth and early nineteenth centuries. One might well have in mind the founder figure of Antoine Laurent de Lavoisier, and his role in establishing modern chemistry as a scientific discipline (though historians now generally agree that chemistry was an established discipline a generation before Lavoisier).[8]

Disciplines create unity. They do so by connecting local scientific subfields, entrusting them with a universal meaning, and at the same time enforcing coherence and convergence. Drawing on Michel Foucault's concept of a regime of truth-dependent social power, Timothy Lenoir regards disciplines as "essential structures for systematizing, organizing, and embodying the social and institutional practices upon which both coherent discourse and the legitimate exercise of power depend."[9] In this view, disciplines are seen as mediators for economies of practice. But how do these mediators come into existence? For Lenoir, the process of discipline formation obeys the rules of invisible market forces, "adjusting relations between producers and consumers of the tools of knowledge production;"[10] and thus he denies the important role traditional histories of disciplines attribute to founder figures. Disciplines are shaped by interacting system effects that make them too complex to be ruled by a small group or even a single founder person. Though this is surely correct, the founder (usually male) plays an important role in creating an identity after the event of discipline building and is usually retroactively designated based on his acceptability as a source of identity. We may see this by thinking about the reverence Otto von Bismarck, the 'founder' of the German Empire in 1870/71, enjoyed in pre-1914 Germany (and even after World War I), thus contributing in an important way to German national identity. This may be compared with the roles Lavoisier, John Dalton, and Justus Liebig had in the identity-building of chemistry in their respective countries. An example from Germany is taken from the 1897 meeting of the *Verein Deutscher Chemiker* (Association of German Chemists). On that occasion, the chairman, industrial chemist and inventor Heinrich Caro, put Liebig on a par with the famous painter Raphael, citing Correggio's phrase *"Anch' Io sono pittore"* and rewording it into: "I am a German chemist, too."[11] In doing so, Caro not only introduced a nationalistic tone completely absent in the Italian original, but also used the phrase to unify the German chemists active in both academia and industry, and to create a common identity among the members of the society over which he presided.

More recently, Mary Jo Nye has developed a schematic taxonomy to explain the process of discipline building in chemistry while drawing on the construction of national identities. Her conception of disciplinary identity includes a genealogical descent connected to a historical mythology, a core literature, codified practices and rituals, a physical and institutional homeland, external recognition, and shared values and unsolved problems.[12] The crossing of boundaries between such well shaped bodies of knowledge appears to provide multiple identities for the scientists involved. Moreover, especially in the twentieth century, the mushrooming of inter- or transdisciplinary fields labeled with the suffix chemistry indicate that the majority of chemists were involved in activities which belonged to the territory of several disciplines. To overcome this seeming paradox, one may draw on Lenoir's proposal to differentiate between discipline-building programs and research programs. Although both are mutually separable resources and are strongly interrelated, scientists active in research do not perceive their goal as creating a discipline. Research is problem-defined and is not confined to a single discipline, while discipline builders use research programs to stabilize their authority and as resources for their institutional goals.[13] Another, tempting, differentiation would be to focus on the separate teaching and research parts of the scientific endeavor. Though disciplines are not restricted to teaching responsibilities, this does constitute an important part of their activities, and it mirrors their development in the nineteenth century. Daryl E. Chubin explicitly assigns to disciplines the role to "form the teaching domain of science, while smaller intellectual units (nestled within and between the disciplines) comprise the research domain." In sociological terminology, these smaller intellectual units are called scientific specialties.[14]

The fact that the number of specialized disciplines in the sciences has increased over time serves as an empirical counter-argument to the positivistic thesis that all of science would be reduced to physics in the long run. For these specialized sciences, reduction to physics – seen as a constraint upon their acceptability – would have the consequence that the more they succeed, the more they tend to disappear.[15] John Dupré, in consequence, has opted for a "promiscuous realism" and has denied that there is a single kind of thing investigated by science: "For if there are numerous distinct ways of classifying objects into real kinds, any one of which schemes of classification could provide the basis for a properly grounded project of scientific inquiry, then there can be no reason to expect a convergence of these projects of inquiry onto one grand theoretical system." A selection of these different ways is influenced heavily by the motivation and goals of the scientific practitioners.[16]

"Through the development and refinement of concepts concerning molecules and their functions, chemistry provides a common resource for experimental science, comparable with the language of quantitative scientific thought provided by mathematics."[17] This statement, taken from Westheimer's 1965 report on U.S. chemistry, places chemistry alongside mathematical techniques such as the differential and integral calculi, Lagrangians and Hamiltonians, matrices and renormalization. It can be compared with Ian Hacking's assessment of mathematics and instruments as unifiers of the sciences, separated by different styles of reasoning.[18] Even if chemistry did not play a role on a par with mathematics and instrumentation, it is clear that chemical concepts and methods were indeed at the core of the development of fields such as genetic engineering and solid-state physics. Throughout this book, science at the research front is shown to be an interconnected patchwork of scientific specialties. We aim to investigate scientific fields that deal with the properties and transformations of materials, as follows from the classic definition of chemistry. Through this, we hope to achieve a better understanding of how the chemical sciences are interwoven throughout the past century's technology, natural sciences, and culture.

An Overview of the Book

Three parts form the scaffolding of the book: *Theoretical Chemistry and Quantum Chemistry, From Radiochemistry to Nuclear Chemistry and Cosmochemistry*, and *Solid State Chemistry and Biotechnology*. While each of these tackles the development of crucial subfields in twentieth century chemistry, they set out to deal with broader issues and topics. In doing so, the authors of the respective chapters connect the chemical sciences to quantum physics and mathematics in part one; to nuclear physics, astronomy, and the geological sciences in part two; and to solid-state physics, biology, and technology in part three. The three parts are preceded by an overview chapter on the history of organic chemistry which focuses on developments in physical-organic chemistry, bio-organic chemistry, and physical instrumentation.

Foremost in the analysis of the authors who deal with theoretical and quantum chemistry is the physical reductionism of chemistry seen from the perspectives of the chemists involved, and the competitive evolution of research schools in the various political systems of the United States, the United Kingdom, France, Italy, and Germany at mid-century. In his introduction, Nikos Psarros from a philosophical perspective differentiates between the conceptual unity of theoretical chemistry as a science and its corresponding disciplinary unity. While drawing on the notion of a theoretical construct, such as the chemical atoms and chemical molecules of nineteenth century chemistry, Psarros compares their primordial state with the quantum mechanical entities developed in the 1930s. Although the theory involved was that of quantum physics, the necessity of the synthesis of substances and empirical testing by chemical means of theoretical predictions made by physicists leads to the methodological autonomy of chemistry.

Ana Simões and Kostas Gavroglu take the re-thinking of reductionism as the starting point of their endeavor to create a framework of five issues in the history of theoretical and quantum chemistry. In their opinion, reductionism might have been a physicist's tool, and not a chemist's. While physicists took reductionism for granted, they could not offer a realization of this program apart for the most simple cases. In contrast, beginning in the early 1900s, chemists strongly resisted a linear reduction of chemistry to physics. This clearly was a long-term trend in the twentieth century, as is seen by a report submitted to the U.S. congress in 1970 about the status of the physical sciences. Though the underlying conceptual unity in the physical sciences enterprise was undisputed, differences in style and approach were decisive. While the physicists were interested in finding "simple" systems with which to test their theoretical predictions, the chemists were more concerned with the huge varieties in the organization of matter. Thus, according to the argument of the scientists who prepared the report, physicists, chemists, and biologists dealt with the conceptually unified structure of matter at different levels of complexity.[19] But the focal point of Simões' and Gavroglu's argument is the uneasy appropriation of mathematics, not physics, by the chemists; and the introduction of mathematical methods and concepts in chemistry has to be seen as relatively independent of the incorporation of physical concepts. Consequently, theirs is a plea for a convergence of three traditions, those of chemistry, physics, and mathematics in the building of theoretical chemistry. The crossing of boundaries between these disciplines became crucial for the emergence of quantum chemistry as an autonomous subdiscipline. Those scientists who first became aware of the fact that sidelines of research in quantum physics constituted the central part of chemistry were among the most successful discipline-builders.

Textbooks played a predominant role in the strategy of building a discourse for quantum chemistry, just as in other disciplines and research specialties.[20] While providing the formalization of the principles, the codification of a curriculum, and the establishment of routines to solve problems, textbooks contributed tremendously to the disciplining of the field. The early textbooks of quantum chemistry, that is in the period of the 1930s and 1940s, explicitly addressed the question of the relative autonomy of quantum chemistry from physics and thus are a valuable source to trace the development of the field in all its diversity. In fact, two of the most influential authors of textbooks, Linus Pauling and George W. Wheland, engaged in a controversy about the ontological status of resonance in particular and the chemical bond in general. For Pauling, the concept of resonance was closely connected to classical structure theory and largely independent of quantum chemistry, while Wheland insisted that resonance was a man-made concept in a more fundamental way than structure theory and was not an intrinsic property of a molecule. With the dissemination of computer techniques in the 1960s, a long-hoped-for dream came true: *ab initio* calculations became possible and endangered the value of semi-empirical methods, where experimental results constrained and directed the course of mathematics in quantum chemistry. Simões and Gavroglu show in detail how the fine balance of the mathematical, experimental, and pictorial concepts prevalent before the advent of modern computing was deeply disturbed and led to a splitting of the quantum chemistry community into two factions, the *ab initio-ists* and the *a posterio-ists*.

While Simões and Gavroglu compare the American, British, and, to a limited extent, the German research schools and traditions in quantum chemistry between the 1930s and 1960s, Andreas Karachalios and Marika Blondel-Mégrelis focus on the Italian and French developments, respectively. The British preferred applied mathematics, the Americans were at ease with the interplay of theory and experiment, and both were very successful in establishing influential research schools that dominated theoretical chemistry in the decades to come. The limited German foray into quantum chemistry before World War II failed precisely because of a sharp division between experiment and theory in the physical communities and the lack of preparedness to understand the quantum mechanical concepts on the part of the chemists. Karachalios' chapter, with its focus on the Bolognese chemist Giovanni Battista Bonino, tells the story of a life in the slow lane, but nevertheless the story of successfully finding a niche in the shadow of the much more influential Anglo-Saxon developments. Crucial for Bonino's orientation and that of his collaborators were the strong interplay of instrumental methods, in Bonino's case infrared and Raman spectroscopy, with theory and mathematical formalisms. Moreover, the career of Bonino was very much influenced by the rise of Italian fascism and he played an important role in Benito Mussolini's science politics and military research during the war. Thus, Karachalios shows the intimate connection of science and politics, especially through his analysis of the German contacts of Bonino, which were of tremendous importance for the knowledge and technology transfer in both directions of the Berlin–Rome axis. Bonino's group contributed in fragmented, but original, ways to quantum and physical organic chemistry. Nevertheless, and surely because of its fragmented character and Italy's isolation during the war, their results were not noted outside Italy.

Karachalios' chapter shows the fine structure and the local character of knowledge production as well as its dependence on information exchange, and the same can be said of the contribution from Blondel-Mégrelis. The career of Jean Barriol after the war and the establishment of the Theoretical Chemistry Laboratory at Nancy extends the story to the period after World War II and expands it to include France. As with Bonino, Barriol regarded the interplay of theory and experiment as of utmost importance for a fruitful development of theoretical chemistry. Although after 1945 the scientific contacts of France with its allies were quite strong, illustrated by the fact that the first conference on quantum chemistry after the war was held in Paris in 1948, the establishment of Barriol's laboratory was closely connected to French traditions in the theoretical aspects of physical chemistry. The provincial Nancy group was able to fill the gap left by the Paris-based *Centre de Chimie Théorique*, which had a much more physical orientation and Nancy was consequently the place where the concepts of the American scientists John Van Vleck and Lars Onsager were merged with the contributions of the French theoretical chemists.

Nuclear structure, nuclear reactions, and radioactive transformations are the subjects of a loosely defined field, mostly termed nuclear chemistry in the decades following World War II, and conducted mainly by scientists trained in chemistry. The only demarcation criterion to nuclear physics seems to be the perception that physicists are involved primarily in instrumentation, while chemists are concerned with "chemical separation, purification, and identification." [21] This traditional argument as raised by the authors of *Chemistry: Opportunities and Needs* is severely questioned by Xavier Roqué in his introductory essay to part two of this volume. Roqué assigns chemists in the post-war period the willingness to manipulate the instruments of high-energy physics and regards this as an important step in the emergence of nuclear chemistry. Moreover, physical evidence had replaced the old-style chemical one, and especially the radiochemists of the inter-war period were early transition figures in the development of today's intensely instrumentalized analytical chemistry. It is evident that the changes that chemistry has undergone in its research methods and procedures from the 1920s and 1930s on do not allow for the traditional demarcation based on methodology. Furthermore, nuclear chemistry includes research on relevant geological and astronomical problems, which makes it difficult to draw boundaries in terms of topics and themes. Consequently, the implicit or explicit argument of the contributors to this section is that disciplinary boundaries do not make sense in the field of radio-, nuclear, geo-, and cosmochemistry. For Helge Kragh these fields are of "completely interdisciplinary nature, . . . thoroughly integrated mixture[s] of elements from all the classical disciplines of science" (see Chapter 9).

According to Roqué, four issues are of general relevance for the approaches taken by the authors: the evaluation of physical evidence in matters chemical; the relationship between the identification of a new element and its manufacture; the difference between natural and artificial elements in the opinions of the scientists involved; and the disciplinary dynamics of the fields under review. With respect to the first point, Roqué argues that radiochemists were the forerunners of the use of physical instruments in chemistry, among others electrometers and ionization chambers, X-ray and electron diffraction devices, and the mass spectrograph. This diffusion of physical methods did not occur without clashes over the reliability and the evidence provided by these techniques. A particular forceful statement was made in 1907 by the British chemist Arthur Smithells, who spoke about the "chemistry of phantoms" with regard to the new science of radioactivity. Chemical isolation and the measurement of atomic weight by traditional methods was still at the core of the argumentation of many chemists, and for this reason alone the "radioactivists" were tempted to produce the newly discovered elements in weighable amounts. Closely intermingled with the scientific evidence were the industrial interests of the actors, which led Marie Curie in the case of radium, Otto Hahn and

Lise Meitner with mesothorium and protactinium, and Ida and Walter Noddack with rhenium to rely on industrial support in their respective endeavors. It is clearly shown that the industrial input was in no way limited to financial help, but in many cases contributed decisively to the scientific success of the research program, by establishing confidence in the existence and properties of the element in question. Drawing on the example of high-energy particle physics, Roqué remarks on the transition of natural to artificial elements. While physicists moved away from the investigation of cosmic rays towards studying the particles produced in accelerators, the chemists introduced artificial elements and isotopes into their research. Some of them never made this transition, be it for lack of resources or for reasons of style. The Noddacks are a good case in point.

It is to the Noddacks' scientific work that Brigitte Van Tiggelen turns in her description of the successful discovery of element 75, rhenium, and the failure to locate element 43, which the Noddacks claimed to have found but could never prove its existence. Rhenium was the last stable element to be discovered and in this case the traditional methods of investigation were to be successful once more. With the help of a manufacturer of electrical equipment, *Siemens und Halske*, and a chemical company, rhenium was soon to be produced in large enough quantities for its use in thermocouples. This transition from a "private" to a "public" element went hand in hand with the recognition of the scientific standing of its discoverers who received the prestigious Liebig medal of the German Chemical Society. Nothing like this ever happened with masurium, the second claim the Noddacks made, but which was soon obliterated. It was eventually found by Carlo Perrier and Emilio Segrè in 1937 when they bombarded molybdenum with deuterons, and was christened technetium in 1947. This marked the end of a research tradition and with it the end of understanding chemistry to be the science of (weighable) matter. The latter opinion was put forward by Ida Noddack in a popular book published during the Third Reich and closely resembled the definition which was used by defendants of the *Deutsche Chemie*, the counterpart of the *Deutsche Physik* in this era. Thus, though Ida Noddack made a striking prediction of nuclear fission as early as 1934, she was dismissed on the grounds of her not understanding the modern developments.

The search for artificial elements and the resulting discovery of nuclear fission is the theme of the chapter by Ruth Lewin Sime. In December 1938, when Enrico Fermi in his Nobel Prize lecture described the finding of new elements heavier than uranium, the so-called transuranium elements, no one in the audience knew that only weeks later Fermi's claims would be eclipsed. The search for the transuranium elements had been initiated by the discovery of artificial radioactivity by Irène and Frédéric Joliot-Curie in 1934. In the following years, their Paris-based research group, the group of Fermi in Italy, and Lise Meitner, Otto Hahn, and Fritz Straßmann at Berlin made serious attempts to expand the periodic table in the region following element 92. Sime thoroughly analyzes the research program of the Berlin group and argues that the research of the physicist Meitner and the chemists Hahn and Straßmann was guided more by co-dependency than by true collaboration. In hindsight, both sides made crucial mistakes: Meitner – with the authority of contemporary physical knowledge – argued that no major changes of the nucleus could occur; Hahn and Straßmann were convinced – with chemical reasoning on their side – that the sought-for transuranium elements would be transition metals. Both opinions proved to be wrong after a change in the experimental method (again initiated by the French group) finally allowed them to establish the existence of barium, a finding of Hahn and Straßmann that was interpreted by Meitner and Otto Frisch with the fission of the uranium nucleus. Already at this time, in early 1939, political events had severely changed the composition of the Berlin group. Meitner was forced to emigrate to Stockholm in mid 1938 and only maintained her intellectual leadership with the help of correspondence and probably one direct discussion with Hahn in November of 1938. Nevertheless, as Sime convincingly argues, Meitner continued to have a decisive influence on the course of events that led to the discovery in late 1938. The separate publication of the discovery – the chemist's side in the German journal *Die Naturwissenschaften* and the physicist's side in the British journal *Nature* – started the distorted understanding of the course of events. One major step in this was the awarding of the Nobel Prize for chemistry to Hahn alone. Symptomatically, the discussion of the history of nuclear fission continues up to the present day, with unsettled controversies. [22]

Helge Kragh, in his chapter, focuses on the formative period of cosmochemistry in the four decades from 1915 to 1955, after giving a brief overview of the nineteenth century history of geochemistry, which he regards as a forerunner of the chemistry of the universe. The starting point of this formative period according to Kragh was the time "when a new generation of scientists attempted to use the new physics to understand how elements were formed and why the stars shine." Though this was based on Einstein's energy-mass relation and the old quantum physics, chemists continued to play a decisive role in this, as Kragh shows with the examples of the eminent physical chemists Walther Nernst, William Harkins, Gilbert N. Lewis, and others. But their work did not contribute to the constitution of a scientific subcommunity. This finally came about with the turn of geochemistry into a mature science with its own visions, problems, and methods; and Kragh identifies the work of the Norwegian scientist Victor Moritz Goldschmidt as the decisive advance. Relying on the masterly use of optical arc-spectrography, Goldschmidt connected nuclear theory with cosmology, mainly through data compilations of the distribution and abundances of elements and isotopes. These data belonged to the empirical basis for both the theories of the origin of the universe and for models of the structure of the nucleus, be it the big-bang, non-equilibrium theory of George Gamow, Ralph Alpher, and Robert Herman, or the nuclear shell model of Maria Goeppert-Mayer, Hans Jensen, Otto Haxel, and Hans Suess. Goldschmidt was very well aware of the pitfalls involved in creating sharp boundaries between disciplines, "cutting through the unity of modern science," as he called it. But soon after his death in 1947, the formative period of geo- and cosmochemistry came to an end with the foundation of the journal *Geochimica et Cosmochimica Acta* in 1950 and the Geochemical Society in 1955. Both scientific and political reasons might have influenced the decision of the geo- and cosmochemical community to choose Goldschmidt as the founding father of this discipline. Goldschmidt's understanding of geo- and cosmochemistry fits very well the understanding of his followers, the "study of the material composition of the universe." Moreover, the Swiss-Norwegian cosmopolitan Jewish scientist with no known political inclination towards the capitalist West or the communist East created no political obstacles. But Kragh leaves no doubt that cosmochemistry is too tightly interwoven with the non-chemical scientific disciplines to be regarded simply as a chemical subdiscipline. He calls for the interdisciplinary study of these interdisciplinary fields.

The same interdisciplinary approach certainly is valid for the themes and topics of the third part of this book, where authors deal with biotechnology, polymers, and materials science. Is technology applied science? Does science constitute a technology driven endeavor? Are both just sides of one and the same coin, interacting as "dancing-partners" and interwoven in the seamless web of post-modern technoscience? These are questions that come readily to mind and for the answers we find valuable hints in the introduction of Peter Morris. In analyzing the history of polymer science, Morris proposes an approach that perfectly suits the slipping gestalt of the mangle of scientific and technological concepts and practice. He argues for an investigation of the discipline-building program, a closer examination of the strategies of the actors involved, the courses that were taught, and the launching of new journals. For this he proposes to study a few key institutions, such as Herman F. Mark's program in polymer science at Brooklyn Polytechnic and the diffusion of it by tracing the careers of early alumni and co-workers. But not always do the best teachers win the prizes. In proposing an early stage of the Matthew-effect in science, Morris focuses our attention on the importance of public relations in science, while drawing on the design of decisive experiments and the importance of clear statements. Shifting to the fourth topic of this part, materials science, Morris assigns to this field the status of a scientific discipline, which had evolved in the period 1965–1990. Established through the merger of metallurgy and polymer science, materials science catered to the needs of the emerging electronics and aerospace industries, and was firmly based in the theories of solid-state physics, developed in the 1940s and 1950s. Securing access to costly instrumentation was a central part of the disciplinary strategy used mainly by metallurgy departments, who also wanted to create a more modern image of their discipline. Morris underlines the importance of chemical firms, who had long before abandoned the classical disciplinary boundaries and were strong supporters of interdisciplinary research.

Nicolas Rasmussen's central argument in his chapter about biotechnology before the "Biotech Revolution" is that of continuity. Through his study of hormone-related research in the United States, he establishes that many of the biotechnological projects underway at present have their roots in the 1930s and 1940s; and he convincingly underscores this with historical vignettes of adrenalin, cortisone, insulin, and sex and plant hormones. Furthermore, the close academic–industrial relationships, so typical for the biotechnology sector after the 1980s, can already be observed vividly in the cases of James Bonner of Caltech, Francis Schmitt of MIT, and Ezra Kraus at the University of Chicago. In Rasmussen's inclusive understanding of biotechnology as "the use of biological science to intervene in life processes on an industrial scale," biotechnology suddenly receives a historical dimension which it lacked when it was seen as just an application of molecular biology. Rasmussen then sets out to explore and explain the origins of the myth of a recent biotech revolution, which he addresses in the interests of both advocates and opponents of the uses of genetic engineering. For businessmen, a claim of novelty polishes the products and enhances their acceptance. For activists against the misuses of the manipulation of the genome, this justifies close observation and serious action. It might be said that historians always are better off with revolutionary concepts than with evolution, and thus join this unholy coalition. But Rasmussen does not quit the field only with the demasking of today's politics. He traces the resurfacing of the academic–industrial relationship in the 1930s and the 1970s, respectively, to the scarcity of funds in the depression years and the drawback of government money in the late 1960s and 1970s. Assisted by favorable patent legislation, scientists in both periods turned to industry with a huge portfolio of usable life science research and this was welcomed by an industry which thirsted for new products and processes as they saw their traditional lines lose their shine, or go into sharp decline. This also helped industry to avoid continuing with older processes that caused environmental problems.

In his study of the history of polymer science in mid-twentieth-century United States and Germany, Yasu Furukawa concentrates on the scientific side of this emerging techno-science. He traces the origins of polymer science to the organic chemistry of 1920s Germany, the rise of polymer physics in 1930s and 1940s America, and the biological nexus after the war. In doing so, he seeks to examine historical key elements, such as concepts and methodology, terminology and pedagogy, textbooks, journals, and scientific debates. The long-standing struggle of the father figure of polymer chemistry, Hermann Staudinger and his colleague and wife Magda Staudinger-Woit, to establish the term macromolecule instead of polymer serves as the Ariadne thread of Furukawa's chapter. Though widely used in Germany, Staudinger's term was less accepted in the English speaking world, where Herman F. Mark's notion of polymer science dominated the field. The success of Mark's industrial and interdisciplinary strategy is well highlighted by the proposal of Robert W. Cairns, vice president of Hercules Company and in 1968 president of the American Chemical Society, that "any student seriously interested in chemistry as a career should be expected to include this [undergraduate] polymer science course in his curriculum. This course should be treated as a special opportunity to counteract the increasing fragmentation of knowledge" (see Chapter 12).

While Mark set out for a successful career in chemical science and industry, his colleague and friend from 1920s Berlin, Michael Polanyi, moved towards the philosophy of science and social studies. Drawing on his research in the studies of surfaces and X-ray diffraction of solids, Polanyi turned his experience into a formulation of a new philosophy of science based on practice, and severely criticized the power of dogma in science. Mary Jo Nye admirably connects Polanyi's scientific work with his philosophical views and thus allows unique insights into the crossing of boundaries between the (supposedly divided) two cultures. Polanyi did not receive the recognition and rewards he deserved because he did not follow the currently accepted dogmas, either in chemistry or in philosophy. Ironically, Polanyi received a belated vindication with solid-state physics now being the most important research field in the physical sciences, while his underlining of scientific practice and tacit knowledge is now one of the cornerstones of modern history and philosophy of science.

An important reason for the growth of solid-state physics to the first rank among the sciences was its basic status in the field of materials science, a founderless science according to Bernadette Bensaude-Vincent. With X-ray diffraction, the new quantum mechanics, and the heuristic of structure-sensitive properties, solid-state physics rocketed materials science to a truly unified techno-science. Bensaude-Vincent follows its development from World War II to 1990, and describes the coming of age of reinforced plastics and their replacement by composite materials as well as the rise of the new scientific paradigm of biomimetics. She ends by addressing the differentiation of scientific disciplines into their teaching and their research parts. With regard to teaching, materials science certainly counts as a discipline, while in research Bensaude-Vincent prefers the notion of a consistent research field. Describing the actual feeling of today's materials scientists, she notes a strong anti-disciplinary tendency, a sort of rebellion against the authoritarian aspects of science. Hopefully, their belief in a promising technology will be matched by their efforts to build a promising future.

The state of a field is best scrutinized in relationship with its neighbors, in order to reflect similarities and differences. Most prominently, the mechanisms and strategies of bridging boundaries between diverse fields deserve our attention. The coexistence of scientific specialties can be shaped either by reductionism and dominance of one field, or by harmonious interrelation, enhancing creative diversity. In part one, unifying forces (such as the use of mathematics in quantum chemistry) are shown both as a means to interrelate chemistry with physics and as a source of debate about the direction of theoretical chemistry. Along with unification, expansion in matters geographical, social, and cultural has been a nearly continuous feature of the development of science in the twentieth century. Expansion does not take place without conflict and losses and, most impressively, the authors of part two analyze the transformation processes that occur when the spheres of activity of chemical research expand to the earth's crust and atmosphere, the universe, and particle accelerators. For a long time, an apparently insurmountable barrier existed between science and technology. The authors of the third part show that this barrier is a fictitious one, generated by the political interests of historical actors. Chemistry in the twentieth century is both a science and a technology, the two characters being inextricably intertwined. The following chapters offer a fascinating perspective on the position of chemistry in the scientific realm of the twentieth century, a realm both decentralized and interrelated, in conflict and in harmony.

Acknowledgments

It is a pleasure to acknowledge the support this book project and the international conference, from which it originated, has received from its very beginning by Christoph Meinel. Celia von Lindern, and Thomas Steinhauser made our conference both efficient and pleasant; and the painstaking work of Christian Forstner helped a lot with the preparation of the final manuscript. For their criticism and invaluable suggestions, I thank heartfully Ralf Hahn, Kristen Haring, Peter J.T. Morris, and Anthony S. Travis. The enthusiasm of Roland Kessinger and Gudrun Walter of Wiley-VCH eventually helped the project to pass the many steps towards completion.

References and Notes

1 National Academy of Sciences, National Research Council, *Chemistry: Opportunities and Needs. A Report on Basic Research in U.S. Chemistry by the Committee for the Survey of Chemistry* (Washington D.C., 1965), 1.

2 P. Oppenheim, H. Putnam, "Unity of science as a working hypothesis," in H. Feigl, M. Scriven, G. Maxwell (eds.), *Concepts, Theories, and the Mind-Body Problem. Minnesota Studies in the Philosophy of Science*, vol. 2 (Minneapolis: University of Minnesota Press, 1958), 3–36.

3 For a description of this discussion see J.K. Stine, "Scientific instrumentation as an element of U.S. science policy: National Science Foundation support of chemistry instrumentation," in R. Bud, S.E. Cozzens (eds.), *Invisible Connections: Instruments, Institutions, and Science* (Bellingham: SPIE Optical Engineering Press, 1992), 238–263.

4 National Academy of Sciences, *Chemistry: Opportunities and Needs*, 103.

5 *The Journal of Chemical Physics* 8 (1940): ii. Managing editor was Harold C. Urey.

6 J.C. Slater, *Introduction to Chemical Physics* (New York: McGraw-Hill, 1939), v, viii.

7 Physics Survey Committee, National Research Council, *Physics in Perspective*, vol. II, part B, *The Interfaces* (Washington D.C.: National Academy of Sciences, 1973), 1012–1014.

8 See M.J. Nye, *From Chemical Philosophy to Theoretical Chemistry. Dynamics of Matter and Dynamics of Disciplines, 1800–1950* (Berkeley: University of California Press, 1993), 32–55, esp. 34–35 and the literature cited therein.

9 T. Lenoir, *Instituting Science. The Cultural Production of Scientific Disciplines* (Stanford: Stanford University Press, 1997), 48–49.

10 Lenoir, *Instituting Science*, 51.

11 A.S. Travis, C. Reinhardt, *Heinrich Caro and the Creation of Modern Chemical Industry* (Dordrecht: Kluwer, 2000), 333.

12 Nye, *From Chemical Philosophy to Theoretical Chemistry*, 19–31.

13 Lenoir, *Instituting Science*, 53–55.

14 D.E. Chubin, "The conceptualization of scientific specialties," *Sociological Quarterly* 17 (1976): 448–476.

15 J.A. Fodor, "Special sciences (or: the disunity of science as a working hypothesis)," *Synthese* 28 (1974): 97–115.

16 J. Dupré, "Metaphysical disorder and scientific disunity," in P. Galison, D.J. Stump (eds.), *The Disunity of Science. Boundaries, Contexts, and Power* (Stanford: Stanford University Press, 1996), 101–17, on 105–106.

17 National Academy of Sciences, *Chemistry: Opportunities and Needs*, 1.

18 I. Hacking, "The disunities of the sciences," in Galison, Stump (eds.), *The Disunity of Science*, 37–74, on 68–69.

19 National Science Board, *The Physical Sciences. Report of the National Science Board Submitted to the Congress* (Washington D.C., 1970), 19–20.

20 For chemistry see Anders Lundgren, Bernadette Bensaude-Vincent (eds.), *Communicating Chemistry. Textbooks and their Audiences, 1789–1939* (Canton: Science History Publications, 2000).

21 National Academy of Sciences, *Chemistry: Opportunities and Needs*, 79.

22 See V. Morgenweck-Lambrinos, M. Trömel, "Lise Meitner, Otto Hahn und die Kernspaltung: eine Legende aus unseren Tagen," *N.T.M.* 8 (2000): 65–76; R.L. Sime, "Whose legend?" *N.T.M.* 8 (2000): 77–84.

1. Research Fields and Boundaries in Twentieth-Century Organic Chemistry

Peter J. T. Morris, Anthony S. Travis, and Carsten Reinhardt

During the nineteenth century, organic chemistry achieved an almost undisputed dominance over all other areas of chemistry. Other sub-disciplines, especially inorganic chemistry, lost much of their former status. The reasons for the advance in organic chemistry had much to do with the tremendous diversity of carbon compounds, the intellectual fascination of structural theory, and the industrial uses of organic chemicals. But at the turn of the century, the hegemony of organic chemists was endangered by the rise of physical chemistry. From then on, organic chemists were forced to adapt to new challenges, often of diverse origins. The quantum physical formulation of the chemical bond, the great impact of physical instrumentation on all aspects of chemical research, and the rising star of molecular biology totally changed the strategies and organization of research in organic chemistry. Organic chemists had to deal with the thought-provoking concepts of physical organic chemistry, the technologies of physical instrumentation, and the wide-ranging pathways of bioorganic chemistry. It is to these fields that we turn in this chapter. We do not intend to provide a complete overview of the historical events. On the contrary, we focus on critical moments and seek out the key decisions that allowed organic chemistry not only to survive, but to thrive in a competitive environment.

1.1 Physical Organic Chemistry

At first sight, physical organic chemistry appears to be somewhat like polymer chemistry. They both involved a fusion of organic and physical chemistry and the development of new techniques. Unlike polymer science, however, physical organic chemistry is now regarded as being firmly within organic chemistry. How did physical organic chemistry develop and why was organic chemistry able to retain it? [1]

Physical organic chemistry is a child of the twentieth century. To be sure, physical aspects of organic chemistry had existed in the nineteenth century, notably the study of optical activity with its associated mutarotation and the concept of the tetrahedral carbon atom. Indeed, the physicist Ludwig Wilhelmy (1812–1864) first developed the concept of reaction velocity in 1850 from his study on the inversion of sugar. Most organic chemists, however, had little interest in physical processes, such as rates of reaction or the pathways through which reactions took place. There was no inclination, for instance, to explore the nature of aromaticity in the late nineteenth century, despite the importance of aromatic compounds in the dye industry.

The first shoots of a new approach to organic chemistry appeared around 1900, with the publications of Arthur Lapworth (1872–1941) on reaction pathways and the papers of Johannes Thiele (1865–1918) on partial valence. Lapworth was also a pioneer in the study of the rate of organic chemical reactions, especially the addition of cyanide ions to ketones. Thiele explained the aromacity of benzene and the phenomenon of 1,4-addition in butadiene in terms of unsatisfied or "partial" valencies associated with double bonds. His ideas influenced the independent Anglo-German chemist and student of Thiele, Bernard Flürscheim (1874–1955), who created a theory of affinity demands to determine the reactive centers in a molecule such as phenol. Flürscheim's theory was championed by the young organic chemist Christopher K. Ingold (1893–1970) in the early 1920s. Lapworth with the support of his colleague Robert Robinson (1886–1975) developed a rival theory of alternating polarities.

Most of this early work was carried out before the electronic theory of valency was introduced in the early 1920s by Gilbert Newton Lewis (1875–1946) in the United States and Nevil V. Sidgwick (1873–1952) in England. With the impetus provided by this new framework, the two opposing groups soon found their chemical battleground in the bromination of diacetylbenzylamine. Flürscheim and Ingold predicted that it would be brominated in the meta position, while Robinson (using an electronic version of the Lapworth-Robinson theory) argued it would be brominated in the ortho and para positions. Robinson was vindicated; Ingold abandoned Flürscheim and reinvented his approach along the same lines followed by Robinson, but employing a clearer terminology. Through this terminology (which itself illustrates the importance of clarity and ready understanding in the formation of new disciplines and their leadership) and his single-minded development of organic chemical kinetics, Ingold wrestled the leadership of physical organic chemistry in Britain from Robinson, Lapworth and other pioneers such as Thomas Martin Lowry (1874–1936) and Kennedy J. P. Orton (1872–1930). [2]

Ingold's partner in this research program at University College, London, was Edward D. Hughes (1906–1963), who had been trained in kinetic chemistry by Herbert B. Watson (1894–1975) and Orton at Bangor. During the 1930s, Ingold and Hughes gradually developed a system of four reaction types (S_N1, S_N2, E1, and E2) and, using reaction kinetics as their main tool, proceeded to categorize most aliphatic organic reactions according to this scheme. Any criticism of their work was met by a mass bombardment of research papers and rhetoric from both Ingold and Hughes. While Ingold was highly successful at creating a new approach to organic chemistry, he made enemies in the process (not least Robinson) and failed to create new centers of physical organic chemistry, with the result that Britain lost its initial lead in this field. In his pugnacity and his failure to establish a research tradition, Ingold bears a strong resemblance to Hermann Staudinger in polymer chemistry. [3]

Four monographs on physical organic chemistry appeared between 1935 and 1941. The first was *Physical Aspects of Organic Chemistry* (1935) by William A. Waters (1903–1985), who had studied under Lowry at Cambridge. [4] The book had been planned by Lowry as early as 1930 in collaboration with Waters and another Cambridge don, Charles P. Snow (1905–1980), but Waters eventually completed the book on his own. After Waters moved to Durham, he had become interested in the chemistry of organic free radicals, a field largely ignored by Ingold. His book contains chapters on dipoles, unsaturation and free radicals and was very much in the Lapworth-Robinson tradition. This was soon followed by Herbert B. Watson's *Modern Theories of Organic Chemistry* (1937). [5] Watson had worked with Orton at Bangor before moving to Cardiff Technical College. He had already supported Ingold and Hughes through his reviews of their research in the Chemical Society's *Annual Reports*. Not surprisingly, his pro-Ingold volume was largely about the "applications of the electronic theory in organic chemistry", though it did contain chapters on free radicals, unsaturation, and even "[t]he factors determining reaction velocity". In a closing footnote, this chapter mentioned Hammett's research on reaction rates. Whereas Ingold and Hughes were principally interested in the mechanisms of organic chemical reactions and used kinetics as a tool, the American physical chemist Louis P. Hammett (1894–1987) of Columbia University was interested in the physical chemistry of organic compounds for its own sake. He not only created the term physical organic chemistry but also extended it beyond mechanistic chemistry into the thermodynamics of transition states. In research which started in 1933, he correlated the kinetics of a reaction with the thermodynamics of a related equilibrium, thereby allowing the effects of substituents to be gauged empirically. The so-called "Hammett" equation for the reaction rates of aromatic compounds was first published in 1937. Hammett published *Physical Organic Chemistry: Reaction Rates, Equilibria and Mechanisms* in 1940. [6] Less of an introductory text than the other two volumes, it was a vehicle for his distinctive approach to mechanistic chemistry, with chapters on equilibrium and reaction rates, although its chapters on structure, acids and bases, and the various reaction types indicate the still immature state of the subject. If Hammett was typical of the growing East Coast school of physical organic chemistry, *The Theory of Organic Chemistry: An Advanced Course* published in 1941 by two professors at the University of California at Berkeley, Gerald Branch (1886–1954) and Melvin Calvin (1911–1997), represented the powerful West Coast school dominated by the theoretical ideas of Lewis and Linus Pauling (1901–1994). [7] Their volume represented a sharp break with the past. Avowedly less elementary than the earlier volumes, it was also less partisan, covering the whole field in a comprehensive manner for the first time. The St Kitts-born Branch was a close friend of Ingold, but *The Theory of Organic Chemistry* was also influenced by Pauling's concept of resonance. Pauling's ideas of a quantum-mechanical resonance between different canonical forms, published by Pauling in 1931 and popularized in his book, *The Nature of the*

Chemical Bond (1939), were often misunderstood by traditional organic chemists who frequently confused this entirely theoretical construct with the physical process of tautomerism. [8] This misapprehension probably increased its popularity rather than the opposite, and the application of resonance to organic chemistry was given a further boost by the publication in 1944 of *Theory of Resonance and its Application to Organic Chemistry* by George W. Wheland (1907–1972). [9]

By the end of World War II, the United States had begun to dominate the field. The reasons for this dominance were varied. Generous funding became available from a variety of sources, including the military. The synthetic rubber research program, the Office of Naval Research, and Du Pont were generous supporters of research before the National Science Foundation began to provide funds in 1955. The Petroleum Research Fund of the American Chemical Society also began to fund chemical research in 1954. Furthermore, American researchers had access to modern physical instrumentation, in a quality and a quantity not available to their British and German counterparts until the 1960s. There was increasing competition among the American universities, a phenomenon largely absent from the state-controlled British and German universities. Hence American universities were more responsive to new ideas, while European universities remained conservative in their teaching of organic chemistry. The American academic system also tended to disperse gifted postgraduates, thereby creating new centers for physical organic chemistry. Consequently, a new generation of talented physical organic chemists sprang up across America in the late 1940s. From Harvard, there was George Wheland, Paul D. Bartlett (1907–1997), William S. Johnson (1913–1995), Donald J. Cram (b.1919), and William von Eggers Doering (b.1917). Doering had synthesized quinine in 1944 with Robert Burns Woodward (1917–1979), who remained a classical organic chemist but strongly supported physical organic chemistry. Andrew Streitwieser (b.1927) studied under Doering at Columbia University. Elliott R. Alexander (1920–1951), Kenneth Wiberg (b.1927), and Jerome Berson (b.1924) also studied at Columbia. At the University of California at Los Angeles, there was William G. Young (1902–1981), Saul Winstein (1912–1969), and John D. Roberts (b.1918). Many of these young academics passed through Harvard on fellowships between 1945 and 1948. There they were inspired by the young Woodward (who was only 28 years old in 1945), Bartlett, and Louis Fieser (1899–1977). The University of Illinois, America's leading (if rather traditional) organic chemistry department, hired Elliott Alexander in 1946. He published *Principles of Ionic Organic Reactions* for traditional Illinois-type organic chemists and undergraduates in 1950, but died in an air crash soon afterwards. [10]

Winstein was the leading figure in post-war physical organic chemistry and he pushed the field to its limits with his concept of non-classical carbonium ions (or carbocations in modern terminology). In particular, starting in 1949, he explained the high reactivity of the norbornyl carbonium ion on the basis of a bridged intermediate in which a remote sigma-bond becomes part of a bridged cation. Such a species was considered bizarre in terms of classical chemical structure, but it could be formulated as a combination of molecular orbitals. Meanwhile, Herbert C. Brown (b.1912) at Purdue University, away from the major centers of physical organic chemistry, had been developing his theory of steric crowding to explain the reactivity of specific compounds. Their rivalry erupted when Brown openly criticized Winstein, claiming that the reactivity of norbornyl chloride could be adequately explained by steric crowding without any recourse to exotic non-classical carbonium ions. In many respects, Brown's conservative position was similar to Staudinger's traditional view of polymers, but the non-classical carbonium ion controversy was not as pivotal as the macromolecular debate. Nevertheless, this "heresy" provoked outrage among the physical organic chemistry community, from the old guard (Ingold and Hughes) and the young Turks (Winstein, Roberts, and Cram) alike. The strength of this outrage had several origins, the defensiveness of a still young sub-discipline, the desire of Winstein and his supporters to promote molecular orbital (MO) theory over the older resonance model,[11] and the veneration with which Winstein was held by most American physical organic chemists. This quasi-religious conflict raged for nearly two decades, Brown being quite impervious to any evidence in favor of non-classical structures, until well after Winstein's early death in 1969. Eventually, NMR data collected at very low temperatures in the 1980s supported a non-classical structure, though Brown characteristically insisted that these data *could* be explained in terms of a very rapid equilibration.[12]

The theoretical chemist Erich Hückel (1896–1980), whose brother Walter (1895–1973) was an organic chemist, attempted to introduce MO theory into organic chemistry in 1931 with limited success.[13] His MO theory of aromaticity (summarized by Hückel's rule) was revived after World War II by Michael Dewar (1918–1997), to explain the structure of colchicine. With the backing of his mentor, Sir Robert Robinson, Dewar published *The Electronic Theory of Organic Chemistry* in 1949, which sought to displace Pauling's concept of resonance with a more modern model that used molecular orbitals to explain, for instance, the Wagner-Meerwein rearrangement. The validity of Hückel's rule was demonstrated by Doering's synthesis of the tropylium cation in 1954 and Ronald Breslow's (b.1931) synthesis of the cyclopropenyl cation in 1957. Meredith Gwynne Evans (1904–1952) at Manchester had suggested in 1938 that the Diels-Alder reaction would have a benzene-like aromatic transition state. Evans was prevented from pursuing this insight by World War II and his early death. His ideas were later taken up and developed by Dewar in terms of MO theory. The idea of using frontier orbital theory to explain some organic reactions was put forward by the chemical engineer Kenichi Fukui (1918–1998) in the 1950s. However, he was away from the main centers of physical organic chemistry and unfortunately published his first work in Japanese. In 1964, Woodward encountered an anomalous ring closure during his synthesis of part of the Vitamin B_{12} molecule and sought to explain the unexpected stereochemistry of the reaction. He soon perceived that MO theory was needed to solve the problem and sought the help of the young theoretical chemist Roald Hoffmann (b.1937). They quickly formulated a set of empirical rules, based on the concept of orbital symmetry, which revolutionized the study of pericyclic reactions.

One striking aspect of physical organic chemistry in the 1950s is the lack of textbooks and periodicals on the subject. Ingold published his magisterial *Structure and Mechanism in Organic Chemistry* in 1953, based on his George Fisher Baker lectures at Cornell University in 1950–51. [14] But, at over 800 pages, it was too long to be a successful textbook and, in any case, Ingold's star was waning. Despite the undoubted expansion of physical organic chemistry, only another four textbooks appeared between 1956 and 1959 and they were not written by the leading figures in the field, who were too busy building up their research schools. [15] Ingold's key paper on organic reaction mechanisms had appeared in *Chemical Reviews* in 1934 and advances were still mostly reported in general review periodicals. A watershed was reached in 1963, when Academic Press in London started the annual *Advances in Physical Organic Chemistry* and Wiley in New York launched a similar annual, *Progress in Physical Organic Chemistry*. Within three years, even the venerable *Journal of the Chemical Society* had given birth to a new "Section B" on physical organic chemistry. The Anglo-American firm of Interscience, already strong in the field of polymer chemistry, created an annual report on *Organic Reaction Mechanisms* in 1965.

Young and Winstein had trained at Caltech under Howard Lucas (1885–1963), who had developed his own electronic theory of organic reaction mechanisms in 1924 and carried out research on olefin addition. His undergraduate textbook, *Organic Chemistry* (1935), was the first to use organic reaction mechanisms. [16] Although James B. Conant (1893–1978) was interested in physical organic chemistry and could be considered one of its founders, his *Chemistry of Organic Compounds* (1933; and revised with Max Tischler in 1939) did not mention organic reaction mechanisms at all. The inclusion of physical organic chemistry in introductory textbooks had to await a new generation of organic chemists who had been trained in this field. The pioneers, both published as *Organic Chemistry* in 1959, were by Robert Thornton Morrison (b.1918) and Robert Neilson Boyd (1914–2000) and by Donald Cram and George Hammond (b.1921). [17] Cram and Hammond had asked John D. Roberts to be their co-author, but Roberts decided to publish his own mechanistic textbook, *Basic Principles of Organic Chemistry*, with his British-born colleague, Marjorie Caserio (née Beckett, b.1929) in 1964. [18] These textbooks had no counterparts in Britain and although some British chemistry departments used the American textbooks, this practice was not widespread. Before 1960, most European textbooks of organic chemistry were in effect either recipe books or encyclopedias. Even in the mid-1960s, the leading British textbooks, notably the one by Ivor L. Finar (1912–1984), were remarkably conservative and they made little use of organic reaction mechanisms. [19] The undergraduate teaching of organic reaction mechanisms was pioneered by Peter Sykes (b.1923) at Cambridge with the support of Alexander (Lord) Todd (1907–1997). His *Guidebook to Mechanism in Organic Chemistry*, first published in 1961, dominated the British undergraduate market for many years. [20] The first British textbook to use organic reaction mechanisms extensively was *Basic Organic Chemistry: A Mechanistic Approach*, published in 1966 by John (the second Baron) Tedder (1926–1994) and Antony Nechvatal (b.1926). [21]

The advance of physical organic chemistry in the United States was mirrored in Britain and Germany, but to a much lesser extent. The Ingold tradition at University College, London, was maintained by John Ridd (b.1927) and there was a strong department at King's College London headed by Victor Gold (1922–1985). Gold's student Donald Bethell (b.1932) set up a research group at Liverpool, and R. O. C. Norman (1932–1993), who had studied under Waters at Oxford, established a school of free-radical chemistry at York. This progress was counterbalanced by an emigration of talented chemists to America, notably the departure of Ingold's son Keith (b.1929) to Canada in 1951, Michael Dewar to Chicago in 1959, and the theoretical chemist John Pople (b.1925) to Pittsburgh in 1964. The physical and institutional rebuilding of chemistry departments in post-war Germany allowed younger chemists such as Rudolf Criegee (1902–1975) and Rolf Huisgen (b.1920) to revive the German tradition of physical organic chemistry established by Arthur Hantzsch (1857–1935), Daniel Vorländer (1867–1941), Thiele, and Hans Meerwein (1879–1965). Significantly, George Olah (b.1927) went to America after he fled from Hungary in 1956 rather than to Germany, despite his admiration for Meerwein. In a move against the trend, the Harvard-educated Paul von Ragué Schleyer (b.1930) emigrated from Princeton to Erlangen in 1976. There was also a strong tradition of physical organic chemistry in the Netherlands which ultimately stemmed from the school of Jacobus Van't Hoff (1852–1911) at Amsterdam.

The introduction of physical organic chemistry met relatively little resistance within organic chemistry departments. This was partly because other organic chemists had started to use mechanistic chemistry in their work and recognized the value of this new field. Its expansion also helped to maintain the prestige of organic chemistry by filling a vacuum left by the collapse of classical structural organic chemistry in the 1960s when it was displaced by physical instrumentation. Physical organic chemistry had a powerful influence on organic synthesis. Not only did it assist the development of new syntheses of natural products through the insights made available by organic reaction mechanisms, but it also stimulated the synthesis of new compounds for the testing of its own theories. Furthermore, even traditional syntheses provided material for research in physical organic chemistry, for instance, the unexpected reaction during the synthesis of Vitamin B_{12} which led to the development of the Woodward-Hoffmann rules. Physical organic chemistry has been retained within organic chemistry, in contrast to polymer chemistry or bioorganic chemistry, but it has transformed traditional areas of organic chemistry at the same time. Modern organic chemistry is not organic chemistry as it would have been understood by a chemist in 1900. In effect, the old name has been retained for a completely new hybrid of organic synthesis and physical organic chemistry.

1.2
Physical Instrumentation and Organic Chemistry

The introduction of instrumental methods of analysis and structure determination during the second half of the twentieth century transformed organic chemistry.[22] Freed from structural studies, some chemists concentrated on organic synthesis, others transferred their attention to biomolecular topics, and yet others switched to physical organic chemistry. Hence one of the effects of the introduction of physical instrumentation was a shift towards other interdisciplinary studies. The passing of classical structural chemistry did not go unmourned. It had created "a great sum of experience,"[23] accumulated over a century of painstaking chemical research. Sir Robert Robinson had assumed, perhaps naively, that such studies would continue even if the structures of organic compounds were already known. Even one of the leaders of the new style of organic chemistry, Carl Djerassi (b.1923), has regretted the loss of the intellectual and creative challenge of the older methods of structure determination.[24]

At first sight, instrumentation, even physical instrumentation within organic chemistry, may not appear to be an interdisciplinary area. However, if by "interdisciplinary" we mean an intellectual zone where scientists from different disciplines meet and interact, no field could be more deserving of the title. Not only did the construction of these instruments draw on new developments in electronics and optics and stimulated further innovation, but the techniques themselves came from outside organic chemistry. Nuclear magnetic resonance and mass spectroscopy, to give just two pertinent examples, crossed over from physics, and organic chemists had to collaborate with chemical physicists to obtain the best results from these new techniques.

There were equally important crossing zones at the borders between industry and academia. Initially, the instrument makers worked with their primary clients, physicists, chemical physicists, and physical chemists interested in fundamental processes. When it became clear that petroleum and petrochemical firms, rather than universities, would be their major market, the instrument manufacturers collaborated with industrial researchers, especially in the period during and immediately after World War II. The early 1950s saw a great rise in funding for academic research from the state, the military and industry, especially in the United States. With the industrial market reaching saturation (or so it seemed at the time), the instrument companies turned to potential new markets in academic organic chemistry, encouraged by the initial efforts of pioneering chemists.

Most of these new instruments had not been built with organic chemists in mind and they had to be assimilated into their new environment. Accordingly, the diffusion of instrumental methods into mainstream organic chemistry was a slow process. The new methods had to gain acceptance from chemists who strongly preferred well-established chemical techniques. Most organic chemists only had a broad understanding of physical chemistry and a shaky grasp of quantum physics and electronics. Furthermore, manufacturers had to overcome the understandable reluctance of chemistry departments to spend most of their annual budgets on a tin-box filled with strange wiring and circuit boards. In the late 1950s, most organic chemists considered the melting point to be the only important physical parameter for most organic compounds and even the dedicated melting-point apparatus was a recent innovation. The polarimeter languished in the darkroom, used mainly by carbohydrate chemists.

To overcome this resistance, instrument companies were forced to use strong promotional methods. They produced attractive trade literature and newsletters, held corporate seminars, hired enthusiastic chemists to spread the word by visiting departments and leading chemists, and encouraged research that employed their instruments. Their target audiences were open-minded graduate students and this technological evangelism was eventually assisted by a new generation of textbooks. The history of the instrumental revolution in organic chemistry is a combination of generational change in science, technological innovation, and marketing strategies.

Instrumental methods are so closely identified with the "tin-box of tricks" that X-ray crystallography is not usually considered to be an instrumental method alongside, say, infrared spectroscopy or nuclear magnetic resonance. Yet, the X-ray camera (whatever form it takes) is as much an instrument as a spectrophotometer. Another reason, perhaps, is that X-ray crystallography came into use earlier than the other methods. It began with the determination by the young Lawrence Bragg (1890–1971) of the structure of the alkali halides in 1913. He obtained the structure of the simple aromatic hydrocarbons, naphthalene, and anthracene, in 1921. While the X-ray crystallography of inorganic ionic compounds flourished, its application to organic chemistry was limited, because it required crystalline compounds that exhibit dielectric behavior. [25]

A major breakthrough was the introduction of the "heavy atom" technique by J. Monteath Robertson (1900–1989) in 1935. In this, once the locations of the "heavy atoms" are established, phase constants for these atoms are used to produce an electron density distribution. From this, the location of the light atoms such as carbon, hydrogen, and oxygen can be found. This led to the first major triumph of X-ray crystallography in 1945, when C. Harold Carlisle (b.1911) and Dorothy Crowfoot [Hodgkin] (1910–1994) at Oxford carried out a full structural determination of cholesteryl iodide, which confirmed the structure first proposed by organic chemists in 1932. About the same time, the Oxford X-ray crystallographers helped to resolve the structure of penicillin. Robinson had postulated two separate rings, while Woodward favored the fused "β-lactam" structure. "[W]orking in a state of much greater ignorance of the chemical nature of the compounds we have had to study than is usual in X-ray analysis," [26] Crowfoot Hodgkin and Charles Bunn (1905–1990), aided by penicillin's large sulfur atom, proved that the β-lactam structure was correct.

By the late 1940s, the X-ray crystallographers had matched the organic chemists, but not surpassed them. The decisive determination was presented by vitamin B_{12}, the anti-anemic factor, which had been isolated independently at Merck and Glaxo in 1948. The organic chemists, led by Todd, clarified several key features of the molecule, but were unable to unravel its complex structure completely. Presented with a crystalline sample by Glaxo, Crowfoot Hodgkin determined the full structure of vitamin B_{12} in 1957, aided by its central cobalt atom. It was the first time that such a complex molecule had been almost entirely elucidated by physical methods and it was followed by the determination of the large biomolecule myoglobin by John Kendrew (1917–1997) at Cambridge, England, in 1960.

X-ray crystallography was revolutionized by the arrival of electronic computers in the 1950s. They enabled the routine determination of bond lengths, bond angles, and spacing between non-bonded atoms. By the 1960s, three-dimensional electron-density distribution patterns, making use of heavy atoms and the technique of isomorphous replacement, enabled the definitive solution to many structural problems.

Although William W. Coblentz (1873–1962) at Cornell University (he later worked at the National Bureau of Standards) had collected the infrared absorption spectra of organic compounds in the early years of the twentieth century, the taking of measurements point-by-point was a lengthy process, and there was no great enthusiasm for this technique from organic chemists. The situation changed with the demands created by the industrial programs set up during World War II, particularly in the petrochemical industry. The example used here is Perkin-Elmer, mainly because its products were more readily available than those of the rival Beckman firm. [27]

Perkin-Elmer was founded in 1937 by an investment banker, Richard S. Perkin (1906–1969), and a publisher, Charles Wesley Elmer (1872–1954), both keen amateur astronomers, to manufacture advanced optical systems. In 1941, the firm opened a factory in Stamford, Connecticut, to meet the demand for military optics, such as tank periscopes. Meanwhile, in the neighboring Stamford central research laboratories of American Cyanamid (opened in 1937), chemists and physicists were advancing their own knowledge of instrumental methods. The two new neighbors teamed up to investigate the applicability of the new instrumentation in chemistry, starting with infrared spectroscopy. This technique was advanced considerably, and more than in any academic laboratory, at the Bound Brook, New Jersey, and Stamford laboratories, by American Cyanamid scientists Edwin Stearns (1911–1992) and R. Bowling Barnes (b.1906). [28] Barnes, who had undertaken research at the University of Berlin in the early 1930s, was head of the Physics Department at Stamford. Both men were present at the October 1943 meeting of the Optical Society of America on infrared spectroscopy and subsequently were among the principal participants in the development and use of spectrophotometers. Numerous publications, for instance in *The Review of Scientific Instruments*, *Journal of Applied Physics*, and *Analytical Chemistry*, documented American Cyanamid's cutting-edge studies, and showed how disciplinary boundaries were continually crossed as optics and electronics interacted with chemistry. [29] Publications by American Cyanamid scientists acknowledged contributions from Perkin-Elmer, and Barnes was co-author of *Infrared Spectroscopy: Industrial Applications, and Bibliography* (1944). The fruit of this collaboration, the Perkin-Elmer Model 12, was commercialized in 1944.

Meanwhile, Arnold Beckman (b.1900), hitherto a manufacturer of electronic pH meters, had joined forces with Robert Brattain – the brother of Walter Brattain of transistor fame – at Shell Research, with the encouragement of the U. S. Government's Rubber Reserve Company. Beckman's first commercial infrared spectrometer, the IR-1, was developed in 1942 and was used by the wartime synthetic rubber research program. However, the classified nature of this and similar work meant that Beckman spectrometers were not generally available until 1945, when the IR-2 was marketed. Meanwhile, in Britain, Adam Hilger and Grubb Parsons independently developed double-beam instruments. The Hilger instrument was closely modeled on the design of MIT physicist Arthur C. Hardy (1895–1977).

After the war, in 1947, Baird Associates in Cambridge, Massachusetts, introduced a double-beam instrument which was the first to present the spectrum in terms of percentage transmission wavelength. These instruments were still expensive and organic chemists needed persuasion to buy them in 1948, Perkin Elmer hired Van Zandt Williams (1916–1966), formerly of American Cyanamid in part to spread the word in chemistry faculties throughout America and Europe. One of his successes was Cambridge, England, where he persuaded Todd to buy one of the new Model 21 double-beam infrared spectrophotometers. The less expensive Perkin-Elmer Model 137 "Infracord," which came out in 1957, brought infrared spectrophotometry within the reach of the ordinary chemistry laboratory. This instrument and its improved successor, the model 157, paved the way for the routine use of spectroscopy in structure determinations. They were also of value in the detection of unknown compounds by comparing the spectra of the compound with atlases of the by then numerous spectra of known compounds.

Although Walter Noel Hartley (1846–1913), a chemist and pioneering ultraviolet spectroscopist, had employed ultraviolet spectroscopy to study the vexed issue of tautomerism in the 1890s, the use of ultraviolet spectroscopy in organic chemistry was very much a product of the twentieth century. In the early years of the century, the London firm of Adam Hilger popularized the use of the ultraviolet quartz spectrograph, in which spectra were recorded photographically. Hardy, at MIT, designed an advanced, but expensive, photoelectric spectrophotometer which was commercialized by General Electric in 1933. These instruments had a limited impact on organic chemistry, but Robert R. Williams (1886–1965) determined the structure of thiamine (vitamin B_1) in 1936 with the help of its ultraviolet spectra.

As in the case of infrared spectroscopy, World War II proved to be a watershed. Beckman brought out the celebrated DU spectrophotometer in 1941, followed by British-made instruments such as the Hilger Uvispeck and Unicam SP 500. The spectra were still obtained by measurement of point-by-point dial readings. Robert C. Hirt (b.1919), at American Cyanamid, Stamford, and colleagues at Bound Brook, contributed particularly to ultraviolet spectrophotometry, which offered more accuracy and precision than infrared for quantitative analysis. Hirt and colleagues modified ultraviolet spectrophotometers, namely the Beckman DU and a Cary machine, for use in identifying individual substances in the UV region. [30] Recording spectrophotometers, with spectra recorded on paper charts, later appeared on the market and were a great improvement, although much more expensive. The Beckman DK-1, which was brought out in 1954, was the first reasonably priced ultraviolet spectrophotometer that was comparable with the GE-Hardy spectrophotometer in its level of automation.

In 1941 and 1942, the young Robert Burns Woodward undertook a careful numerical analysis of published spectral data for various steroidal ketones containing double bonds. These ketones characteristically absorbed ultraviolet light strongly around 230–250 nm. From this data, Woodward drew up general rules that related these absorption maxima to the arrangement of the double bonds and the substituents attached to the ketones. Using these rules and with his typical self-confidence, Woodward declared that some of the accepted structures were incorrect.[31] This was not just a theoretical breakthrough. It was also the first systematic application of instrumentation, apart from the polarimeter, to a major area of natural product chemistry and foreshadowed the changes that were to take place in organic chemistry.

The mass spectrometer produces positive ions from a sample and uses a strong magnetic field to resolve them into a series of beams recorded on photographic plates or by electronic detectors. The beams are presented as a series of peaks representing mass to charge ratios. Francis William Aston (1877–1945) and Arthur J. Dempster (1886–1950) independently made the first mass spectrographs around 1919. Alfred O.C. Nier (1911–1994) developed the first high resolution mass spectrometer at the University of Minnesota in the late 1930s.[32]

Mass spectroscopy entered the arena of organic chemistry during wartime research in the strategic petroleum and synthetic rubber industries. Instrument makers, for example Consolidated Electrodynamics Corporation (CEC) of Pasadena, California, developed the mass spectrometer as a highly reliable and precise instrument suitable for organic analysis.[33] Many spectra and fragmentation patterns were obtained and published, mainly through the efforts of the Hydrocarbon Research Group of the American Petroleum Institute. Some of the chemists who were trained in the early use of the technique were later to undertake research on highly complex organic compounds. The extension of mass spectrometry to structural organic chemistry was not, however, immediate. Most organic chemists regarded mass spectrometers as very expensive and elaborate instruments that were difficult to handle. There was also the problem of the industrial approach, in which mass spectra were used for analysis of mixtures of known structure.[34] Furthermore, the fragmentation patterns of hydrocarbons did not appear to show any promise for correlation with structures.

Fred W. McLafferty (b.1923) of Dow Chemicals (later at Cornell University) overcame the "terrible" reputation of mass spectroscopy when used in analysis of organic compounds caused by what he called the "random rearrangements" of hydrocarbons. However, "specific rearrangements" (directed by functional groups in the molecules) did provide critical insights into mechanisms, and therefore helped to elucidate structures. With the help of the rationalization of these rearrangements, the fragmentation of organic molecules in the mass spectrometer – formerly seen as a problematic disorder – now brought together physical organic chemistry, mass spectroscopy, and natural product chemistry.

An early pioneer in the application of mass spectroscopy to natural products was Ivor Reed at the University of Glasgow, who used fragmentation analysis in 1956 to determine the structure of the side-chains of various steroids. Other researchers included Einar Stenhagen's group at the University of Göteborg (Sweden). They employed the mass spectrometers designed and built by Ragnar Ryhage from the Karolinska Institute in Stockholm mainly for the elucidation of the structures of long-chain fatty acids and esters. Another early group that engaged in systematic studies of the structures of natural products was Klaus Biemann's (b.1926) at MIT

(Cambridge, Mass.) with a focus on alkaloids, amino acids, and peptides.[35] Biemann had a strong background in synthetic organic chemistry and worked on the synthesis and structure of natural products. In the early 1960s, he used mass spectroscopy to determine the structure of complex alkaloids, especially indole alkaloids. By early 1964, Biemann had acquired a high-resolution mass spectrometer (first used in organic chemistry by John Beynon (b.1923) at ICI in the 1950s). This enabled the entire mass spectrum to be displayed on a single photographic plate. Using an IBM 7094 computer, Biemann used this data to calculate the exact molecular mass of each fragment. This very powerful technique determined the structure of compounds for which only minute samples were available.

The application of mass spectroscopy to natural product chemistry in the 1960s was of tremendous value in rapid unambiguous identification and structural elucidation. Djerassi and his group at Stanford did much to establish mass spectroscopy as a tool for the organic chemists, based largely on the assumption that most of them would not measure the spectra, but would interpret them. Djerassi and colleagues regarded the mechanistic approach as the best from a pedagogical perspective. They published a series of books that dealt with natural products such as alkaloids, steroids, terpenoids, and sugars.[36] Djerassi was also well aware of the possible pitfalls of assigning specific "mechanisms" to fragmentations of molecules. At the end of the 1960s, his investigations, which were based mainly on high resolution work and isotopic labeling techniques, led to confirmation or revision of many proposed fragmentation mechanisms. Significantly, the success of mass spectroscopy in chemical research during the 1960s derived from research formulated in the language of physical organic chemistry. This strategy of pushing the physical organic chemistry approach was determined by cognitive and pedagogical considerations, due to the relatively marginal role of natural product chemistry in the United States (as compared with organic synthesis and especially physical organic chemistry).

The nuclear magnetic resonance (NMR) effect was first reported in 1946 by Felix Bloch (1905–1983) at Stanford and, independently, by Edward Purcell (1912–1997) at Harvard University. James Arnold (b.1920), a postgraduate at Stanford, obtained the first NMR spectra in 1951. He later worked for the Californian firm Varian Associates. These showed separate resonances for protons located at different positions in the molecule: the field-dependent chemical shifts. Herbert S. Gutowsky (1919–2000) at Illinois, who studied spin-spin coupling, was one of the first to introduce NMR into organic chemistry in the early 1950s. Bloch had the bright idea of the spinning tube to reduce sample inhomogeneities in 1954 while he was stirring a cup of tea.

The first commercial NMR spectrometer was marketed by Varian in 1952. Operating at 30 MHz, it was purchased by large companies but was too expensive for academic use. At a cost of $ 26 000, John D. Roberts obtained an NMR machine, the Varian 40 MHz instrument that was "the first commercial NMR spectrometer to be sited in a university. If it was not the first piece of such equipment, I'm sure it was the first to be put under the jurisdiction of an organic chemist."[37] In the late 1950s, the use of NMR in organic chemistry was heavily promoted by Varian. James

Shoolery (b.1925), who was employed by the firm, went round organic chemistry departments and showed how it could be used to solve structural problems, as well as carrying out his own research on the structure of steroids. In contrast to physical organic chemistry, NMR studies were widespread geographically, but did not inspire much interest from most organic chemists.

As IR spectroscopy had already demonstrated, a relatively cheap and easy-to-use instrument was required if NMR was to be widely adopted. The breakthrough came with the introduction of the A-60 by Varian in 1961 and the arrival of a new generation of organic chemists with a stronger grasp of physics. The interpretation of NMR spectra was greatly improved by the growing availability of stronger magnetic fields. Higher field strengths produced greater chemical shift separations, which allowed chemists to distinguish between peaks created by spin-spin coupling and peaks from wholly different protons. By the end of the 1960s, both the powerful superconducting magnets and the more sophisticated Fourier-Transform instruments were commercially available. With advances in resolution and sensitivity, chemists were able to correlate spin-spin coupling constants with physical features of molecules. Frank Anet (b.1926) and A. J. R. (Tony) Bourn introduced the Nuclear Overhauser Effect (discovered in 1953) into chemical NMR in 1965. This was useful for the study of conformations, as in the case of carbohydrates, because it provided information about the positions of protons in space. Raymond Andrew (b.1921) developed the technique of "magic-angle" spinning in 1971 to overcome the problem of dipole-dipole coupling in solid state NMR. The 1980s saw a second wave of innovations in NMR, involving the study of NMR spectra in two and three dimensions, using a combination of ^{1}H (proton) and ^{13}C NMR for structural work and, to a lesser extent, nitrogen, fluorine, and phosphorus NMR. [38]

NMR has become hugely popular, in part because it is essentially a highly sensitive probe inside the molecule itself, but largely because it is so versatile. Comparing the results of ^{1}H and ^{13}C NMR usually allows the entire molecular structure of a compound to be determined from a tiny sample and without any crystallization or other working up. This is a chemist's dream come true and, not surprisingly, has led to an increasing reliance on the results of NMR measurements on their own without the back-up of a full X-ray crystallographic determination. While NMR is usually reliable, this does mean that there is a degree of uncertainty about many modern structures that are proposed in the literature. At the same time, these very same features of NMR make it a vital tool for the physical organic chemist, particularly for studying organic reaction mechanisms. It played a decisive role, for instance, in settling the protracted non-classical carbonium ion controversy (see the section above on physical organic chemistry). With the development of two- and three-dimensional NMR in the 1980s and 1990s, it has dominated much instrumental work in molecular biology and is used to study the all-important conformations of large proteins. Moving from the laboratory into the hospital, NMR (under the guise of Magnetic Resonance Imaging) has enabled us to see the soft tissues inside the body.

This transformation of organic chemistry was reinforced by changes in the hitherto conservative world of science publishing, with the arrival of more aggressive publishers who used new processes such as color printing. With the encouragement of Bill Benjamin of McGraw Hill, John D. Roberts published *Nuclear Magnetic Resonance* in 1959 and the more specialized (and by Roberts' own admission, less read) *Introduction to the Analysis of Spin-Spin Splitting in High-Resolution Nuclear Magnetic Resonance Spectra* two years later. [39] The best known (if not often actually read) monograph on NMR, *High Resolution Nuclear Magnetic Spectroscopy* by John Pople, William Schneider (b.1915), and Harold Bernstein (b.1914), was also published by McGraw-Hill in 1959. [40] Biemann brought out his textbook on *Mass Spectroscopy* in 1962 with McGraw-Hill [41] and Gordon Barrow (b.1923) published his *Introduction to Molecular Spectroscopy* in the same year with the same publisher. [42]

Relatively few organic chemists were willing to invest time in reading in-depth accounts of one technique and broader surveys of the whole field played a decisive role. The first volume of *Determination of Organic Structures by Physical Methods*, edited by Ernest Braude (1922–1956) and Frederick Nachod (1913–1992), had an immediate impact when it was published in 1955. [43] *Elucidation of Structures by Physical and Chemical Methods*, edited by Kenneth Bentley (b.1925), was also popular when it came out in 1963. [44] As well as introducing physical organic chemistry, Roberts and Caserio also emphasized the role of instrumentation in organic chemistry in *Basic Principles of Organic Chemistry* (1964). As in the case of physical organic chemistry, British textbooks were more conservative and their American counterparts were considered too basic for the more specialized British degree. Oliver and Boyd in Edinburgh published *Physical Methods in Organic Chemistry*, edited by J. C. Peter Schwarz (b.1927), in the same year. [45] In some respects this was an old-fashioned book and, in 1966, McGraw-Hill brought out the more modern *Spectroscopic Methods in Organic Chemistry* written by two Cambridge dons, Dudley Williams (b.1937), who had worked with Djerassi, and Ian Fleming (b.1935). [46] Like its counterpart in physical organic chemistry, Peter Sykes's *Guidebook to Mechanism in Organic Chemistry*, it has endured, with a fifth edition in 1995.

The introduction of electronic instrumentation after 1940 was nothing less than a scientific and technological revolution. It has led to the near-total displacement of classical "wet and dry" methods in organic structure elucidation. The routine of one type in chemistry was transformed into routine of another type, with major implications for organic chemistry and organic chemists. Organic chemists who might have spent their careers carrying out chemical degradations on, say strychnine, now used NMR to study peptide chains. Whether the latter can really be called organic chemistry is perhaps a moot point. The instrument companies themselves have diversified into the more profitable biomolecular and biomedical fields. This was demonstrated in June 1999, when Perkin Elmer (now PE Corporation), one of the founders of this revolution, switched entirely from chemistry into automatic gene sequencing and sold its analytical instrument division to EG&G, Inc. PE Biosystems Group/Applied Biosystems Inc. high-speed DNA sequencers mapped the genetic code announced to the world by the PE subsidiary Celera Genomics Corporation in June 2000.

1.3 Bioorganic Chemistry

There has been a long-standing interaction between biology and chemistry, dating back to the Chemical Revolution in the 1780s, giving rise to a stream of sub-disciplines in the nineteenth century, including organic chemistry, medical chemistry, the closely related physiological chemistry, and biological chemistry. However, the most important interaction was the development of biochemistry in the early years of the twentieth century.[47] Although chemists played an important role in the creation of biochemistry, notably the work of Emil Fischer (1852–1919) on peptides, biochemistry drifted away from chemistry in the 1920s and became an independent discipline. In the 1960s, however, organic chemistry faced a crisis on two fronts. It was threatened by the growing importance of biochemistry and molecular biology, following the discovery of the double helix in 1953. At the same time, the classical chemistry of natural products was on the point of collapse, following the introduction of physical instrumentation into structural organic chemistry.

A group of organic chemists with a strong interest in the new field of physical organic chemistry attempted to solve this crisis by creating the new discipline of bioorganic chemistry. They were convinced that the reactions observed in biochemistry were in principle identical to those in organic chemistry; and thus they realized the opportunity for an expansion of their field. Two journals were created in 1971 to establish this field: *Bioorganic Chemistry* and *Progress in Bioorganic Chemistry*. In the preface to the first volume of *Progress in Bioorganic Chemistry*, the editors declared that:

> Bioorganic chemistry is a new discipline emerging from the interaction of biochemistry and physical organic chemistry... As with all interdisciplinary sciences, bioorganic chemistry uses many of the methods and techniques of the disciplines from which it is derived; many of its protagonists qualify themselves as physical organic chemists, enzymologists, biochemists or kineticists. It is, however, a new science of its own by the criterion of having developed its own goals, concepts and methods. The principal goal of bioorganic chemistry can be defined as the understanding of biological reactions at the level of organic reaction mechanisms, that is, the identification of the basic parameters which govern these reactions, the formulation of quantitative theories describing them, and the elucidation of relationships between the reactivity and the structures of the molecules participating in the process.[48]

In practice, bioorganic chemistry proved harder to define and for the most part, it has been overshadowed by molecular biology. It was mainly favored by organic chemists with biological interests, notably Eugene van Tamelen (b.1925) at Stanford. The use of the term bioorganic chemistry declined in the 1980s, but has been revived in recent years. The current "Information for Authors" on the website of *Bioorganic Chemistry* defines its subject area as "research [that] either use[s] the principles and techniques of organic and physical organic chemistry in attempting to solve some problem of relevance to biology or... chemical studies that are inspired by some biological observation." [49] A historical overview of the origins of bioorganic chemistry will aid our understanding of how perceptions changed and paradigms shifted, according to the expanding and prevailing state of knowledge, and the backgrounds and interests of participants.

Arising from within physical organic chemistry, bioorganic chemistry shared its intellectual origins. There was a similar focus on mechanisms, structure, and function. However, in bioorganic chemistry, the term mechanism covered both organic reaction mechanisms and the mechanisms of biogenesis, the pathway by which a particular compound was made in nature. Here one sees the influence of Robert Robinson who was a pioneer in both fields. One of Robinson's protégés, and one of the founders of bioorganic chemistry, Alexander (Lord) Todd, emphasized the importance of Robinson's "rationalization of structural relations in the alkaloid field in terms of biogenesis," and his synthesis of tropinone, an alkaloid related to the atropine group of the deadly nightshade, under physiological conditions. His work had a deep impact on the thinking of organic chemists in terms of biogenetical reasoning. [50] This may – for good reasons – be regarded as a classical example of the establishment of a founder figure, designated more than 60 years after the event.

The study of these biogenetic pathways was much assisted by the use of isotopic labeling, and Harold Urey (1893–1987) at Columbia and Martin D. Kamen (b.1913) at Berkeley were both proto-bioorganic chemists. In more recent years, NMR has come to play an important role in both mechanism and structure studies (see section on physical instrumentation). The concept of a relationship between the structure of a compound and its biochemical functioning goes back to Emil Fischer's model of a "lock and key", first formulated in 1894, but many years were to elapse before bioorganic chemists were able to show how the "lock" and "key" fitted together.

The origins of bioorganic chemistry can be traced to the work of German organic chemists on biological molecules around the beginning of the twentieth century. Albrecht Kossel (1853–1927), who had trained in medicine but always did his research in organic chemistry, laid the foundations of nucleotide chemistry in the 1880s and 1890s. He realized that nucleic acids were a combination of purines and pyrimidines with sugars and, although he identified some of the bases, he was unable to determine the structure of the sugars. Emil Fischer turned from the study of carbohydrates to polypeptides and proteins around 1900. He was able to make the first synthetic polypeptides, but his failure to synthesize proteins led him to doubt the existence of very large molecules. Richard Willstätter (1872–1942) was initially a conventional (if brilliant) organic chemist, but as a result of his work on the structure of chlorophyll, he switched to the study of enzyme kinetics in the 1910s. He rejected the idea that enzymes were proteins and insisted that they were small active molecules carried on the surface of larger proteins.

Unfortunately, the German chemists were unable to bridge the gap between organic chemistry and biochemistry; and they also fell behind in the new field of physical organic chemistry. Nevertheless, bioorganic chemistry developed in American and British research groups was sympathetic to the German tradition. These groups were also led by organic or physical chemists interested in physical organic chemistry, kinetics and biogenetic ideas. Between the wars, they typically carried out research on proteins, enzymes, and nucleotides. At Columbia University in New York in the 1930s, there were two independent groups interested in bioorganic chemistry. In the chemistry department, there was the group led by John M. Nelson (1876–1965), who was an inspired teacher and also encouraged chemical physicists, such as Harold Urey, to become involved in this field. The significance of the Professor of Biological Chemistry, Hans Thacher Clarke (1887–1972), is more controversial. He had an unusual background, studying under William Ramsay in London and under Fischer in Berlin before emigrating to the United States to become head of the organic chemicals division at Eastman Kodak. Clarke was highly regarded by James B. Conant, but he was described as a poor lecturer. [51] Crucially, he was both pro-German and pro-Semitic, welcoming exiled Jews from Nazi Germany into his group, notably Konrad Bloch. At Yale, the two chemistry professors Henry L. Wheeler (1867–1914) and his successor Treat B. Johnson (1875–1947) carried out research on nucleic acids and proteins. The physical chemist William M. Clark (1884–1964) was put in charge of biochemistry at Johns Hopkins University at Baltimore in 1927 and he carried out important research on the metalloporphyrins. Clark also appointed a pioneering physical organic chemist, Leslie Hellerman (1896–1971), who was interested in enzyme kinetics, to his group. However, since Clark's group remained small, its impact was limited.

In the immediate post-1945 period, there was a strong growth of bioorganic chemistry on the West Coast of America, in parallel with the development of physical organic chemistry. At Caltech, Linus Pauling, in collaboration with Robert B. Corey, became increasingly interested in the structure of biological molecules and in 1951 they postulated the α-helix structure for proteins. At Berkeley, a bioorganic group was set up within the Radiation Laboratory under the direction of Melvin Calvin (see below). Much later, in 1960, William S. Johnson, a physical organic chemist with strong biological interests, moved from Wisconsin to Stanford and recruited the leading bioorganic chemist Eugene van Tamelen from Wisconsin in 1962. [52]

The situation in Britain was even more diffuse. Robinson, the founder figure of bioorganic chemistry in Britain, took his interest in biogenesis, protein, and enzyme chemistry with him to Oxford in 1929. His main contribution to bioorganic chemistry, however, were his students, in particular Alexander (Lord) Todd and John W. ("Kappa") Cornforth (b.1917). After working on vitamins at Edinburgh University under another proto-bioorganic chemist George Barger (1878–1947), at the Lister Institute in London, and Manchester University, Todd became Professor of Chemistry at Cambridge in 1944. He worked on the structure of vitamin B_{12}, coenzymes, and nucleotides and was awarded the Nobel Prize in 1957. Todd encouraged the use of organic reaction mechanisms and kinetics in organic chemistry, and his friendship with Woodward and Djerassi created an important bridge between British and American chemistry. His students included James Baddiley (b.1918) and George Kenner (1922–1978).

Cornforth by contrast worked at the National Institute for Medical Research at Mill Hill. This was a non-academic research center with a reputation for brilliance and independent thinking. It was established by Sir Henry Dale (1875–1968), a pioneering pharmacologist and future father-in-law of Alexander Todd, in 1914. In its early years, the chemical side was dominated by the steroid chemists Otto Rosenheim (1871–1955) and Harold King (1887–1956). Dale was succeeded in 1942 by (Sir) Charles R. Harington (1897–1972), a brilliant biochemist, who was determined to make the NIMR one of best research institutes in the world. The researchers had no security of tenure and had to make their own equipment for the most part, but in return they were given complete freedom to pursue their own ideas in a very informal atmosphere. Thus James Lovelock, who was hired as an organic chemist (he had studied under Todd at Manchester) became a medical researcher and the inventor of the electron capture detector. Archer Martin (b.1910) developed paper and gas chromatography at the NIMR in the 1940s and 1950s. Cornforth worked with the biochemist George J. Popják (b.1914) on the biosynthesis of cholesterol.

Another independent research institute was the Carlsberg Laboratory in Copenhagen founded in 1875 under the leadership of Johan Kjeldahl (1849–1900). Given Carlsberg's dependence on an enzyme in its beer production, there was an early focus on enzyme kinetics. The laboratory was a pioneer in the application of physical chemistry to biological problems and its second director Søren Sørensen (1868–1939) introduced the concept of pH in 1909. Kai Linderstrøm-Lang (1896–1959) continued the research on enzymes and proteins in the 1920s and 1930s.

There are many similarities between the NIMR and the Rockefeller Institute for Medical Research founded in New York in 1901. The Rockefeller played a key role in the development of modern bioorganic chemistry (and also molecular biology). It was pro-European in its outlook, strongly influenced by organic and physical chemistry, and shared with the NIMR a belief in the freedom of research. The first chemical research leader was Phoebus A. Levene (1869–1940), a Russian emigrant who had studied (briefly) under Kossel and Fischer.[53] Although his education background was largely medical, during his long career at the Rockefeller he concentrated on the organic chemistry of the nucleic acids and succeeded where Kossel failed, showing that the sugars involved were ribose and deoxyribose. His achievements were however overshadowed by his failure to determine the structure of DNA. The Rockefeller also combined biology with physical chemistry, setting up one group under Duncan MacInnes (1885–1965) in 1926, and Leonor Michaelis's (1875–1949) famous group was established three years later. Wendell M. Stanley (1904–1971) joined the Rockefeller in 1931 and within a few years had crystallized the tobacco mosaic virus, showing that viruses were essentially chemical compounds. Unfortunately, he thought that they were simple proteins (rather than a compound of a protein and RNA). In 1934, the institute received a major impetus with the arrival of Max Bergmann (1886–1944), a leading student of Fischer who had fled Nazi Germany. Bergmann pioneered the use of proteolytic enzymes to study proteins and his students William H. Stein (1911–1980) and Stanford Moore

(1913–1982) used this approach to determine the structure of the enzyme ribonuclease A in 1963.

Of all the researchers at Rockefeller in the 1920s and 1930s, perhaps John H. Northrop (1891–1987), who had studied under Nelson at Columbia, was the closest to being a bioorganic chemist. In 1920, he had isolated pepsin but, failing to crystallize it, he abandoned this line of research for the time being. Six years later, James B. Sumner (1887–1955) at Cornell crystallized urease, but Willstätter and his supporters objected to Sumner's categorization of urease as a protein. Northrop decided to crystallize pepsin and then subject it to a battery of tests to show that it was a pure protein but with the same activity as the enzyme.[54] By 1929, he had crystallized pepsin and began his tests, which ranged from repeated crystallization, rate of diffusion, and rate of sedimentation in the ultracentrifuge. From the point of view of bioorganic chemistry, the most interesting of Northrop's work was the measurement of the rate of inactivation with alkali and the rate of hydrolysis with acid. His group also measured the solubility curves of pepsin in different salt solutions and its denaturation by ultraviolet light using the absorption maximum of pepsin to measure its concentration. All these tests showed that the crystalline pepsin was a pure compound possessing all the activity of the enzyme, thus demolishing Willstätter's concept of enzymes being small molecules, despite a last ditch attempt by his supporters to show pepsin's activity could be transferred to melon seed protein. Subsequently, Northrop's co-workers crystallized trypsin, chymotrypsin, ribonuclease, deoxyribonuclease, hexokinase, and pyrophosphatase, and demonstrated in the same manner that they were also pure proteins.

One of the most striking successes in the course of events that may be qualified as belonging to bioorganic chemistry was the elucidation of the carbon pathway in photosynthesis by Melvin Calvin and his collaborators in the years from 1946 to 1956.[55] Moreover, the paths that led to Calvin's discovery shed light on the interdisciplinary nature of bioorganic research. Calvin found himself "working at some time or another in atomic physics, physical chemistry, organic synthesis, cellular biochemistry, neurochemistry, plant molecular biology, chemical evolution and organic geochemistry, biophysics and animal behavior."[56] What held this impressive range of scientific activities together? Calvin's interest in coordination chemistry, which began in the mid 1930s with a project on phthalocyanine dyes in Michael Polanyi's (1891–1976) group at the University of Manchester,[57] led him via chlorophyll to the study of photosynthesis and guided him also through a large part of his other research topics. In the late 1930s and early 1940s, at the University of California at Berkeley, these topics were mainly theoretical aspects of organic structure, and a major outcome was his authoring, together with Branch, of *The Theory of Organic Chemistry* (see Section 1.1).

Through war work on extraction methods for plutonium and uranium, Calvin became acquainted with Ernest O. Lawrence (1901–1958), head of the Radiation Laboratory at Berkeley which played an important role in the initial production of plutonium. On his suggestion, a bioorganic group was founded at the Radiation Laboratory under Calvin's direction. This may have been the first use of the term bioorganic in an official way. The initial task for Calvin's group was to use the supply of radioactive carbon-14 for the study of organic reaction mechanisms and its potential use in cancer therapy. Thus from its very beginning, Calvin's group was a truly interdisciplinary one, bridging the gap from atomic physics to medicinal chemistry. [58]

Calvin literally inherited a vial of barium carbon-14-carbonate from one of its discoverers, Samuel Ruben (1913–1943). Together with Kamen, Ruben had already set out to investigate the first steps of photosynthesis, but the minute amounts of carbon-14 at their disposal in 1940 were of no use for that. In 1945, when nuclear reactors produced larger amounts of this long-lived isotope, the situation had changed. It proved to be the ideal technique for tracing the path of carbon dioxide to the carbohydrates produced by green plants. At the beginning of their work, Calvin and his colleagues used classical chemical methods for the isolation and identification of the unknown compounds. But they soon turned to the new paper chromatography recently developed by Archer Martin and Richard Synge (1914–1994). Combining this with the radioisotopic tracer method, they were able to identify even minuscule amounts of the carbohydrates they were interested in, using also fluorescence and ultraviolet absorption spectroscopy. [59]

Photosynthesis is comprised of two distinct phases, the dark reactions and the light reactions. This had been assumed already in 1906 by research of Frederic Frost Blackman (1866–1947) and subsequent interpretation by Otto Warburg (1883–1970). Though Calvin did work on both dark and light reactions, he was most successful with the elucidation of the reactions occurring in the dark, which bring about the synthesis of carbohydrates from carbon dioxide and water. The most pressing problem at the beginning of Calvin's research was the question about the primary product of the assimilation of carbon dioxide in the plant. Calvin and his team were able to establish that this assimilation product was a compound long known to biochemists, phosphoglyceric acid, a degradation product of sugar. In a year-long search, they identified the nature of the acceptor molecule of carbon dioxide in the plant as a phosphate of a five-carbon sugar, ribulose. Starting with phosphoglyceric acid, the plant synthesizes the sugars fructose and glucose, which are the building blocks of more complex carbohydrates. In a cyclic reaction pathway, the sugar ribulose is also synthesized and converted into phosphoglyceric acid and this starts the photosynthetic carbon reduction cycle again.

The energy-rich compounds which drive the dark reactions are derived from the light reaction of photosynthesis, which takes place in the chloroplasts of green plants. The mechanism for the formation of adenosin triphosphate (the most important of these compounds) was discovered by Peter Mitchell (1920–1992) in the early 1960s. The structure of the membrane-bound photosynthetic reaction center was established through X-ray crystallography by Hartmut Michel (b.1948), Johann Deisenhofer (b.1943), and Robert Huber (b.1937) between 1982 and 1985. In a typical endeavor of bioorganic chemistry, Calvin's group in the 1970s set out to mimic nature by synthetic chloroplasts or artificial photosynthetic systems. [60]

Originally, the staff of the bioorganic chemistry group of the Radiation Laboratory at Berkeley was joined by six senior chemists and decisions were made informally during weekly meetings. In its early stage, the group consisted of sub-groups dealing with carbon-14 synthetic chemistry, animal biochemistry, and photosynthesis itself. Later the group's name was changed to the Laboratory of Chemical Biodynamics of the University of California at Berkeley; and it was joined by members of the faculties of chemistry and botany. The strong interdisciplinary nature of its research was always mirrored by the openness of the laboratory space, and a very informal discussion culture. In 1964, the laboratory moved to a building specially designed for the purpose of free flow of information and ideas. All working areas in the circular building (now the Melvin Calvin Laboratory) radiate outward from the center occupied by a coffee table. The funding agencies show the multifarious aspects of Calvin's work. Among others, they include the Atomic Energy Commission, the National Institutes of Health, and private sources such as the Charles Kettering Foundation. [61]

A multitude of scientists, disciplines, and techniques were involved in the extended investigation of the biological synthesis of cholesterol. Cholesterol, first found in the eighteenth century in gallstones, is the precursor of all steroidal hormones in the body, and infamously known too as being involved in atheriosclerosis. Based on X-ray analysis of a related compound (ergosterol) by John Desmond ("Sage") Bernal (1901–1971), Otto Rosenheim, and Harold King (and independently Heinrich Wieland (1877–1957) and Elisabeth Dane (1903–1984)) in 1932 formulated its structure. Its biosynthesis remained unclear until the mid 1950s. In 1926, Harold J. Channon (1897–1979) had shown that if squalene (a triterpene first isolated in 1916 from shark liver oil) was fed to rats, this led to an increase in the content of cholesterol in rat tissue. Thus, a first hint was given for the synthesis of a sterol (which contains three six-membered rings and one five-membered) from a long-chain hydrocarbon. After the structure of squalene had been proved through its synthesis by Paul Karrer (1889–1971) in 1931, Robinson pointed out that the carbon skeleton of cholesterol was very similar to that of squalene, also proposing a mechanism for its conversion in the tetracyclic sterol.

But as Robinson pointed out in 1955, "the comparison of structures *per se* gives no information about the details of the mechanisms of biosynthetic reactions. It is the task of the biochemist to determine these by appropriate experiment." [62]

The appropriate experiment for the study of the pathway leading to cholesterol came into existence with the radioisotopic tracer technique developed by George de Hevesy (1885–1960) in the 1920s and 1930s. This method was used to great benefit by Rudolph Schoenheimer (1898–1941) at Columbia University in New York. After the carbon-14 isotope had become available, Konrad Bloch (1912–2000) in Schoenheimer's group and later at the University of Chicago was able to show that all of the carbon atoms contained in cholesterol were supplied by acetic acid (which contains only two carbon atoms). In subsequent work, Bloch and co-workers showed how acetic acid was converted into the 30-carbon squalene. In 1953, Woodward suggested a steroidal intermediate, lanosterol, in the folding process leading to squalene and this was proved by isotopic studies undertaken by Bloch. Finally, an unexpected precursor of squalene, the six carbon mevalonic acid, was identified in 1956 by Karl Folkers during vitamin research unrelated to cholesterol. [63]

The stereospecifity of the reactions sequences was investigated by John W.

Cornforth and George J. Popják at the National Institute for Medical Research, and from 1962 on at Shell's Milstead Laboratory of Chemical Enzymology. They used specifically labeled mevalonic acid molecules to establish the molecular rearrangement in detail. Because the synthesis of squalene needs 14 steps, each governed by stereospecific enzymes, there are 2^{14} (16 384) different routes; and only one leads to squalene.[64] Cornforth's research was impressive proof both of the astonishing specificity of enzymatic action and of the power of the meticulously used tracer technique in bioorganic chemistry. In the 1960s and 1970s, William S. Johnson synthesized two- and three-ring steroidal compounds from polyenes, using acetal initiators, thus mimicking the biological pathway.

In his presentation speech for the Nobel Prize Winner in chemistry for 1957, Alexander Todd, Arne Fredga, member of the Nobel Committee for Chemistry of the Royal Swedish Academy of Science, described Todd's work on nucleotides, chemical compounds that had been first found in the nuclei of cells. In the middle of the Cold War, Fredga felt the necessity to make clear that Todd's work on nucleotides had nothing to do with atomic nuclei, nuclear fission, and the hydrogen bomb.[65] Todd's prize-winning work was the elucidation of the structure of the building blocks of deoxyribonucleic acid (DNA) and ribonucleic acid (RNA) and their artificial synthesis. Combined with proteins, nucleotides are also contained in virus molecules and as coenzymes play a crucial role in metabolic processes. The energy-rich compound adenosine triphosphate (ATP) driving the photosynthetic cycle is one of many examples for this. What scientists hoped for, but nobody knew for sure at that moment, was the tremendous impact the unraveling of the structure of the building molecules of the hereditary material would have on the transformation of science and society.

According to Todd, the definition of organic chemistry was best given by the leading chemist of the first half of the nineteenth century, Jöns Jacob Berzelius (1779–1848), who saw organic chemistry as "the chemistry of substances found in living matter."[66] This seeming anachronism nicely underscores the transformation of organic chemistry in the second half of the nineteenth and the first half of the twentieth centuries, when most of organic chemistry dealt with artificial synthesis of carbon-containing compounds. In the 1950s, when the intellectual roots of bioorganic chemistry were formed, this approach was already too restrictive, at least in Todd's eyes. Moreover, though trained in synthetic organic chemistry, Todd took up Robinson's notion of synthesis as being an equal counterpart of degradation reactions in order to unravel the structure of an unknown molecule. In addition, artificial synthesis of biologically active compounds could lead to further insights in their functions and significance.

Todd's research provided the knowledge of how the nucleotides are linked in DNA and RNA; and thus it belongs to the basis of James D. Watson's (b.1928) and Francis H. Crick's (b.1916) work on the structure of DNA, the double helix. In addition, his work on coenzymes opened new vistas for the understanding and manipulation of many biochemical processes and thus contributed fundamentally to concepts that soon would become of use in medicinal chemistry.

By crossing the boundary into medicinal chemistry, bioorganic chemistry merged academic and industrial interests in a way not unlike that achieved a century earlier. Indeed, several chemical manufacturers divested themselves of traditional lines and concentrated on developments in life sciences, with the main emphasis on animal and human health and agricultural products. Moreover, the technical and scientific sides of chemistry cannot be separated easily in the twentieth century.[67] A particular striking example is the history of steroids, such as cortisone, with the major participation of Edward C. Kendall (1886–1972, who cooperated with Merck), Woodward (who consulted for Monsanto), and Djerassi (who was employed by Syntex). In addition, the chemistry of steroids was shaped both by fundamental research such as the conformational analysis by Derek Barton (1918–1998) and Odd Hassel (1897–1981) and by tactical, industrial syntheses of steroidal compounds not found in nature.[68]

One of the major goals of bioorganic chemistry is the elucidation of the synthetic pathways of compounds in living systems. With the help of radioisotopic methods, chemists were able to test the theories of biogenesis. Important contributions, such as the work of Konrad Bloch and Feodor Lynen (1911–1979) on the mechanisms and regulations of the cholesterol and fatty acid metabolisms, were rewarded with the Nobel Prize. Furthermore, they revealed an underlying order in nature, much to the pleasure and excitement of the scientists involved. The natural order was recreated in the laboratory of the bioorganic chemists through the synthesis of natural products and systematic testing of biochemical pathways.[69] This attempted unification of the natural and the artificial world certainly underlined the scientific realism upheld by many scientists at mid-century.

A second aim of many bioorganic chemists was the conceptual unification of the two mother disciplines, biology and organic chemistry. The major obstacle was the seeming incompatibility of the two central theories in their respective fields, evolutionary theory in biology and structural theory in organic chemistry. This incompatibility, according to the organic chemists Steven A. Benner (b.1954) and Andrew D. Ellington (b.1959), was firmly rooted in different research traditions, uses of language, and problem-solving strategies. Furthermore, chemists with their preference for rigorous, serious laboratory experiments considered evolutionary biology as a "soft" and intellectually disappointing field. But in 1990, with the help of the concepts and techniques of bioorganic chemistry and the bountiful data containing structural information of biological macromolecules, this obstacle seemed to be surmountable. Therefore, the evolutionary picture of macromolecular chemistry was, according to Benner and Ellington, close at hand and could serve as a coherent model for the behavior of biomolecules and the design of experiments. Although bioorganic chemists did not succeed in their attempts of complete conceptual unification of biology and chemistry, the strategy was useful for protein engineering, the prediction of tertiary structures of proteins, and the organization and interpretation of sequence data.[70]

The foundation journals *Bioorganic Chemistry* and *Progress in Bioorganic Chemistry* were joined two decades later by two journals whose titles demonstrated the direction in which the discipline was heading: *Bioorganic and Medicinal Letters* (1991) and *Bioorganic and Medicinal Chemistry* (1993). Understanding of structure-function relations, metabolic processes, molecular and cellular recognition, and the reproduction of life was achieved with the help of modern techniques for structure elucidation and mechanistic considerations. The manipulation of biological macromolecules and their design based on biomimetic principles became the basic tools for this endeavor. The wheel had turned full circle and bioorganic chemistry returned to its nineteenth-century roots in medicinal chemistry. At the same time, it has partly filled the gap left by the disappearance of traditional structure studies in organic chemistry.

1.4
Conclusion

Historically, the three scientific specialities dealt with in this chapter, physical organic chemistry, physical instrumentation, and bioorganic chemistry were strongly interrelated, and there was a steady flow of concepts, methods, ideas, and people between them. The emergence of physical instrumentation, NMR in particular, had a powerful influence on the development of physical organic chemistry and bioorganic chemistry. Furthermore, bioorganic chemistry grew out of physical organic chemistry and the closely related field of enzyme kinetics. Between them, these three specialities have transformed organic chemistry. The introduction of physical instrumentation in the 1950s and 1960s destroyed the traditional field of structure studies, which had hitherto been the mainstay of organic chemistry. One of the effects of the instrumental revolution in organic chemistry was a shift towards interdisciplinary studies. Freed from time-consuming structural determinations by degradation and synthesis, organic chemists could concentrate on biomolecular themes, and reaction mechanisms. Physical organic chemistry remained within organic chemistry, and at the same time, it has transformed the older discipline completely. By partly filling the vacuum left by the collapse of classical structural organic chemistry, the expansion of physical organic chemistry maintained the prestige of organic chemistry. In effect, the old name has been retained for a completely new hybrid of organic synthesis, physical organic chemistry, and bioorganic chemistry.

References and Notes

1 For general accounts see W. H. Brock, *The Fontana History of Chemistry* (London: Fontana Press, 1992), 508–569, and references therein; M. J. Nye, *From Chemical Philosophy to Theoretical Chemistry. Dynamics of Matter and Dynamics of Disciplines* (Berkeley: University of California Press, 1993), 139–223; R. E. Kohler, "The Lewis-Langmuir theory of valence and the chemical community, 1920–1928," *Historical Studies in the Physical Sciences* 6 (1975): 431–468; papers by C. A. Russell, M. D. Saltzman, and J. Shorter in a special "Robert Robinson" issue of *Natural Product Reports* 4 (1987): 1–88; M. D. Saltzman, "The development of physical organic chemistry in the United States and the United Kingdom: 1919–1939, parallels and contrasts"

Journal of Chemical Education 63 (1986): 588–593; L. Gortler, "The physical organic chemistry community in the United States, 1925–50," *Journal of Chemical Education* 62 (1985): 753–757; T.M. Brown, A.T. Dronsfield, and P.J.T. Morris, "Who really invented the curly arrow?" *Education in Chemistry*, in press; T.M. Brown, A.T. Dronsfield, and P.J.T. Morris, "A reaction which changed organic chemistry" *Education in Chemistry*, forthcoming. For first hand accounts, see the autobiographies of: D.J. Cram, *From Design to Discovery* (Washington D.C.: American Chemical Society, 1990); M.J.S. Dewar, *A Semiempirical Life* (Washington D.C.: American Chemical Society, 1992); E. Havinga, *Enjoying Organic Chemistry, 1927–1987* (Washington D.C.: American Chemical Society, 1990); R. Huisgen, *The Adventure Playground of Mechanisms and Novel Reactions* (Washington D.C.: American Chemical Society, 1994); W.S. Johnson, *A Fifty-Year Love Affair with Organic Chemistry* (Washington D.C.: American Chemical Society, 1998); J.D. Roberts, *The Right Place at the Right Time* (Washington D.C.: American Chemical Society, 1990); A. Streitwieser, *A Lifetime of Synergy with Theory and Experiment* (Washington D.C.: American Chemical Society, 1996); C. Walling, *Fifty Years of Free Radicals* (Washington D.C.: American Chemical Society, 1995).

2 On Ingold see Û‰K.T. Leffek, *Sir Christopher Ingold – A Major Prophet of Organic Chemistry* (British Columbia: Nova Lion Press, 1996); Û‰M.D. Saltzman, "The Robinson-Ingold controversy. Precedence in the electronic theory of organic reactions" *Journal of Chemical Education* 57 (1980): 753–757; Nye, *From Chemical Philosophy to Theoretical Chemistry*, 196–223.

3 For Staudinger and the origins of polymer chemistry see the chapter of Yasu Furukawa in this volume.

4 W.A. Waters, *Physical Aspects of Organic Chemistry* (London: G. Routledge & Sons, 1935).

5 H.B. Watson, *Modern Theories of Organic Chemistry* (Oxford: The Clarendon Press, 1937).

6 L.P. Hammett, *Physical Organic Chemistry: Reaction Rates, Equilibria and Mechanisms* (New York ; London: McGraw-Hill, 1940).

7 G.E.K. Branch, M. Calvin, *The Theory of Organic Chemistry: An Advanced Course* (New York: Prentice-Hall, Inc., 1941).

8 L.C. Pauling, *The Nature of the Chemical Bond and the Structure of Molecules and Crystals* (Ithaca: Cornell University Press, 1939)

9 G.W. Wheland, *Theory of Resonance and Its Application to Organic Chemistry* (New York: J. Wiley and Sons; London: Chapman and Hall, 1944). The Pauling-Wheland controversy on the ontological status of resonance is described in the contribution of K. Gavroglu and A. Simões in this volume.

10 E.R. Alexander, *Principles of Ionic Organic Reactions* (New York: J. Wiley and Sons, Inc; London: Chapman and Hall, 1950).

11 A brief account of the earlier history of the molecular orbital theory and the resonance theory of chemical bonding is given in the contribution of K. Gavroglu and A. Simões in this volume.

12 See S. Weininger, "'What's in a name?' From designation to denunciation – the nonclassical cation controversy," *Bulletin for the History of Chemistry* 25 (2000): 123–131 and references therein; C. Walling, "An innocent bystander looks at the 2-norbornyl cation," *Accounts of Chemical Research* 16 (1983): 448–454.

13 J.A. Berson, "Erich Hückel, pioneer of organic quantum chemistry: reflections on theory and experiment," *Angewandte Chemie International Edition in English* 35 (1996): 2750–2764; A. Karachalios, "Die Entstehung und Entwicklung der Quantenchemie in Deutschland," *Mitteilungen der Gesellschaft Deutscher Chemiker Fachgruppe Geschichte der Chemie* 13 (1997): 163–179.

14 C.K. Ingold, *Structure and Mechanism in Organic Chemistry* (London: Bell; Ithaca: Cornell University Press, 1953).

15 *Physical Organic Chemistry* (1956) by Jack Hine; *Organic Chemistry: Electronic Theory and Reaction Mechanism* (1956) by Rowland I. Reed, a pioneer mass spectroscopist at the University of Glasgow; *Electronic Theories of Organic Chemistry: An Introductory Treatment* (1958) by J.W. Baker, best known for the Baker-Nathan effect; and *Mechanism and Structure in Organic Chemistry* (1959) by Edwin S. Gould, who had taken his Ph.D. at the University of California at Los Angeles.

16 H. J. Lucas, *Organic Chemistry* (New York: American Book Company, 1935).

17 R. T. Morrison, R. N. Boyd, *Organic Chemistry* (Boston: Allyn and Bacon, 1959); D. J. Cram, G. S. Hammond, *Organic Chemistry* (New York: McGraw-Hill, 1959).

18 J. D. Roberts, M. C. Caserio, *Basic Principles of Organic Chemistry* (New York : W. A. Benjamin, 1964).

19 I. L. Finar, *Organic Chemistry*, vol. 1, *Fundamental Principles*, vol. 2, *Stereochemistry and the Chemistry of Natural Products* (First Edition, London: Longmans, 1951, 1956; Sixth Edition, vol. 1, Fifth Edition, vol. 2, London: Longmans, 1973, 1975).

20 P. Sykes, *Guidebook to Mechanism in Organic Chemistry* (London: Longmans, 1961).

21 J. M. Tedder, A. Nechvatal, *Basic Organic Chemistry: A Mechanistic Approach* (London: Wiley, 1966); four more volumes (with different co-authors) were published up to 1975.

22 P. J. T. Morris, A. S. Travis, "The role of physical instrumentation in structural organic chemistry," in J. Krige, D. Pestre (eds.), *Science in the Twentieth Century* (Amsterdam: Harwood Academic Publishers, 1997), 715–739; D. Baird, "Analytical chemistry and the 'big' scientific instrumentation revolution," *Annals of Science* 50 (1993): 267–90; D. S. Tarbell, A. T. Tarbell, *Essays on the History of Organic Chemistry in the United States, 1875–1955* (Nashville: Folio Publishers, 1986), 335–352; H. A. Laitinen, G. W. Ewing (eds.), *A History of Analytical Chemistry* (Washington D. C.: Division of Analytical Chemistry, American Chemical Society, 1977); M. E. Bowden, T. Benfey, *Robert Burns Woodward and the Art of Organic Synthesis* (Philadelphia: Beckman Center for the History of Chemistry, 1992); J. K. Stine, "Scientific instrumentation as an element of U. S. science policy: National Science Foundation support for chemistry instrumentation," in R. Bud and S. E. Cozzens (eds.), *Invisible Connections: Instruments, Institutions, and Science* (Bellingham: SPIE Optical Engineering Press, 1992), 238–263. For first hand accounts, see the autobiographies of: A. J. Birch, *To See the Obvious* (Washington D. C.: American Chemical Society, 1995); C. Djerassi, *Steroids Made It Possible* (Washington D. C.: American Chemical Society, 1990); R. U. Lemieux, *Explorations with Sugars: How Sweet It Was* (Washington D. C.: American Chemical Society, 1990); J. D. Roberts, *The Right Place at the Right Time* (Washington D. C.: American Chemical Society, 1990).

23 J. Hunt, *The Ascent of Everest* (London: Hodder and Stoughton, 1953), 7.

24 C. Djerassi, *The Pill, Pygmy Chimps, and Degas' Horse: The Autobiography of Carl Djerassi* (New York: Basic Books, 1992), 104.

25 L. Bragg in D. C. Phillips and H. Lipson (eds.), *The Development of X-ray Analysis* (London: G. Bell & Sons, 1975), 176–193; J. P. Glusker, "Brief history of chemical crystallography. II: organic compounds," in J. Lima-de Faria, *Historical Atlas of Crystallography* (Dordrecht: International Union of Crystallography, 1990); P. P. Ewald (ed.), *Fifty Years of X-Ray Diffraction* (Utrecht: International Union of Crystallography, 1962).

26 D. Crowfoot Hodgkin, "The X-ray analysis of complicated molecules," *Science* 150 (1965): 979–988, on 983.

27 See papers on the history of infrared spectroscopy by F. A. Miller, "The infrastructure of IR spectrometry. Reminiscences of pioneers and early commercial IR instruments, " *Analytical Chemistry* 64 (1992): 824A-831A; P. A. Wilks, Jr., "The evolution of commercial IR spectrometers and the people who made it happen," *ibid.*, 833A-838A; P. R. Griffiths, "Strong-men, Connesmen and Block-Busters or how Mertz raised the Hertz," [sic] *ibid.*, 869A-875A; N. Sheppard, "The U. K.'s contributions to IR spectroscopic instrumentation: from wartime fuel research to a major technique for chemical analysis," *ibid.*, 877A-883A; Y. M. Rabkin, "Technological innovation in science: the adoption of infrared spectroscopy by chemists," *Isis* 78 (1988): 31–54; R. N. Jones, "Analytical applications of vibrational spectroscopy: a historical review," in J. R. Durig (ed.), *Chemical, Biological and Industrial Applications of Infrared Spectroscopy* (Chichester: Wiley, 1985), 1–50.

28 See for example E. I. Stearns, "Applications of ultraviolet and visible spectrophotometric data," in M. G. Mellon (ed.), *Analytical Absorption Spectroscopy: Absorptimetry and Colorimetry* (New York: Wiley, 1950), 306–438. For Barnes and the American

Cyanamid-Perkin-Elmer connection see Rabkin's important study, "Technological innovation in science," on 38 and 44.

29 Typical review articles by American Cyanamid Stamford chemists include R. C. Gore, "Infrared spectroscopy," *Analytical Chemistry* 30 (1958): 570–579, and R. C. Hirt, "Ultraviolet spectrophotometry," *Analytical Chemistry* 30 (1958): 589–593.

30 R. C. Hirt and F. T. King, "Use of micrometer baly cells with Beckman and Cary ultraviolet spectrophotometers," *Analytical Chemistry* 24 (1952): 1545–1548. For American Cyanamid Stamford see R. P. Chapman, "Organisation and functions of an analytical and testing group," *Chemical Industries* 65 (1949): 718–721.

31 R. B. Woodward, "Structure and the absorption spectra of unsaturated ketones" *Journal of the American Chemical Society* 63 (1941): 1125.

32 J. H. Beynon, R. P. Morgan, "The development of mass spectrometry. An historical account," *International Journal of Mass Spectrometry and Ion Processes* 27 (1978): 1–30; H. Remane, "Zur Entwicklung der Massenspektroskopie von den Anfängen bis zur Strukturaufklärung organischer Verbindungen," *N. T. M.* 24,2 (1987), 93–106.

33 See H. W. Washburn, H. F. Wiley, S. M. Rock, "The mass spectrometer as an analytical tool," *Industrial and Engineering Chemistry (Analytical Edition)* 15 (1943): 541–47.

34 K. Biemann, Applications of mass spectrometry, in K. W. Bentley (ed.), *Elucidation of Structures by Physical and Chemical Methods*, Part one (New York: Interscience Publishers, 1963), 259–316, on 260.

35 K. Biemann, "The Massachusetts Institute of Technology mass spectrometry school," *Journal of the American Society for Mass Spectrometry* 5 (1994): 332–338; K. Biemann, "The coming of age of mass spectrometry in peptide and protein chemistry," *Protein Science* 4 (1995): 1920–1927.

36 H. Budzikiewicz, C. Djerassi, D. H. Williams, *Interpretation of Mass Spectra of Organic Compounds* (San Francisco: Holden-Day, 1964); *Structure Elucidation of Natural Products by Mass Spectrometry*, 2 vols (San Francisco: Holden-Day, 1964).

37 J. D. Roberts, *The Right Place at The Right Time*, 154.

38 T. Lenoir, C. Lécuyer, "Instrument makers and discipline builders: the case of nuclear magnetic resonance," *Perspectives on Science* 3 (1995): 276–345; E. D. Becker (ed.), *Encyclopedia of Nuclear Magnetic Resonance*, volume one, *Historical Perspectives* (Chichester: Wiley, 1996); J. Feeney, "Development of high resolution NMR spectroscopy as a structural tool," in Bud, Cozzens, *Invisible Connections*, 199–216.

39 J. D. Roberts, *Nuclear Magnetic Resonance: Applications to Organic Chemistry* (New York: McGraw-Hill, 1959); *id.*, *Introduction to the Analysis of Spin-Spin Splitting in High-Resolution Nuclear Magnetic Resonance Spectra* (New York: W. A. Benjamin, 1961).

40 J. A. Pople, W. G. Schneider, H. J. Bernstein, *High Resolution Nuclear Magnetic Spectroscopy* (New York: McGraw-Hill, 1959).

41 K. Biemann, *Mass Spectrometry. Organic Chemical Applications* (New York: McGraw-Hill, 1962).

42 G. M. Barrow, *Introduction to Molecular Spectroscopy* (New York: McGraw-Hill, 1962).

43 E. A. Braude, F. C. Nachod, *Determination of Organic Structures by Physical Methods* (volume 1, New York: Academic Press, 1955; volume 2, New York: Academic Press, 1962).

44 K. W. Bentley (ed.), *Elucidation of Structures by Physical and Chemical Methods*, volume 11 of A. Weissberger (ed.), *Technique of Organic Chemistry*, in two parts (New York, London: Interscience, 1963).

45 J. C. P. Schwarz (ed.), *Physical Methods in Organic Chemistry* (Edinburgh: Oliver & Boyd, 1964).

46 D. H. Williams, I. Fleming, *Spectroscopic Methods in Organic Chemistry* (London: McGraw-Hill, 1966).

47 J. S. Fruton, *Proteins, Enzymes, Genes: The Interplay of Chemistry and Biology* (New Haven and London: Yale University Press, 1999); R. Olby, *The Path to the Double Helix: The Discovery of DNA*, enlarged edition (New York: Dover Publications, 1994); R. E. Kohler, *From Medical Chemistry to Biochemistry: The Making of a Biomedical Discipline* (Cambridge: Cambridge University Press, 1982).

48 E. T. Kaiser, F. J. Kézdy (eds.), *Progress in Bioorganic Chemistry*, vol. 1 (New York: Wiley-Interscience 1971), vii.

49 http://academicpress.com/www/journal/bh/bhifa.htm. Accessed on 13 November 2000.

50 A. R. Todd, "Dedication," *Bioorganic Chemistry* 1 (1971): vii.

51 J. S. Fruton, *Proteins, Enzymes, Genes*, 94.

52 For Johnson see his autobiography, *A Fifty-Year Love Affair with Organic Chemistry* (Washington D. C.: American Chemical Society, 1998).

53 G. K. Hunter, "Phoebus Levene and the tetranucleotide structure of nucleic acids," *Ambix* 46 (1999): 73–103.

54 R. M. Herriott, "John Howard Northrop," in *Biographical Memoirs of the National Academy of Sciences of the United States of America* (Washington D. C.: National Academy Press, 1994), 431–433; J. H. Northrop, *Crystalline Enzymes: The Chemistry of Pepsin, Trypsin, and Bacteriophage* (New York: Columbia University Press, 1929), 26–38.

55 On Calvin see G. B. Kauffman, I. Mayo, "Multidisciplinary scientist–Melvin Calvin. His life and work," *Journal of Chemical Education* 73 (1996): 412–416.

56 Melvin Calvin, *Following the Trail of Light. A Scientific Odyssey* (Washington D. C.: American Chemical Society, 1992), 134.

57 On Polanyi see the contribution of Mary Jo Nye in this volume.

58 Calvin, *Following the Trail of Light*, 51–56.

59 M. Calvin, "The path of carbon in photosynthesis," in Nobel Foundation, *Nobel Lectures. Chemistry 1942–1962* (Amsterdam: Elsevier, 1964), 619–644.

60 Calvin, *Following the Trail of Light*, 96–101.

61 Calvin, *Following the Trail of Light*, 102–110.

62 R. Robinson, *The Structural Relations of Natural Products* (Oxford: Clarendon Press, 1955), 1, cited after J. S. Fruton, *Molecules and Life. Historical Essays on the Interplay of Chemistry and Biology* (New York: Wiley-Interscience, 1972), 470.

63 See Fruton, *Molecules and Life*, 470–474.

64 A. Fredga, "The Nobel Prize for Chemistry," in T. Frängsmyr, S. Forsén (eds.), *Nobel Lectures. Chemistry 1971–1980* (Singapore: World Scientific, 1993), 181–183; J. W. Cornforth, "Asymmetry and enzyme action," in Frängsmyr, Forsén, *Nobel Lectures*, 187–198.

65 A. Fredga, "Chemistry 1957," in Nobel Foundation, *Nobel Lectures. Chemistry 1942–1962* (Amsterdam: Elsevier, 1964), 519–521. See also Todd's autobiography, *A Time to Remember: The Autobiography of a Chemist* (Cambridge: Cambridge University Press, 1983).

66 Alexander Todd, "Synthesis in the study of nucleotides," in Nobel Foundation, *Nobel Lectures. Chemistry 1942–1962* (Amsterdam: Elsevier, 1964), 522–536, on 522.

67 See, with special emphasis on cortisone and other biomolecules, the paper of Nicolas Rasmussen in this volume.

68 For a discussion of the terms *fundamental* and *tactical* research in the context of the history of steroid chemistry see L. Slater, "Industry and academy: The synthesis of steroids," *Historical Studies in the Physical Sciences* 30 (2000): 443–480.

69 Slater, "Industry and academy," 479–480.

70 S. A. Benner, A. D. Ellington, "Evolution and structure theory: the frontier between chemistry and biology," in H. Dugas (ed.), *Bioorganic Chemistry Frontiers*, vol. 1 (Berlin: Springer, 1990), 1–70.

Part I
Theoretical Chemistry and Quantum Chemistry

2.
Theoretical Quantum Chemistry as Science and Discipline: Some Philosophical Remarks on a Historical Issue

Nikos Psarros

2.1
The Quarrel of the Faculties

At first glance, the methodological differences between a philosophical and a historical reflection on science appear to be overwhelming and insurmountable. There is, it seems, no possibility for communication between a historian and a philosopher of science. Some people would be content with such a state of affairs according to the maxim: Do not meddle with others' business. The claims are marked out and everybody has to dig in his own place for gold. However, I think that – notwithstanding academic freedom – as members of the academic republic we have neither the right nor the possibility to act in this manner.

We do not have the right to do so, because as scholars and teachers we are committed to a public educational and orienting function the aim of which is to produce a coherent view of the world. By saying this, I do not put myself in the tradition of Unified Science. Nevertheless, I think that our scientific knowledge about the world should be systematized in such a manner that enables the recognition and evaluation of the relative position of its faculties, even by the non-professional. I think this is the only way to render scientific knowledge useable by the non-scientific public. As scholars and academic teachers, we are committed to educate people who can be integrated into a highly co-operative society of specialists. How can we fulfil this duty successfully, unless we co-operate and communicate across the boundaries of our faculties? We can refer to and defend our academic freedom only as long as we use it with sanity, that is (among others), as long as we try to fulfil our academic and educational responsibilities. I cannot imagine that society would tolerate for long mutually tacit scholars who do not display a trace of understanding each other.

We do not have the possibility to refrain from interdisciplinary communication for another reason. In the normal mode of practicing our own particular science, we already use terms, concepts, and statements, and we also take points of view that stem from other scientific provinces. Doing this, we automatically participate in discourses about validity and standards of adequacy that take place in the fields where these notions originate. This means in our case that both the philosopher and the historian of science may – or sometimes must – formulate and defend a historical program and a philosophical idea, respectively.

On the background of those reflections, I would like to address some systematic and philosophical aspects in the papers of the first part of this book. I would like to use a principle of characterization that can be described as situating the papers within the framework of the history of *science*, or within the frame of the history of the *discipline* of theoretical quantum chemistry. Here a first terminological distinction has to be made, namely that between science and discipline. I understand *science* as an ensemble of true statements about a set of objects that are arranged in a logical system and of methods for their confirmation. The objects are constituted within the frame of methodological rules, the adequacy of which has to be justified by recourse to pre-scientific practice, rooted in the every day life world. [1] The term *discipline* refers to the implementation of a scientific practice in society. It includes the various forms of its institutional organization for the purposes of teaching, or of research and application of scientific knowledge, including public relations and strategies for fund raising. It is not necessary that every science has its disciplinary implementation in society, nor that the scientific and disciplinary history of a field of knowledge coincide. In some cases, the emergence of the corresponding discipline occurred decades or centuries after a particular scientific field had been established, normally within a precursory institutional tradition. According to this distinction, the paper by Kostas Gavroglu and Ana Simões clearly addresses issues that belong to the history of theoretical quantum chemistry as a science. In contrast, Andreas Karachalios's and Marika Blondel-Mégrelis's papers address both scientific and disciplinary aspects, combined with biographical studies and studies of individual research styles.

2.2 Theoretical Quantum Chemistry: Establishing a New Science in the Twentieth Century

Simões and Gavroglu investigate the impact of the following systematic points on the development of theoretical quantum chemistry in the second and third quarters of the past century. The first point is the importance of the interpretation of quantum mechanical and quantum chemical concepts as *theoretical concepts* for the proper understanding of the possibilities and the confines of this science. The second point is the *methodological autonomy* of chemistry. Their historical investigations show that the majority of chemists – even that of the theoretical quantum chemists – clearly recognized this. This majority was immune against the reductionist temptations which came mainly from the direction of the physicists and the physics-dominated philosophy of science.

Typical theoretical concepts occurring in chemical and quantum chemical theories are the various "corpuscles" – that is "atoms", "molecules", "ions", and "electrons" – and "orbital", "spin", "chemical bond", and "electric charge". The expressions atom, molecule, electron, and ion refer to particles that are thought to be the building blocks of what is macroscopically perceived as substantially homogeneous things, i. e., as the constitutive parts of so-called chemical substances. These kinds of entities, which are treated somehow as material ones, are called *theoretical constructs.* [2] On the other hand, orbital, spin, chemical bond, and charge refer to the *theoretical properties* of the constructs (e. g., their energetic states). Thus, the macroscopically detectable properties of chemical substances are thought to be the result of the combined effects of the theoretical constructs and their properties. It is a widespread opinion to ascribe theoretical constructs with the same mode of existence that applies to microscopic material objects like dust grains or bacteria. There is, however, a fundamental difference in the ontic status between those two groups of objects, since microscopic objects are always "epistemically primordial." That is, when we make a microscopic object visible with the aid of a microscope or an other device, we can talk about it and its material properties (shape, volume, etc.) without *taking into consideration the mode of function of the visualizing or magnifying device.* On the contrary, the visualization of a corpuscular theoretical construct always requires a theory of function of the visualizing device, in which the construct under consideration represents a constitutive part. In other words, the "existence" of theoretical constructs is always dependent on theories that explain the function of devices, or the occurrence of certain phenomena. The meaning of the concepts that refer to them is given only within the realms of the corresponding theories. Theoretical concepts are – as it is said – *implicitly defined.* Correspondingly, theoretical constructs are not in the same way existing as bodily objects, no matter how small they may be.

Theories do not always necessitate the introduction of theoretical constructs with respect to concepts. Newtonian mechanics, classical thermodynamics, parts of chemical kinetics, and the old theory of chemical affinities are examples of this. Sometimes, however, phenomena, which are explained by construct-free theories, should or must be integrated in a more general theory. This may be the case either because the new theory is more promising than the precursory ones, or because the observed phenomena occur together, as in the case of the so-called *Law of Integer Gas Volume Ratios,* which was established by Joseph Gay-Lussac in the early nineteenth century. This fundamental law of chemistry states that when gaseous substances react to form gaseous products, the ratios of the reacting volumes have small integer values. Gaseous chlorine and hydrogen react for example to gaseous hydrogen chloride, their gas volumes displaying a ratio of 1:1:2. Here a chemical phenomenon (the reaction) and a physical one (the change in volume) occur together after the mixing and igniting of the two gases. In order to explain the coincidence that is also observed in other gas reactions, Amadeo Avogadro introduced in 1811 [3] the theoretical constructs of the *chemical atom* and of the *chemical molecule,* i. e., the idea that given volumes of a gas consist of an amount of particles that split into simpler particles during the course of the reaction. Simple particles of each gas combine to build the more complex particles of the resulting substance. These constructs were later utilized for the explanation of the reactions of non-volatile substances via the theoretical concepts of *molecular weight* and *mole.* Their success in explaining and predicting other chemical phenomena and in integrating them in one chemical theory resulted in their stabilization in the chemical language and today in their unanimous acceptance by the chemists.

Similarly, quantum mechanical constructs and their properties have been introduced in order to integrate spectroscopic, steric, and chemical bond phenomena into a pre-existing chemical theory which, however, *already used* the concepts of atom, molecule, and electron. Thus, the latter can be regarded as primordial in respect to the former. The question about the "existence" of quantum mechanical and quantum chemical entities is resolved in the same manner as the question about the existence of atoms and molecules: All of them are theoretical constructs. Asserting their "existence" means that we can formulate true statements about them, or better, true statements that contain the corresponding theoretical concepts. Nevertheless, the confirmation of such statements takes place in a different manner than the confirmation of statements that do not contain any theoretical concept or statements that do not refer to the corresponding theoretical constructs. Thus, Coulson's statement that "the chemical bond is a concept of our imagination" should be agreed to, adding however, that this applies also to the atoms and to all the other denizens of the chemical world. It is here perhaps worth noting that the German chemist Alwin Mittasch took a similar position in his 1937 published paper "Über Fiktionen in der Chemie". [4]

In respect to the methodological autonomy of chemistry, Simões's and Gavroglu's paper also shows clearly that the chemists of the era under investigation regarded their science as methodologically autonomous and not reducible to physics. There are two reasons in favor of this attitude: First, it should be considered that regardless of the particular methods applied for the investigation of chemical phenomena, the objects of the investigations are always chemical, namely substances. They have to be produced by chemical means involving chemical expertise. Second, the theoretical predictions of quantum mechanics concerning chemical phenomena have to be tested in the chemical laboratory. Thus, the experimental ability of the chemists in realizing the predicted quantum chemical phenomena is a *sine qua non* for the deployment of the expertise of the physicists and mathematicians who deliver their theoretical treatments.

2.3
Giovanni Battista Bonino:
Pioneer of the New Science and Founder of a New Discipline in Italy

Turning to Andreas Karachalios' study, I would like first to draw attention to the fact that Giovanni Battista Bonino shared the same attitude towards the methodological autonomy of chemistry as his Anglo-American colleagues. Second, I would like to emphasize that Karachalios's paper has the merit of opening a new field of historical research, namely the Italian chemistry of the period between the wars. Besides its historical focus, Karachalios's paper also addresses a systematic point, namely the question of the justification of the use of certain theoretical constructs and the rejection of others. This problem can also be stated as the question of how arbitrariness in the choice of theoretical constructs can be avoided. The answer is that theoretical constructs in scientific theories have to comply with three sets of constraints. First, they must not conflict with the frame that is given by the norms and the basic laws that govern the field under investigation. For example, they must not violate the fundamental norms of preservation of mass and energy, or the second law of thermodynamics, etc. Second, they must fit into hitherto known effects and phenomena, in our case isomerism or line spectra. Third, they must enable the successful explanation of known phenomena and the prediction of new ones.

Therefore it becomes obvious that the introduction of a given theoretical construct is the more justified, the higher the predictive and explanatory power of the theory is, in which it occurs. Bonino's theoretical and experimental work between 1920 and 1940 was devoted to the investigation of the applicability and the adequacy of the new theoretical constructs and concepts of theoretical quantum mechanics on problems of physical and physical organic chemistry. Another important point addressed in Karachalios' paper is the study of the interaction of epistemic, social, personal, political, and economic aspects that shaped Bonino's scientific biography, his academic profile, and the disciplinary embodiment of the new science in Italy's academia. Philosophers of science tend to concentrate their attention on epistemic aspects and to neglect the disciplinary, social, and biographical ones. It is, however, very important if the social, ethical, political, and economical impacts of the scientific enterprise on society as a whole are also investigated. In this sense, Karachalios gives an integrated picture of the situation by investigating both the scientific and the disciplinary sides of his object. In his paper, he describes how the political and social frame of fascist Italy provided a fertile ground both for the development of the new discipline and for shaping the career of an intelligent and ambitious young scientist who early recognized the political and scientific opportunities of the new disciplinary niche in fascist Italy.

2.4
Jean Barriol: The French Version

While Bonino's career and Italian theoretical quantum chemistry were embedded in the political and social frame of fascist Italy, the protagonist of the last story and his institute – Jean Barriol and the Theoretical Chemistry Laboratory in Nancy – are of a more "idiographic" character in a Windelbandian terminology. Nevertheless, Marika Blondel-Mégrelis succeeded in chapter 5 by showing the individual and social aspects that formed this remarkable biography. In spite of the fact that both men were contemporaries and had similar scientific interests, there are striking differences between them. Both managed to become appointed to academic chairs and institutes that were created especially for them. Both were the protagonists of the disciplinary embodiment of the new science in their countries. However, Bonino's career was that of a university-educated practitioner who discovered his interest for theory because of concrete application problems he was working on.

Being initially a pure chemical experimenter, he had to acquaint himself later with the mathematics necessary for the comprehension of quantum mechanics. Barriol, on the other hand, enjoyed a theoretical and mathematical education from the beginning. His approach to chemical problems was always from a mathematical point of view, being convinced that chemical problems were as accessible for mathematical treatment as physical ones. Bonino pursued his academic career before and during the war up to the top of the Italo-German scientific community, being always in close touch with the political elites of his country. Blondel-Mégrelis's picture of Barriol is in contrast more that of an enlightened, but somehow apolitical theoretical scientist who had to make his way in a disinterested academic environment. Additionally, belonging to one of the nations that lost the first battles of World War II, he spent a considerable time in scientific isolation as a prisoner of war. This separated him from academic discourse, but gave him the opportunity to deepen his studies of the new science of quantum chemistry and to create his own understanding of it. It is thus not surprising that the institutionalization of theoretical quantum chemistry in France in the form of Barriol's laboratory in Nancy took place later than in Italy, in England, or in the USA. It is worth noting, however, that Barriol's attitude against the methodical and epistemic autonomy of chemistry was in accord with that of Bonino's and with that of the majority of the European and American Theoretical Chemists. Compared with the overall view of theoretical quantum chemistry presented in chapter 3, chapters 4 and 5 concentrate on one individual person or a single institution. Such studies should be regarded as equally important for the following reasons: First, they provide and preserve important factual knowledge about individuals and single institutions and make it available to other scholars for comparative studies. Second, they help to preserve the shape of a particular national scientific community. Finally, they can be regarded as pieces of a project that I would call *Creating a Map of Knowledge of Chemistry in the Twentieth Century*; and thus they are a contribution to the cultural heritage of mankind.

References and Notes

1 N. Psarros, *Die Chemie und ihre Methoden* (Weinheim: Wiley-VCH, 1999), 11 ff.

2 D. Hartmann, *Naturwissenschaftliche Theorien* (Mannheim: Bibliographisches Institut, 1993), 110 ff.

3 A. Avogadro, "Essay d' une maniére de déterminer les masses rélatives des molécules élémentaires des corps, et les proportions selon lesquelles elles entrent dans les combinaisons," *Journal de Physique* 73 (1811): 58–76, reprinted and translated in: Ostwalds Klassiker der exakten Wissenschaften, Bd. 8 (Leipzig, 1921; Reprint Leipzig 1983), 3–22.

4 A. Mittasch, "Über Fiktionen in der Chemie," *Angewandte Chemie* 50 (1937): 423–434.

3.
Issues in the History of Theoretical and Quantum Chemistry, 1927–1960

Ana Simões and Kostas Gavroglu

Contrary to what is sometimes supposed, the theoretical chemist is not a mathematician, thinking mathematically, but a chemist, thinking chemically. [1]

3.1
Introduction

In this paper we discuss a number of issues which manifest the theoretical particularity of quantum chemistry and which are usually not discussed in an explicit manner either in the historical or in the philosophical studies related to quantum chemistry. We shall focus on five issues: the re-thinking of the problem of reductionism, the discourse of quantum chemistry as a confluence of the traditions of physics, chemistry, *and* mathematics, the role of textbooks in consolidating this discourse, the ontological status of resonance, and the more general problem of the status of the chemical bond. Finally, we shall briefly discuss the impact of large scale computing.

3.2
Re-thinking Reductionism or the Chemists' Uneasy Relation with Mathematics

The question of reductionism has been reigning supreme in any discussions concerning the philosophical, theoretical, methodological and many times historical aspects of chemistry. In 1929, Paul A. M. Dirac, after he had successfully incorporated the spin quantum number into the newly developing quantum mechanics, expressed what every physicist felt to be true and what every chemist was afraid that it might be true.

> The underlying physical laws necessary for the mathematical theory of a large part of physics and the whole of chemistry are [now] completely known, and the difficulty is only that the exact application of these laws leads to equations much too complicated to be soluble. [2]

The interesting aspect of this dictum was that although it was trivially true, it was, at the time, of no practical help nor of any consequence for the chemists. In fact the very success of the Heitler-London paper – which could be taken as the first instantiation of Dirac's program – was a strong indication that what prevented chemistry from being reduced to physics was mathematics – or rather, the lack of it. Dirac's 1929 pronouncement encapsulated what was already part of the physicists' culture for many decades. And, with Dirac's specific contributions to the development of quantum mechanics, it became possible to articulate this reductionist program. After the Heitler-London paper, chemistry could be perceived as being different manifestations of spin, and spin, after all, was under the jurisdiction of the physicists. And though physicists felt that the new quantum mechanics had also taken care of chemistry, the chemists themselves did not appear to have panicked that their identity was being transformed and they were being turned into physicists. Nor did they feel that their very existence was being threatened, though it appeared that what they had been doing could now be done much better by the physicists. The appropriation of quantum mechanics, the attempts to overcome cultural resistances within the chemical community on how to appropriate quantum mechanics, and the different views on how to form the appropriate discourse are some of the issues related to the problematic of reductionism which we have already studied. [3]

Here we want to raise a different issue and investigate whether reductionism may be a misplaced category if one wants to discuss a number of questions for chemistry. Perhaps reductionism is a physicist's analytical tool and not a chemist's. Might it be the case that the whole notion of reductionism expresses a trend that is dear to the physicist's own culture rather than that of the chemists? Though physicists took for granted the reduction of chemistry to physics and did little about it, the chemists did not have the luxury of waiting for history to fulfill such an agenda. For reductionism may have been a program, but it was nearly impossible to realize it because, as became evident right at the beginning, one could not deal analytically with any of the other elements except hydrogen and helium, even in grossly approximate terms.

Are there any other dimensions to reductionism, whose discussion may be considered more fruitful in addressing the same set of problems? What we would like to discuss is the (uneasy) relationship of chemists and mathematics and argue that the chemists' relationship with the appropriation of mathematics into their culture was far more complex and difficult than their appropriation of physics. And though the two cannot be considered as totally independent of each other, it can in fact be argued that chemists were more resistant in accepting the use of mathematics rather than the physical concepts, and the physical techniques.

Like all forms and expressions of appropriation, opinions differed among the members of the chemical community. Some pushed quite strongly for introducing mathematics into chemistry. The chemist Edward Frankland predicted that the future of chemistry was to lay in its alliance with mathematics. The chemist Paul Schützenberger believed that mathematics would become an instrument as useful to the chemist as the balance. Jacobus H. Van't Hoff could not have been more mathematical in his systematic study of chemical thermodynamics. Wilhelm Ostwald's extensive use of mathematics would have been much more influential had it been not undermined by his insistence on energetics. Gilbert N. Lewis was not less skilled in mathematics. Even Joseph Larmor and Joseph J. Thomson before him tried to propose a mathematical framework for dealing with chemical problems. But resistance to such programs came from different quarters.

As early as 1884, one of the pillars of the British chemical establishment and a person who was very sympathetic to the physicists meddling into the chemists' affairs, Henry E. Roscoe was still not sure how successful mathematics would be for chemistry.

> One of the noteworthy features of chemical progress is the interest taken by physicists in fundamental questions of our science. We all remember Sir William Thomson's interesting speculations, founded upon physical phenomena, respecting the probable size of the atom. Also Helmholtz's about the relation between electricity and chemical energy. A further subject of interest to chemists is the theory of vortex-ring constitution of matter first proposed by WT and lately worked out from a chemical point of view by J. J. Thomson ... *How far this mathematical expression of chemical theory may prove consistent with the facts remains to be seen.* [4]

In 1906, Ostwald had indicated that he was having second thoughts about his denial of the existence of atoms. The British chemists did not waste any time to settle their accounts with the person who – when he gave the Faraday Lecture in London 2 years earlier – had put them in such a difficult situation, trying to convince them to abandon the atomic notions and lure them into the vague promises of energetics. Without even waiting for Ostwald's official declaration that he did, indeed, believe in atoms, Arthur Smithells, the forceful spokesman of British chemistry, at the 1907 meeting of the British Association for the Advancement of Science, took it upon himself to deliver the pangeric of the victorious.

Smithells, initially, expressed his excitement about the state of chemistry. Though the discovery of radioactivity did mark a new epoch in the history of chemistry, and radium was in a way an embarrassment, since it was elementary and it also broke into elementary substances, there was not enough evidence to warrant any unsettlement of "the scientific articles of the chemists' faith."

> The perplexities of the chemists are not due to the new ideas being presented, but to the invasion of chemistry by mathematics ... With radioactivity, in relation to the ponderable, we seem almost to be creating a chemistry of phantoms ... this reduction in the amount of experimental materials, associated as it is with the exuberance of mathematical speculation of the most bewildering kind concerning the nature, or perhaps I should say the want of nature, of matter, is calculated to perturb a solid and earthly philosopher whose business has hitherto been confined to comparatively gross quantities of materials and to a restricted number of crude mechanical ideas. [5]

He said that as a representative of the chemists he wanted to make some points, because in recent times, even before the advent of radium, a good deal had happened which had given chemists occasion to ask themselves whether chemistry was not beginning to drift away from them. In the past years, the most important developments had been on the physical side, and one great chemist remarked to him that he was feeling "submerged and perishing in the great tide of physical chemistry, which was rolling up into our laboratories."

> It is precisely such men who should be preserved to chemistry. Though chemistry and physics meet and blend there is an essential difference between the genius of physics and the genius of chemistry. Apart from his manipulative skills, the latter is not given to elaborate theories and is usually averse to speculation; nor has he usually an aptitude in mathematics. Such the normal chemist is, or was, and I hope he always may be – naked perhaps in some respects, but unashamed. There seems to be a solicitude in some quarters to make a chemist more than a chemist, a solicitude which, if granted will make him something less than one .... The most important undertaking in the education of the chemists is to be trained in the act of exact experiment and that his experimental conscience should acquire a finer edge. *Chemistry should not be invaded by mathematical theorists.* [6]

Smithells's conservative backlash was complemented in two years by Henry Armstrong's aggressive stand. Obviously to prepare his audience about the spirit of the things to come, Armstrong started by quoting a professor of world-wide reputation – whom he preferred to be anonymous – who had said that a "man's opinions are of much more value than his arguments." And in the longest address to the chemical section of the BAAS for over 30 years, the audience was, indeed, treated to the man's opinion, including a long tirade of why women should not be made Fellows of the Chemical Society, since to encourage such a move will "inevitably lead women to neglect their womanhood". Concerning his views about the energeticists, Armstrong boasted that, even though his attitude was one of "complete antagonism towards the speculations of the Ostwald school", he was nevertheless the first English chemist to publicly remark that Ostwald's investigations were of the highest importance. But now Ostwald had changed his mind, and Armstrong warned his fellow chemists in a most dogmatic manner about the dangers of dogmatism. He reminded his audience how Ostwald had:

> charged his test tubes with ink instead of chemical agents and by means of a too facile pen he has enticed chemists the world over into becoming adherents of the cult [of his school] – a cult the advance of which may well be ranked with that of Christian science, so implicit has been the faith of its adherents in the doctrines laid down for them, so extreme and narrow the views of its advocates ... The lesson we shall have learnt will be of no slight import ... if it serve to bring home to us the danger of uncontrolled literary propagandism in science, if it but cause us always to be on our guard against the intrusion of authority and of dogmatism in our speculations. [7]

But that was not the end. There was one more account to be settled. Armstrong appealed to the physicists to make themselves more acquainted with the methods of the chemists and to stop speculating unnecessarily.

> Now that physical inquiry is largely chemical, now that physicists are regular excursionists into our territory, it is essential that our methods and our criteria be understood by them. I make this remark advisedly, as it appears to me that, of late years, while affecting almost to dictate a policy to us, physicists have taken less and less pain to make themselves acquainted with the subject matter of chemistry, especially with our methods of arriving at the root conceptions of structure and the properties as conditioned by structure. It is a serious matter that chemistry should be so neglected by physicists. [8]

Though Armstrong's views may be taken to express the chemists' assertiveness, it is interesting to note the following. When Armstrong died in 1937, Ernest Rutherford wrote to Nevil V. Sidgwick and said that he had been told that "Armstrong had never got beyond arithmetic, and that even algebraic symbols were Greek to him. This may account for his attitude to all mathematical theory." [9]

This uneasy relationship between chemists and mathematics can, also, be traced during the emergence of quantum chemistry. All those who were directly involved in the development of quantum mechanics were confronted with the evaluation of the relations of chemistry to physics and by extension to mathematics. In 1928, in a review paper written for *Chemical Reviews*, John Van Vleck outlined the problems faced by the new science of "mathematical chemistry". Most were eager to point to the subsidiary role of mathematics. Linus Pauling managed to present a coherent treatment of the chemical bond which was appealing to the chemists because of its frequent reliance on the "chemists' intuition" and the use of a lot of existing experimental data to be able to explain or predict other experimental data. Though it was repeatedly stressed that the understanding of the nature of the chemical bond was possible only because of the developments due to quantum mechanics, his use of detailed mathematical formulations was reduced to a bare minimum.

Some years later, Charles A. Coulson argued that the "splendid and elegant elucidation" of so large a part of chemistry by quantum mechanics forbade chemists to be happy with an electronic theory of valence couched in pre-quantum mechanical terms. In 1952, he was careful to stress in his book *Valence* that quantum chemistry should be understandable by a chemist with no mathematical training. The presentation of the principles of quantum mechanics was reduced to two introductory chapters, and in many instances mathematical results were illustrated or complemented by the extensive use of visual representations, an implicit acknowledgment that visualizability, instead of elaborate mathematics, still remained one of the constitutive features of chemistry.

Hugh C. Longuet-Higgins, one of Coulson's students, went even further in assessing the complex relation of quantum chemistry to mathematics. He turned the whole argument upside down. He did not consider that there was a danger that quantum chemistry might be subsumed under mathematics, and boasted that the time had come for chemists to teach mathematics to the mathematicians. He introduced the paper "An application of chemistry to mathematics" with the bold statement:

> I imagine that the title of this paper will shock many of the readers of this Journal. It is generally taken for granted, at least by mathematicians, that in the hierarchy of the exact sciences mathematics holds first place, with physics second and chemistry an insignificant third. Organic chemistry is considered at best a practical necessity and at worst a rather noisome branch of cookery. In this paper I hope to show that pure mathematics is occasionally enriched not only by the fruits of physics, but also by those of chemistry, and to establish this thesis by proving a mathematical theorem of some intrinsic interest which was, in fact, suggested by an empirical generalization in organic chemistry. [10]

The problem was the solution of Schrödinger's equation under certain simplifying assumptions, and specifically the task of obtaining expressions for the electronic energy and electron densities by recourse to the theory of complex variables for benzenoid hydrocarbons, a rather special but very important class of molecules. Such is the case of naphthalene, which Coulson and his group had defined as an alternant hydrocarbon, and for which experiment suggested that the effect on the electron density around atom *r* due to the effect of a perturbation at atom *s* is of one sense if *r* and *s* belonged to the same system and of the opposite sense if they belonged to different systems. It was this experimental result that Longuet-Higgins showed to imply the validity of a certain mathematical theorem. He concluded by pointing that the discovery of many other theorems, with an intrinsic interest from the purely mathematical point of view, was prompted by chemical laws. He hoped then that "the more trained mathematicians will come to recognize theoretical chemistry as a subject not altogether unworthy of their professional attention." [11]

We have noted these cases not in order to make any conclusive argument about the relationship of chemists to mathematics, but rather as indicative instances of a trend among chemists which has not been properly discussed up to now: as chemists were discussing the appropriation of physics into their own culture, there was a parallel and relatively independent discussion among them concerning their appropriation of mathematics.

3.3
Convergence of Diverging Traditions: Physics, Chemistry, *and* Mathematics

When referring to the different approaches to the question of atomic bonding, nearly all textbooks and research papers project two such methods: the Heitler-London-Slater-Pauling valence bond method and the Hund-Mulliken method of molecular orbitals. Elsewhere we have argued that the views of these protagonists about theory building and the role of theory in chemistry form a set of criteria that justifies a different classification: the Heitler-London approach versus the Pauling-Mulliken approach. Walter Heitler and Fritz London shared, in effect, Dirac's reductionist view: the underlying laws governing the behavior of electrons were known; and hence to do chemistry meant to deal with equations which were in principle soluble, even though in practice they may only produce approximate solutions. Pauling and Robert S. Mulliken thought differently on how the newly developed quantum mechanics could, in practice, be applied to problems of chemistry and, more specifically, to the problem of the chemical bond. They felt that a reductionist agenda was, in practice, useless to the chemist, and by making ample use of semi-empirical methods they developed their respective approaches, whose only criterion for acceptability was their practical success. And, most significantly, they both shared a common outlook on how to construct their theoretical schemata, on the character of the constitutive features of their theories, on what the relation of physics to chemistry should be and on the discourse they developed to legitimize their respective theories.

Let us now turn to a number of issues associated with the theoretical outlook shared by Pauling and Mulliken. Pauling's valence bond and Mulliken's molecular orbital approaches were not simply two practical methods to solve valence problems. They were part of two different conceptual schemata, which can be explained in terms of two different legacies – that of physics in the case of Mulliken and that of chemistry in the case of Pauling. Their contributions were simultaneously the culmination of two different research traditions and the beginning of a new discipline and a new practice. Their research programs evolved from two different research programs developed in the context of the old quantum theory. Pauling was eager to establish a continuity between his contributions and Lewis's program for the explanation of the covalent bond developed in the context of the work of the community of physical chemists. In contrast, Mulliken's work on band spectra structure can be seen as an instantiation of the research agenda carried out by the American molecular physics community. Pauling's research program was presented as an extension of the classical structure theory whereas Mulliken's agenda was presented in sharp contrast with them. [12]

Extending Heitler and London's work but demarcating himself from their methodological orientation, Pauling outlined a chemical theory based on the concept of resonance. The appropriation of resonance from the quantum mechanical context in which it was used by the physicists Heitler and London, to a new context in which it served a strictly chemical end, played a fundamental role in the formulation of a number of new concepts, from the hybridization of bond orbitals, to one-electron and three-electron bonds, the discussion of the partial ionic character of covalent bonds in heteropolar molecules, and the idea of resonance among several hypothetical bond structures. In certain aromatic compounds such as benzene, Pauling suggested that the wave function should be written as a superposition of wave functions associated with the different valence bond structures which chemists had introduced to represent all its properties. The new concept explained in "an almost magical way" [13] the many puzzles that had plagued organic chemistry, establishing the connecting link between Pauling's new valence theory and the classical structural theory developed throughout the second half of the nineteenth century.

The most characteristic feature of Pauling's approach, which became known as the valence bond approach (VB) is that it considers the combining atoms as units. The molecule is therefore supposed to be formed by bringing together two or more atoms that are then allowed to interact. Pauling's ontological commitments were associated from the start to a number of rules enabling him to get numerical values of bond energies and bond angles. This quantitative dimension of the VB approach was not matched for a while by the molecular orbital approach (MO), developed by Mulliken and others, in which the molecules are taken as the main building blocks. It is assumed that only nuclei (or nuclei plus inner electrons) are brought together into their positions, and only afterwards are the remaining electrons – the valence electrons – allowed to be fed into what were called molecular orbitals. According to Mulliken, the first method followed the "ideology of chemistry" [14] whereas the latter departed from it.

The clarification of the relations between electronic states and band spectra structure led Mulliken to dispense altogether with classical valence theory and to propose an entirely different approach to the question of molecule formation and chemical bonding. Mulliken rejected the accepted notion of chemical structure and proposed to analyze the phenomena of molecule formation in terms of the electronic structure of molecules. Reasoning by analogy with Bohr's building-up principle for atoms, Mulliken considered that molecules were formed by feeding electrons into molecular orbitals, that is, into orbitals that encircled two or more nuclei. Electrons were delocalized in the sense that there was a non-zero probability of finding them near more than one nucleus. The assignment of quantum numbers to electrons in molecules, and the classification of molecular orbitals, was achieved by exploring the relations to the united-atom description and the separated atom description. New auxiliary concepts were introduced such as promoted and unpromoted electrons, bonding, non-bonding, and anti-bonding electrons, and varying bonding power of electrons. In 1929, John Lennard-Jones introduced the physical simplification of representing molecular orbitals as linear combination of atomic orbitals (LCAO), a step that was fundamental to the mathematical development of MO theory.

Many reasons contributed to the successful way in which quantum chemistry developed in the United States. Mulliken summarized them well when he called himself a middleman between theory and experiment, and between physics and chemistry. A particular kind of institutional atmosphere accounted for the appearance of this new type of scientist, whose definition as a chemist or physicist was in many instances a matter of chance, personal preferences, or institutional affiliation. The institutional ties between chemistry and physics were stronger in the United States than in Europe. At universities like Berkeley and Caltech, chemistry students were often learning as much physics as chemistry and thus were more apt to learn and accept quantum mechanics than their European counterparts. Pauling's knowledge of physics was impressive and Mulliken was an expert on the quantum theory of molecules. Besides Berkeley, Caltech, Harvard, and MIT, more universities were promoting the cooperation between their physics and chemistry departments. Examples were Princeton, Chicago, Michigan, Minnesota, and Wisconsin. But before Mulliken, Pauling, John Slater, and Van Vleck, the preceding generation of chemists and physicists – chemists like Lewis and Arthur A. Noyes, Richard C. Tolman, and William Harkins, and physicists like Edwin C. Kemble and Raymond T. Birge – planted the seeds which blossomed into quantum chemistry.

The ability of the particular scientists to be at ease with both theory and experiment might well account for the successful development of quantum chemistry. Mulliken started as an experimentalist but shifted into theory, owing to the delay in getting the high-resolution spectrograph he had been promised when he moved to Chicago in 1928. Pauling's determinations of crystal structures were instrumental as a source of practical information on bond angles and bond lengths to be used in his future, more theoretical endeavors.

An altogether different situation occurred in Europe, specifically in Germany. There was a sharp division between theory and experiment in the German physical community. As to the German chemists, they were in general ill-prepared to cope with the challenges of quantum mechanics. One example was Kasimir Fajans, a professor of Physical Chemistry in Munich when the young Pauling was there in 1925–1927. Fajans was already one of the leading physical chemists. Many years later, in 1987, Pauling remembered that Fajans' inability to get a good grasp of quantum mechanics was a problem that bothered him for the rest of his life.[15]

In Germany, by the early 1930s, chemistry and physics were well established disciplines, entertaining few disciplinary, methodological, or institutional ties to each other. Therefore, scientists whose profile could favor an attack on chemical problems using the tools of the newly developed quantum mechanics were hard to find. Several German physicists, but not chemists, were interested in applications to chemistry and contributed initially to the field; but they were unable in the long run to carry out their research programs. Such were the cases of Heitler, London, Friedrich Hund, and Max Born. An exceptional case in the German context was the physicist Erich Hückel. He was able to overcome his deficient chemical background by taking advantage of his brother Walter Hückel's expertise in organic chemistry, which probably helped him in asking the pertinent questions in organic chemistry to be answered in the framework of quantum mechanics.[16] Erich Hückel developed a reductionist program in which the facts of organic chemistry were to be interpreted by taking seriously the peculiar theoretical features of quantum mechanics. Its non-visualizability was seen as forbidding Pauling's description of the structure of benzene by means of resonance among several fictitious valence bond structures. By 1937, Hückel abandoned the field, unable to challenge a scientific establishment in which German physicists were not yet ready to accept research on the quantum mechanical properties of the chemical bond as a topic of research for physicists, and German chemists did not consider quantum chemistry a field of chemistry.

The genesis and development of quantum chemistry as an autonomous subdiscipline owed much to those scientists who were able to realize that "what had started as an extra bit of physics was going to become a central part of chemistry". Those that manage to escape successfully from the "thought forms of the physicist"[17] by implicitly or explicitly addressing issues such as the role of theory in chemistry, and the methodological status of empirical observations helped to create a new space for chemists to go about practicing their discipline. The ability to "cross boundaries" between disciplines was perhaps the most striking and permanent characteristic of those who consistently contributed to the development of quantum chemistry. Moving at ease between physics, chemistry, and mathematics became a prerequisite to be successful in borrowing techniques, appropriating concepts, devising new calculational methods, and developing legitimizing strategies.

In the mid- and late 1930s when quantum chemistry was already delineated as a distinct sub-discipline, there was in Britain a group of people whose contributions to the further entrenchment of the disciplinary boundaries of quantum chemistry proved rather decisive. Lennard-Jones, Douglas R. Hartree, and Coulson were the best known members of this group. Coulson was the most vocal and the person in whose work we find all those trends that have characterized the "British approach" to quantum chemistry. If the "German approach" inaugurated by London, Heitler, Hund, and E. Hückel was emphasizing the application of first principles of quantum mechanics to chemistry, and if the "American approach" of Pauling, Mulliken, Van Vleck, and Slater was characterized by a pragmatism combined with a creative disregard towards the strict obeisance to the first principles of quantum mechanics, the British succeeded in enlarging the domain of applied mathematics so as to include techniques derived from their discussion of problems of quantum chemistry.

Both Cambridge, where Coulson studied and completed his doctorate, and Oxford, where Coulson became professor of Applied Mathematics and then professor of Theoretical Chemistry, had researchers who were particularly receptive to the new possibilities offered by the new quantum mechanics for chemistry. Two in particular, Ralph H. Fowler at Cambridge and Sidgwick at Oxford, were quite decisive in creating a milieu where these possibilities were actively sought. When the new quantum mechanics was first formulated, Fowler was 37 and Sidgwick was 53 years old, and they immediately became enthusiastic converts to the new ideas. Their subsequent work was not directly related to the developments of quantum mechanics, but Sidgwick through his book *Some Physical Properties of the Covalent Link in Chemistry* (1933), his annual reports, and his presidential addresses became one of the most effective propagandists of the immense usefulness of resonance for chemistry. Fowler, on the other hand, was himself one of the very expressions of the Cambridge tradition of mathematical physics. Two of the students Fowler supervised became professors at Cambridge in 1932, the same year he himself was appointed professor of Mathematical Physics. Dirac became the Lucasian Professor in Natural Philosophy and Lennard-Jones the first Professor in Theoretical Chemistry.

Furthermore, the supervisor and his two students appeared to share similar views on the relations of the new quantum mechanics to chemistry. At the end of 1929 Fowler, who was one of the editors for Cambridge University Press, asked London whether he would be interested to write a book on "the foundations of chemistry in quantum mechanics". [18] Dirac during the same year had, as we saw, expressed his view about chemistry, which would permanently mark the physicists' culture. By 1931, Fowler was already expressing a subtler view of the whole problem. In a report delivered at the Centenary Meeting of the British Association for the Advancement of Science, he expressed the view that now the chemical theory of valence had shown that it was no longer independent from physical theory, but just a beautiful part of a simple self-consistent whole, that of non-relativistic quantum mechanics. He felt that he had sufficient chemical appreciation to claim that quantum mechanics is glorified by the successes in theoretical chemistry, rather than saying that the recent developments had shown that "there is some sense in valencies." [19] He still believed, though, that a full quantum mechanical explanation of the valence rules of the quantum chemist was to be reached in the near future.

Lennard-Jones in an article in *Nature* in 1931 which also echoed the views he expressed in lectures at the Physical and also Mathematical Societies in London, considered the connection between the pairing of electrons with the "valency rules of the chemist" as a consequence of the same "mathematical and physical principles which have been formulated for other branches of physics." [20] He was convinced that the general principles behind the different forces were understood and that such insights may come to be regarded as one of the greatest achievements of the present formulation of quantum mechanics. What was now required were mathematical techniques capable of applying them to particular cases.

The year 1932, when they were all appointed professors, was also the year Coulson started his doctorate, first as a student of Fowler, and then he was nominally supervised by Lennard-Jones. But Coulson's researches, though they were deeply grounded in this Cambridge tradition, showed a characteristic resistance against being lured by the excesses of this program. Coulson, the mathematical physicist, would "refuse" to become the long hand of physics in chemistry. His works and the evidence in the archival material show that Coulson was progressively displaying an increased sensitivity to the needs of the chemists rather than taking a physicist's patronizing view. It was he who legitimized the use of such heavy – by the chemists' criteria – mathematics in chemistry and managed to have a rather wide recognition by the chemical community.

The British quantum chemists perceived the problems of quantum chemistry first and foremost as calculational problems and, by devising novel calculational methods, they tried to bring quantum chemistry within the realm of applied mathematics. It may not have been as exciting an undertaking as the Germans' or the Americans', but it was surely a particularly effective one. In that specific context, the demand for more rigor was not primarily a demand for a rethinking of the conceptual framework, but rather it was a demand for developing as well as legitimizing formal (mathematical) techniques and methods to be used in chemical problems. For the members of this group, and for Coulson in particular, the demand to make a discipline more rigorous meant to have more mathematical techniques that will be inimical to the discipline itself, and that meant to get involved with (applied) mathematics.

One of the most intriguing aspects of the initial phase of quantum chemistry was the formulation of a host of concepts devised to conceal the impotence of mathematics to produce exact solutions and to cater to a community for whom visualization was one of the necessary ingredients of their practice. Exchange integral, hybridization, directed valence, bonding, and antibonding orbitals, and, above all, resonance were attempts to mellow the blow felt when chemists realized that, perhaps, chemistry is a purely quantum phenomenon, since it is so dependent on spin. And we know that the explanatory strength of quantum theory has been a factor undermining the perennially difficult process of pictorial representation.

For a long time, it appeared that in quantum chemistry the four procedures – conceptual, mathematical, experimental, and pictorial – were complementary to each other, each having a relative autonomy and at the same time each coming to the rescue of the whole enterprise whenever the other(s) were reaching their limits. With the extensive development of numerical techniques, one gets the feeling that something had been changing. It is rather intriguing to pursue the question whether the development of calculational techniques progressively "de-conceptualized" quantum chemistry. This is not to imply that the work of the British was devoid of any new and novel concepts. The question to be discussed is to understand the characteristics of their work as they were developing the mathematical techniques for chemistry. Or to put it another way, understanding the particularities of the British quantum chemists entails the understanding of their overall outlook to reformulate the problems of quantum chemistry as problems of applied mathematics.[21] Is it the case that in their attempts to build theoretical schemata, their specific methodological choices and ontological commitments led to their becoming less dependent on concepts and more on mathematics – and then more dependent on calculating machines?

3.4
The Role of Textbooks in Building a Discourse for Quantum Chemistry

Textbooks have always played a rather dominant role in the early stages of the formation of sub-disciplines: by formalizing the "principles" of the sub-discipline, making explicit the solutions to hitherto unsolved problems, reviewing the state of the field, codifying what there is to be taught, and giving background information for non-experts to learn about the field, the early textbooks in a sub-discipline's history contribute to the legitimation and institutionalization of the field.

The development of quantum chemistry has been no exception. Is quantum chemistry an application or use of quantum mechanics for chemical problems? Is quantum chemistry the totality of chemical problems formulated in the language of physics and which could be dealt by a straightforward application of quantum mechanics with, of course, the ensuing conceptual readjustments? Or is it the case that chemical problems could be dealt with only through an intricate process of appropriation of quantum mechanics by the chemists' culture? By attempting to provide an answer to these seemingly pedantic questions and often implicitly posed questions, various textbooks attempted to define the status of quantum chemistry, that is, to define its degree of autonomy with respect to both physics and chemistry as well as the extent of its non-reducibility to physics. And even though these issues were being discussed in the research papers, the meetings, and the conferences, the early textbooks of quantum chemistry became equally decisive in articulating the constitutive aspects of quantum chemistry.

Textbooks in general are – necessarily – a-historical and only in a very few instances do we find a mention and, in even fewer cases, a discussion of some of the disputes in a discipline's early history. Interestingly, the early textbooks of quantum chemistry could also be read as polemic or partisan texts: by proposing and arguing in favor of particular (ontological) hypotheses and approximation methods, each one of them adopts a particular viewpoint on how to answer the question of whether quantum chemistry is an application or use of quantum mechanics for chemical problems.

Early textbooks in a discipline's history could also be viewed as a genre for consolidating a consensus as to the language to be used and the practice to be adopted. In the case of quantum chemistry, such an agenda revolved around the question of whether chemists should start diverging from the accepted norms of their disciplinary culture where chemistry is not thought of as a mathematical science, or whether they should continue to be faithful to such a culture and appropriate the right dose of quantum mechanics for their own purposes. The dilemma, then, of whether chemists should apply quantum mechanics to chemical problems or use quantum mechanics in chemistry, and the ensuing issues as to the extent of mathematics to be introduced, was really a dilemma concerning the status of quantum chemistry: the question, that is, about the extent of its *relative* autonomy with respect to physics.

We will now give a few examples. *The Electronic Theory of Valence*, the book the English chemist Sidgwick published in 1927, the year often considered to be the birth date of quantum chemistry, announced a new era, sensing the promises that lay in the road ahead. His next textbook *Some Physical Properties of the Covalent Link in Chemistry* (1933), born out of a series of lectures delivered in the USA, went further in assessing the methodological guidelines to be followed by the new discipline. By 1939, the Americans had imposed their agenda. Coincidentally, this was the year of the publication of two textbooks – Pauling's *The Nature of the Chemical Bond*, and John Slater's *Introduction to Chemical Physics*. These two articulate writers aimed – by adopting different viewpoints – at educating an audience of both students and professionals in the ways of the new discipline. Pauling, the chemist, proceeded to a reform of the whole of chemistry from the standpoint of quantum chemistry. Slater, the physicist, saw the beginnings of quantum chemistry, which he christened chemical physics, as heralding the unification of physics and chemistry. Both books reflected the tendency to impose a new (sub)discipline by establishing a new language, a new practice, a new theoretical agenda, and a concomitant methodology, and finally by securing an audience. That was no longer the case with two other textbooks. The organic chemist George W. Wheland, one of Pauling's former students, contributed more than any other chemist towards the extension of the scope of the theory of resonance to organic chemistry. He adopted Pauling's research agenda and pushed it ahead by arguing that it is possible for organic chemists to use quantum chemistry without having to turn their discipline into a fully mathematized science. In fact, his book *The Theory of Resonance and its Application to Organic Chemistry* published in 1944 was to play a prominent role in the education of organic chemists. In 1952, the textbook *Valence* written by Coulson reflected a growing awareness on the part of some chemists that Pauling's viewpoint had been strongly overrated. Coulson's book was the first serious and successful attempt to replace *The Nature of the Chemical Bond*, with important repercussions in the teaching of quantum chemistry.

Whether we look at textbooks written during the 1930s, the 1940s, or even later on, we conclude that many of the textbooks dealt with the implications of quantum mechanics to chemistry, however, taking different views on the matter. In some cases, they provided qualitative discussions of the applications of quantum theory to chemistry, particularly to chemical bonds, avoiding as much as possible the mathematical structure of the theory. In other cases, they presented quantum mechanics with full consideration of its mathematical methods and different degrees of emphasis on topics of chemical interest. In still other cases, they attempted to combine the advantages of both approaches. In some instances, these different strategies reflect implicit or even explicit views about the autonomy of quantum chemistry, that is, about the hypothetical reduction of chemistry to physics and, interestingly, textbooks written in later years continued to display a similar ambivalence towards the kind of mathematical details to be introduced in quantum chemistry.

It is, thus, interesting to note that from the very first days when it became possible to expand the domain of quantum mechanics to chemical problems until the period when there was a consensus among the chemists of the relative merits and shortcomings of a number of approaches, the problem of reductionism was at the forefront of pressing questions for many chemists. A number of leaders in the field had no qualms in declaring that quantum chemistry was a branch of physics, others, by emphasizing the qualitative arguments so prevalent in chemical thinking, were attempting to define a framework where quantum chemistry will develop a relatively autonomous status with respect to physics. In many textbooks this problematic was expressed by the authors' dilemmas as to whether quantum chemistry will be an application of quantum mechanics to chemical problems or whether quantum chemistry would be able to articulate its language by the successful appropriation of quantum mechanics by chemistry. It is this subtle differentiation between the approaches which led to the writing of a number of pedagogically effective and ideologically diverse textbooks in quantum chemistry.

3.5
The Ontological Status of Resonance

Pauling's resonance theory raised questions as to the ontological status of theoretical entities very similar to the *problématique* associated with discussions about scientific realism. Differences in the assessment of the methodological and ontological status of resonance were the object of a dispute between Pauling and Wheland, who worked towards the extension of resonance theory to organic molecules. Wheland, in his book *The Theory of Resonance and Its Applications to Organic Molecules* dedicated to Pauling, argued that resonance was a "man-made-concept" [22] in a more fundamental way than in most other physical theories. This was his way to counter the widespread view that resonance was "a real phenomenon with real physical significance," which he classified as one example of the nonsense organic chemists were prone to.

> What I had in mind was, rather, that resonance is not an intrinsic property of a molecule that is described as a resonance hybrid, but is instead something deliberately added by the chemist or the physicist who is talking about the molecule. In anthropomorphic terms, I might say that the molecule does not know about resonance in the same sense in which it knows about its weight, energy, size, shape, and other properties that have what I call real physical significance. Similarly... a hybrid molecule does not know how its total energy is divided between bond energy and resonance energy. Even the double bond in ethylene seems to me less "man-made" than the resonance in benzene. The statement that the ethylene contains a double bond can be regarded as an indirect and approximate description of such real properties as interatomic distance, force constant, charge distribution, chemical reactivity, and the like; on the other hand, the statement that benzene is a hybrid of the two Kekulé structures does not describe the properties of the molecule so much as the mental processes of the person who makes the statement. Consequently, an ethylene molecule could be said to know about its double bond, whereas a benzene molecule cannot be said, with the same justification, to know about its resonance ... Resonance is not something that the hybrid *does*, or that could be "seen" with sufficiently sensitive apparatus, but is instead a description of the way that the physicist or chemist has arbitrarily chosen for the approximate specification of the true state of affairs. [23]

Pauling could not disagree more. For him, the double bond in ethylene was as "man-made" as resonance in benzene. Pauling summarized their divergent viewpoints by saying that Wheland seemed to believe that there was a "quantitative difference" in the man-made character of resonance theory when compared to ordinary structure theory – but he could not find such a difference. He asserted that Wheland made a disservice to resonance theory by overemphasizing its "man-made character." [24] Wheland conceded that resonance theory and classical structural theory were qualitatively alike, but he still defended, contrary to Pauling, that there was a "quantitative difference" between the two. He viewed his disagreement with Pauling as a result of different value-judgements on what he classified as philosophical, rather than scientific matters.

Nevertheless, acknowledging or denying the existence of differences between resonance theory and classical structural theory was dependent on their different assessments of the role of alternative methods to study molecular structure. Wheland equated resonance theory to the valence bond method and viewed them as alternatives to the molecular orbital method. Pauling conceded that the valence bond method could be compared with the molecular orbital method, but not with the resonance theory that was largely independent of the valence bond method. For Pauling the theory of resonance was not merely a computational scheme. It was an extension of the classical structure theory, and as such it shared with its predecessor the same conceptual framework. If one accepted the concepts and ideas of classical structure theory one had to accept the theory of resonance. And, how could one reject their common conceptual base if they had been largely induced from experiment?

> I think that the theory of resonance is independent of the valence-bond method of approximate solution of the Schrödinger wave equation for molecules. I think that it was an accident in the development of the sciences of physics and chemistry that resonance theory was not completely formulated before quantum mechanics. It was, of course, partially formulated before quantum mechanics was discovered; and the aspects of resonance theory that were introduced after quantum mechanics, and as a result of quantum mechanical argument, might well have been induced from chemical facts a number of years earlier. [25]

This discussion with Wheland prompted Pauling to make his position about these issues public. More than the question of the artificiality of the resonance concept, to which he alluded briefly in his Nobel lecture, [26] he wanted once and for all to state as clearly as possible his views on theory building. A new version of the arguments brought about in the discussion with Wheland appeared in *Perspectives in Organic Chemistry* [27] and later on in the third edition of *The Nature of the Chemical Bond.* [28] In the preface, Pauling pointed out that the theory of resonance involves "the same amounts of idealization and arbitrariness as the classical valence-bond theory". Pauling added a whole section in the new edition to discuss this question. His manifesto was called "The Nature of the Theory of Resonance." There, he argued that the objection concerning the artificiality of concepts applied equally to resonance theory as to classical structure theory. To abandon the resonance theory was tantamount to abandoning the classical structure theory of organic chemistry. Were chemists willing to do that? According to Pauling, chemists should keep both theories because they were chemical theories and as such possessed "an essentially empirical (inductive) basis".

> I feel that the greatest advantage of the theory of resonance, as compared with other ways (such as the molecular-orbital method) of discussing the structure of molecules for which a single valence-bond structure is not enough, is that it makes use of structural elements with which the chemist is familiar. The theory should not be assessed as inadequate because of its occasional unskillful application. It becomes more and more powerful, just as does classical structure theory, as the chemist develops a better and better chemical intuition about it ... The theory of resonance in chemistry is an essentially qualitative theory, which, like the classical structure theory, depends for its successful application largely upon a chemical feeling that is developed through practice. [29]

In 1947, Coulson wrote an article in a semi-popular magazine on what he thought was resonance:

> Is resonance a *real* phenomenon? The answer is quite definitely no. We cannot say that the molecule has either one or the other structure or even that it oscillates between them ... Putting it in mathematical terms, there is just one full, complete and proper solution of the Schrödinger wave equation which describes the motion of the electrons. Resonance is merely a way of dissecting this solution: or, indeed, since the full solution is too complicated to work out in detail, resonance is one way – and then not the only way – of describing the approximate solution. It is a "calculus", if by calculus we mean a method of calculation; but it has no physical reality. It has grown up because chemists have become used to the idea of localized electron pair bonds that they are loath to abandon it, and prefer to speak of a superposition of definite structures, each of which contains familiar single or double bonds and can be easily visualizable. [30]

The question as to the ontological status of resonance was not an issue that was confined to this exchange between Pauling and Wheland. Pauling's theory of resonance was viciously attacked in 1951 by a group of chemists in the Soviet Union in their Report of the Commission of the Institute of Organic Chemistry of the Academy of Sciences. [31] They themselves, stressed that their main objection was methodological. They could not accept that by starting from conditions and structures that did not correspond to reality one could be led to meaningful results. Of course, they discussed analytically the work of Aleksandr M. Butlerov who in 1861 had proposed a materialist conception of chemical structure: this was the distribution of the action of the chemical force, known as affinity, by which atoms are united into molecules. He insisted that any derived formula should express a real substance, a real situation. According to the report, Pauling was moving along different directions. For him a chemical bond between atoms existed if the forces acting between them were such as to lead to the formation of an aggregate with sufficient stability to make it convenient for the chemist to consider it as an independent molecular species. To these chemists Pauling's operational definition was totally unacceptable.

> In this treatment the objective criterion of reality of the molecule and of the chemical bond vanishes. Since the definition of the molecule and the chemical bond given by Pauling is methodologically incorrect, it naturally leads, when logically developed, to absurd results. [32]

It is interesting to note the initiative of the New York Chapter of the National Council of Arts, Sciences and Professions to organize a meeting on the subject. It was proposed that the meeting have the form of a debate where N. D. Sokolov from Moscow, Coulson, and Pauling would each contribute a paper and there would follow a discussion of the points raised in the communications. Coulson felt that the best way would be for Sokolov and Pauling to present their viewpoints and that he would make a series of comments. Each party would be asked to provide answers to the following questions: What is the resonance theory? What is the evidence in proof or disproof of the resonance theory? Is the convenience of the theory a proof or a corroboration of the theory? Is the resonance theory essentially a theory with physical meaning, or a mathematical technique or both? Has the resonance theory a basis in related sciences, such as physics? Is the resonance theory applicable in all aspects of chemical valence or is it in conflict? [33] The meeting did not take place basically because of the unwillingness of the Soviets, but the points that each party would have had to address were indicative of the uncertainties involved as to the methodological significance and ontological status of resonance in quantum chemistry.

3.6 The Status of the Chemical Bond

Coulson called the chemical bond a "concept of the imagination", and used it to illustrate the role and status of concepts within quantum chemistry. According to Coulson, all chemistry rests on the idea of a chemical bond, and every generation of chemists has tried, in its own way, to describe what is a bond. The different descriptions that have been given show how greatly our understanding of the "real essence of chemistry" [34] has developed in the past since Frankland or Kekulé. For nearly one hundred years chemists noticed the characteristic affinities of one substance for another. Lewis had suggested that this affinity is related to the disposition of two electrons, but "remember, no one has ever seen an electron!". [35] Since 1927, applied mathematicians have been able to handle the differential equations of wave mechanics, although they soon faced the embarrassment of not being able to solve exactly Schrödinger's wave equation, which describes the behavior of the wave function that carries the answer to every chemical question we can ask. The quantum mechanical underpinning of Lewis's description showed next that the shared electrons have their spins pointing in opposite, or anti-parallel, directions, but "remember, no one can ever measure the spin of a particular electron!" [36] However, everyone was captivated by "the simplicity of the idea." [37] Then the distribution in space of these electrons is described analytically with closer and closer degrees of precision, but "remember, there is no way of distinguishing experimentally the density distribution of one electron from another!" [38]

Concepts like hybridization, covalent and ionic structures, resonance, and fractional bond orders have been introduced in the process, and Coulson was rather uneasy that none of these concepts could be linked to a directly measurable quantity. Nevertheless, "chemical knowledge and, perhaps even more, chemical intuition, find their full expression and their proper setting within the mathematical framework that has now been devised." [39] The importance of conceptual insightfulness together with the usefulness and truthfulness of concepts is stressed again and again in Coulson's writings. As "concepts of the imagination" they have not necessarily to be real.

> It does not require our friends the logical positivists to give us pause. (...) I described a bond, a normal simple chemical bond; and I gave many details of its character (and could have given many more). Sometimes it seems to me that a bond between two atoms has become so real, so tangible, so friendly that I can almost see it. And then I awake with a little shock: for a chemical bond is not a real thing: it does not exist: no-one has ever seen it, no-one ever can. It is a figment of our own imagination. [40]

In the inaugural lecture as professor of Applied Mathematics, Coulson had already made the same point, perhaps even with more poise:

> Dare we make a lesser claim than this for the modern description of a chemical bond? For all these *concepts of the imagination* give us such understanding and feeling for the thing that sometimes it seems to me that a chemical bond is so real, so huge, so life-like that I can almost see it. Then I wake with a shock to the realization that neither I nor anyone else will ever see one: a chemical bond does not exist: it is a figment of imagination which we have invented – it is most useful, most satisfying, but (though perhaps in this building [the Physical Chemistry Laboratory, Oxford] I should be careful with what I say) no more real than the square root of −1! [41]

What is going to happen in the future to the idea of a bond? Coulson gave two possible answers to this interesting question. The work of the next years will have to be more concerned with refining and perhaps simplifying the sort of description already worked out. [42] In a symposium commemorating the 50 years of valence theory which took place in 1970, Coulson went much further.

> So to the question: has the chemical bond now done its job? Have we grown to that degree of knowledge and that power of calculation that we do not need it? Certainly in the more elaborate of the calculations that I have referred to, the authors seldom if ever use the word "bond." This a tantalizing question. And only a little can be said by way of comment. Chemistry is concerned to explain, to give us insight, and a sense of understanding. Its concepts operate at an appropriate depth, and are designed for the kind of explanation required and given. If the level of enquiry deepens, as a result of our better understanding, then some of the older concepts no longer keep their relevance. No one talks much now about the polarization of non-bonding electrons, of dynamic oscillation, or of bond fixation. From its very nature a bond is a statement about two electrons, so that if the behavior of these two electrons is significantly dependent upon, or correlated with, other electrons, our idea of a bond separate from, and independent of, other bonds must be modified. In the beautiful density diagrams of today the simple bond has got lost. It is as if we had outgrown the early clothes in which, as children, we could be dressed, and now needed something bigger. But whether that 'something bigger' that should replace the chemical bond, will come to us or not is a subject, not for this Symposium, but for another one to be held in another 50 years time, and bearing for its title: The Changing Role of Chemical Theory. [43]

3.7
The Impact of Computers in Quantum Chemistry: the Split of the Community

Let us now come to our last point. Since we want to talk until about the end of the 1960s when it was still possible to go a long way in both physics and chemistry without the use of computers, we shall be pointing out to the beginnings of the diverging trends among the chemists and the effects they had in the community, rather than discussing the totally new practice which was consolidated when computers started more or less to dictate to the theoretical chemists the kinds of problems they would work on and the ways to deal with these problems.

The introduction and growing dissemination of digital computers in quantum chemistry opened the way for the calculation of ever more difficult molecular integrals and made it possible to seriously consider the delineation of an extensive program of "completely theoretical" (*ab initio*) calculations. It was, in a way, an old dream come true. These calculations contrasted with those "semi-empirical" calculations, in which the impossible analytical calculation of certain parameters was substituted by the introduction of their values as given by experimental determinations. Semi-empirical calculations had become one of the constitutive aspects of quantum chemistry since its early days, and had contributed decisively to the articulation of its partial autonomy in relation to physics. What were the implications of *ab initio* calculations for quantum chemistry? It soon became clear that quantum chemists gave different answers to the former question, and that there was the danger of an irreversible splitting of the quantum chemical community reflecting divergent and irreconcilable attitudes towards the outcome of the use of computers.

The Conference on Molecular Quantum Mechanics held at Boulder, Colorado, in June 1960, was the first major meeting of its kind since the 1951 Shelter Island Conference. It was also the first meeting where the many theoretical chemists started realizing that there were deep – and perhaps irreconcilable – divisions in the community of quantum chemists among those who continued the semi-empirical calculations with the use of computers and the *ab initio-ists*. Coulson, again, emerges as one of the more perceptive observers of this situation and in the after dinner speech he delivered one finds Coulson not preaching tolerance but advocating partisanship. [44]

In discussing the major conclusions from the Conference he noted that "the whole group of theoretical chemists is on the point of splitting into parts... almost alien to each other." [45] The splitting was the result of the different views concerning the large-scale use of electronic computers – but there could even be a deeper reason than that. During the week of the conference, he had heard more than once the phrase "Oh, but you're not doing quantum chemistry." The occasions which gave rise to such assessments were the computational techniques presented for calculating energy values for atomic helium and molecular hydrogen, the calculations of a "highly empirical" kind to estimate energy levels and charge distributions of heteronuclear aromatic molecules and, the tabulation and interpretation of barriers to internal rotation in substituted ethane type molecules. His view was that these three situations represented quite distinct aspects of what used to be called quantum chemistry, since they differed considerably in their underlying assumptions. But each group thought that what the others did was not quantum chemistry. "The situation is indeed serious. For my own part, I am very far from laughing at it, and I want us to look at as openly and as dispassionately as possible. The questions that we are really asking concern the very nature of quantum chemistry, what relation it has to experiment, what function we expect it to fulfill, what kind of questions we would like it to answer. I believe we are divided in our own answers to these questions." [46]

The splitting, he thought, in the community resulted from the antagonism of two extreme groups. The first group possessed great computational skills and advocated that there are a number of problems that a dispute can only settle by computation since experiments are too difficult. To many people, this group of chemists appeared to be moving away from the conventional concepts of chemistry, such as bonds, orbitals, and overlapping hybrids "as to carry the work itself out of the sphere of real quantum chemistry." [47] On the other extreme were calculations with very rough approximations for biological molecules. These calculations give quite interesting results, but the approximations put forward would be greatly upsetting to the people who extensively used computers.

"Where, in all this, does "real" quantum chemistry lie?" Coulson wondered. The possibilities offered by the electronic computers enabled one to distinguish three levels of activity – a distinction with which most of the exponents of computing at the conference agreed.

Firstly, there are the molecules or atomic systems of 1–6 electrons, for which one could effectively calculate energies as accurately as they can be measured. Secondly, the all too realistic prospects for faster computers allowed to extend the range of molecules for which it would become possible to have effectively exact solutions to those with 6–20 electrons. Nevertheless, accurate results for these cases were achieved at the expense of visualizability. Coulson thought it was not very probable – and also not particularly desirable – to deal in such a manner with molecules of more than 20 electrons. There was such a deep distinction between those chemists whose main interest laid in the 1–20 range, and consequently thought in terms of full electronic computation, and those who did not think in these terms that the two groups deserved distinct names – Group I (the electronic computors or *ab initio-ists* as some would call them) and Group II (the non-electronic computors or *a posterior-ists*).

But he thought that it would be an oversimplification to think that the difference is only a difference having to do with the use of electronic computers. In their desire for complete accuracy, group I appeared to be prepared to "abandon all conventional chemical concepts and simple pictorial quality in their results." Against this, the exponents of group II argued that chemistry is an experimental subject, whose results are built into a pattern around quite elementary concepts. He did not make any effort to conceal that his sympathies lay with the latter and re-emphasized that the role of quantum chemistry is to understand these concepts and show what are the essential features in chemical behavior. Nevertheless, he was also aware that none of these concepts could be made rigorous.

> Chemistry itself operates at a particular level of depth. At that depth certain concepts have significance and – if the word may be allowed – reality. To go deeper than this is to be led to physics and elaborate calculation. To go less deep is to be in a field akin to biology. Once this is recognized, it is not difficult to see that there is a perfectly sound basis for all three comments about "not doing quantum chemistry" that I reported earlier. [48]

Coulson felt that it would be a great disaster if quantum chemistry were limited to either the "very deep" or the "shallow" level. And certainly it would be a serious loss if it did not maintain a close link with experiment and with conventional thought forms of chemistry. He felt strongly that there was a danger that group I people will forget that chemistry is associated with the real world. He ended in a pessimistic mood.

> Mathematically a bond is an impossible concept for group I. It is not surprising that it is practically never used by them. Yet the existence of bond properties is basic to all chemistry... It is not surprising that the orientations of these two groups of quantum chemists are so different that cross fertilization has now become much less frequent than in earlier days... Many members of group I do not realize what is happening to them; and members of both groups display an undesirable lack of sympathy for each other's work. [49]

In a way Coulson's work contributed decisively in making the chemists' nightmare come true: Dirac's pronouncement of 1929 could, in fact, be realized and chemistry was, in fact, physics. Though his work accelerated the *ab initio-ist* culture of theoretical chemistry, Coulson himself appears to be entrenched in the more traditional culture. He was deeply committed to the view that theoretical chemistry was first and foremost an enterprise whereby mathematical notions, numerical methods, experimental measurements, pictorial representations and, above all, chemical concepts, constituted an undivided whole. There was a fine balance among all these aspects, a balance that could not be articulated in any distinct way and yet, Coulson felt, it was the distinctive feature of theoretical chemistry itself.

References and Notes

The following abbreviations are used:
LP – Fritz London Papers, Duke University
PP – Ava Helen and Linus Pauling Papers, Kerr Library Special Collections, Oregon State University
SP – Nevil V. Sidgwick Papers, Lincoln College, Oxford University.

1 C. A. Coulson, *Valence* (Oxford University Press, 1952), preface, on v.

2 P. A. M. Dirac, "Quantum mechanics of many electron systems," *Proceedings of the Royal Society of London* A123 (1929): 714–733, on 714.

3 A. Simões, *Converging Trajectories, Diverging Traditions: Chemical Bond, Valence, Quantum Mechanics and Chemistry, 1927–1937*, Ph. D. Dissertation, University of Maryland at College Park, 1993; K. Gavroglu, A. Simões, "The Americans, the Germans and the beginnings of quantum chemistry," *Historical Studies in the Physical and Biological Sciences* 25:1 (1994): 47–110; K. Gavroglu, *Fritz London. A Scientific Biography* (Cambridge: Cambridge University Press, 1995).

4 H. Roscoe, "Presidential address to Section

B-Chemistry," *Proceedings of the British Association for the Advancement of Science* (1884): 659–669, on 666. Italics ours.

5 A. Smithells, "Presidential address to Section B-Chemistry," *Proceedings of the British Association for the Advancement of Science* (1907): 469–479, on 477.

6 Ibid., 478. Italics ours.

7 H. Armstrong, "Presidential address to Section B-Chemistry," *Proceedings of the British Association for the Advancement of Science* (1909): 420–454, on 423.

8 Ibid., 424.

9 SP. V-Correspondence. Item 71. Rutherford. Letter Rutherford to Sidgwick, 26 July 1937.

10 H.C. Longuet-Higgins, "An application of chemistry to mathematics," *Scientific Journal of the Royal College of Science* 23 (1953): 99–106, on 99.

11 Longuet-Higgins, "An application of chemistry to mathematics," 106.

12 A. Assmus, "The molecular tradition in early quantum theory," *Historical Studies in the Physical and Biological Sciences* 22 (1992): 209–231; A. Assmus, "The americanization of molecular physics," *Historical Studies in the Physical and Biological Sciences* 23 (1993): 1–33; A. Simões, K. Gavroglu, "Different legacies and common aims: Robert Mulliken, Linus Pauling and the origins of quantum chemistry," in J.-L. Calais and E.S. Kryachko (eds.), *Conceptual Perspectives in Quantum Chemistry* (Dordrecht: Kluwer, 1997), 383–413.

13 PP, Box 242, Popular Scientific Lectures 1925–1955, "Resonance and organic chemistry," 1941.

14 R.S. Mulliken, "Electronic structures of polyatomic molecules and valence. VI. On the method of molecular orbitals," *Journal of Chemical Physics* 3 (1935): 375–378.

15 Letter Pauling to Reymond Holmen, March 1987, as quoted in R. Holmen "Kasimir Fajans" *Bulletin for the History of Chemistry* 6 (1990): 7–15.

16 Walter Hückel's *Theoretische Grundlagen der organischen Chemie* (1931) included quantum interpretations and was very influential when eventually translated into English. H. Kragh, "The young Erich Hückel: His scientific work until 1925," invited paper given at the Erich Hückel Festkolloquium at the Philipps-Universität, Marburg, 28 October 1996; J.A. Berson, "Erich Hückel, pioneer of organic quantum chemistry: Reflections on theory and experiment," *Angewandte Chemie International Edition in English* 35 (1996): 2750–2764; A. Karachalios, "Die Entstehung und Entwicklung der Quantenchemie in Deutschland," *Mitteilungen der Gesellschaft Deutscher Chemiker Fachgruppe Geschichte der Chemie* 13 (1997): 163–179.

17 C.A. Coulson, "Recent developments in valence theory," *Pure and Applied Chemistry* 24 (1970): 257–287, on 259.

18 LP, Letter D. Whyte to London, 14 December 1929; Letter Fowler to London, 14 January 1930.

19 R.H. Fowler, "A report on homopolar valency and its mechanical interpretation," in *Chemistry at the Centennary Meeting of the British Association for the Advancement of Science* (Cambridge: W. Heffer and Sons Ltd, 1932), 226–246, on 226.

20 J.E. Lennard-Jones, "The nature of cohesion," *Nature* 128 (1931): 462–463, on 462.

21 A. Simões, K. Gavroglu, "Quantum chemistry *qua* applied mathematics. The work of Charles Coulson (1910–1974)," *Historical Studies in the Physical and Biological Sciences* 29 (1999): 363–406.

22 G.W. Wheland, *The Theory of Resonance and Its Applications to Organic Molecules* (New York: John Wiley & Sons, 1944).

23 PP, Box 115, Letter Wheland to Pauling, 20 January 1956.

24 PP, Box 115, Letter Pauling to Wheland, 26 January 1956.

25 PP, Box 115, Letter Pauling to Wheland, 8 February 1956.

26 L. Pauling, "Modern structural chemistry. Nobel lecture, December 11, 1954," in *Nobel Lectures in Chemistry 1942–1962* (Amsterdam: Elsevier Publishing Company, 1964), 134–148.

27 L. Pauling, "The nature of the theory of resonance," in A. Todd (ed.), *Perspectives in Organic Chemistry, Dedicated to Sir Robert Robinson* (New York: Interscience publishers, 1956), 1–8.

28 L. Pauling, *The Nature of the Chemical Bond and the Structure of Molecules and Crystals. An Introduction to Modern Structural Chemistry* (New York: Cornell University Press, 1967), third edition, 215–220.

29 L. Pauling, "The nature of the theory of

resonance," in *Perspectives in Organic Chemistry*, 6–7; L. Pauling, *The Nature of the Chemical Bond*, 219–220.

30 C. A.Coulson, "The meaning of resonance in quantum chemistry," *Endeavour* 6 (1947): 42–47, on 47.

31 D. N. Kursanov, M. G. Gonikberg, B. Dubinin, M. I. Kabachnik, E. D. Kaveraneva, E. N. Prilezhaeva, N. D. Sokolov, R.Kh. Freidlina, "The present state of the chemical structural theory." Translation in English by I. S. Bengelsdorf, published in *Journal of Chemical Education* (January 1952): 2–13; V. M. Tatevskii, M. I. Shakhparanov, "About a machistic theory in chemistry and its propagandists." Translation by I. S. Bengelsdorf in *Journal of Chemical Education* (January 1952): 13–14. It was not, of course, the case that such sentiments were shared by all the chemists of the community. Characteristic of the differences is the editorial note to the first article where it is stressed that any particular way of dealing with chemical phenomena should not be excluded on *a priori* grounds, but should be first closely studied. In the same article Ya. Syrkin and M. Dyatkina were also attacked. They were the authors of the excellent book *The Chemical Bond and the Structure of Molecules* and had translated Pauling's book into Russian. See also I. Moyer Hunsberger, "Theoretical chemistry in Russia," *Journal of Chemical Education* (October 1954): 504–514.

32 D. N. Kursanov et al., "The present state of the chemical structural theory." Translation in English by I. S. Bengelsdorf published in *Journal of Chemical Education* (January 1952): 2–13, on 5.

33 PP, Box 261. Letter M. V. King to Pauling, 23 January 1953; letter Coulson to Pauling, 7 October 1953; letter Coulson to King, 18 January 1954; letter King to Pauling, 9 February 1954.

34 C. A. Coulson, "What is a chemical bond?" *Scientific Journal of the Royal College of Science* 21 (1952): 11–29, on 11.

35 C. A. Coulson, *The Spirit of Applied Mathematics* (Oxford: Clarendon Press, 1953), 20–21.

36 Ibid.

37 C. A. Coulson, "Recent developments in valence theory," Symposium: Fifty Years of Valence Theory, *Pure and Applied Chemistry* 24 (1970): 257–287, on 287.

38 Coulson, *The Spirit of Applied Mathematics*, 20–21.

39 Coulson, "What is a Chemical Bond?" 13.

40 C. A. Coulson, "The contributions of wave mechanics to chemistry," *Journal of the Chemical Society* (1955): 2069–2084, on 2084.

41 Coulson, *The Spirit of Applied Mathematics*, 20–21. Italics ours.

42 Coulson, "What is a chemical bond?" 12.

43 Coulson, "Recent developments in valence theory," 287.

44 C. A. Coulson, "Present state of molecular structure calculations," Conference on Molecular Quantum Mechanics, University of Colorado at Boulder, June 21–27, 1960, *Reviews of Modern Physics 32* (1960): 170–177.

45 Ibid., on 172.

46 Ibid.

47 Ibid.

48 Ibid., on 174

49 Ibid.

4.
Giovanni Battista Bonino and the Making of Quantum Chemistry in Italy in the 1930s

Andreas Karachalios

4.1
Introduction

In this paper, I will consider the emergence of a new scientific discipline, quantum chemistry, in Italy. This requires taking developments in three fields into account: physical organic chemistry, quantum physics, and mathematics. Quantum chemistry thus emerged as a new field with interdisciplinary character, an area of research founded on physical knowledge applied to chemical problems. Moreover, the term interdisciplinary indicates an orientation toward scientific research distinguished by the crossing of boundaries between established disciplines, an interaction with methodological, logical, and conceptual implications.

The leading person in Italy who conducted this process was Giovanni Battista Bonino (1899–1985), founder of a research school in physical chemistry at the University of Bologna during the 1930s. [1] As will be shown, Bonino's views about theory-building and the role of experiment in chemistry were decisive for the making of this new discipline. Moreover, I will emphasize the role of infrared and Raman spectroscopy as well as the importance of the mathematical formalism, in particular group theory, for the emergence of quantum chemistry in Italy. Recent historiography of science has shown how instruments and instrumental techniques were crucial for the development of the chemical and physical sciences. [2] In accord with this, we shall see that for the emergence of quantum chemistry a significant function was carried out by the contacts between Italian, Austrian, and German scientists – primarily chemists, physicists, and physical chemists – at the beginning of the 1930s. In particular, these relations involved the importation of certain laboratory techniques into Italy. Then, I will also show how this interest in laboratory techniques was connected with Bonino's very specific theoretical interests at the frontiers of Italian quantum chemistry.

Since Bonino's personality was greatly influenced by the political climate in fascist Italy, one must consider him and his research group in relation to the social, political, and cultural environment of that time. Thus, Bonino's scientific politics and the emergence of the new discipline of quantum chemistry were embedded in a larger political context that deeply influenced the scientific culture of the period. Among the significant political events were the establishment of the Rome–Berlin axis in October 1936, the cultural agreement between Rome and Berlin during the autumn of 1938, the steel pact of May 1939, and the wartime alliance between Nazi Germany and fascist Italy, beginning in March 1940. These events inevitably marked scientific relations between the two countries during the late 1930s.

4.2 Early Career

Bonino was born on 3 May 1899, in Genoa, Italy. He attended school and later the university at the city of his birth. It was during his school days that Bonino became interested in science, although at the time he felt a strong attraction for languages and humanistic fields, such as philosophy. His scientific interests, however, already pointed him toward the study of chemistry, and he entered the University of Genoa with that purpose, taking his degree *summa cum laude* on July 18, 1920. [3] Alongside his principal field of study, Bonino chose theoretical physics as his secondary field. In his "Tesina di fisica teorica", a short research paper, he dealt with the vaporization of liquids. [4] For this work, Bonino consulted, among other sources, the French translation of Max Planck's lectures on thermodynamics. Shortly afterwards, Bonino presented his thesis work in analytical chemistry. [5] This was the principal topic of his studies, which had a more experimental rather than a theoretical emphasis. Bonino used purely chemical methods to study the quantitative relations of the action of $K_4[Fe(CN)_6]$ on AgCl in the presence of AgBr and AgI. His quantitative and qualitative analysis attested to the presence of the complex compound $Ag_3K[Fe(CN)_6]$. Bonino did not confine himself to studying the formation of this new compound, but also sought to explain its constitution. Thus he elucidated his findings by means of Alfred Werner's coordination theory and its applications to the complex compounds of cyanides. This approach had been realized earlier by the Italian chemist Arturo Miolati, one of Werner's pupils. [6] In his thesis Bonino also attempted to explain the number of valence units acting on the iron atom in accordance with the electroatomic theory of Johannes Stark. [7] With this study, Bonino showed that he possessed an excellent knowledge of physics and chemistry along with the ability to present his ideas clearly and to make creative use of modern theories of the chemical bond.

Immediately after graduating, he began work at the chemical company Andrea Bevilacqua in Genoa. Here he proved to be an imaginative theorist as well as a meticulous experimenter, combining techniques in pure as well as applied chemistry. Among many other tasks, Bonino was called upon to solve problems involving the theoretical interpretation of experimental and observational evidence. For this reason, Bonino turned in 1921 to Mario Betti, a professor at the new "*Istituto di Chimica Farmaceutica*" at the University of Genoa. [8] Betti was well known from numerous publications, in which he had been using physical methods to approach chemical problems. Especially well known were his studies concerning the relationship between the chemical constitution and optical properties of organic compounds. This early contact with Betti came at a crucial moment in Bonino's academic career, which began at the "*Istituto di Chimica Farmaceutica*", where he worked first as a technician and the following year as Betti's assistant. [9] Using an infrared spectrograph, Bonino began investigating the relationships between physical properties and chemical constitution of organic compounds. The instrument he employed for this research was the "Ultraspiegelspektrometer" first introduced by Heinrich Rubens; this apparatus was made available by the chemical company where he worked. [10]

Although this spectroscopic work was at the cutting edge of research, Bonino also realized that his knowledge of physics and mathematics left a lot to be desired. For this reason, he worked hard to overcome these deficiencies and to catch up with the most recent developments in physics. Among the most important works which he studied at this time were *Atombau und Spektrallinien* by Arnold Sommerfeld as well as Stark's *Prinzipien der Atomdynamik*. [11] These books played a pivotal role in Bonino's education in quantum theory and spectroscopic analysis. Two other books, William Coblentz's *Investigations of Infrared Spectra* and Victor Henri's *Etudes de Photochimie*, were also very important for furthering his knowledge of the experimental side of this new field. [12] Coblentz's study contained over one hundred spectra of organic compounds and provided the first experimental evidence for the link between molecular structure and the spectral characteristics of chemical substances. His analyses of the intensity of absorption in relation to the constitution of the molecule were, however, solely of a qualitative nature. Henri was the first to study spectroscopic bands in a quantitative manner; he also made an attempt to consider the intensity of the bands at 3.4 microns as an additive property of the groups CH, CH_2, and CH_3 in the molecule.

4.3 Bonino and the Beginning of Infrared Spectroscopy in Italy

In the meantime, Bonino's mentor, Betti, was transferred from Genoa to the chair in general chemistry at Bologna. Bonino followed him and, after having received his *Libera Docenza per titoli in Chimica Fisica*, was appointed in 1924 as assistant professor of physical chemistry at the University of Bologna. Three years later he became professor of physical chemistry at the Superior King School of Industrial Chemistry in Bologna, and in 1930 he accepted a new chair on physical chemistry at the university there.

Bonino worked intensely at developing his own research agenda, based on a mature understanding of theoretical and experimental techniques. A series of reports on this work appeared, beginning with a preliminary paper in 1923 entitled "Studi di spettrochimica nell'ultrarosso". This was the first of 14 papers that came out in the next 3 years in the *Gazzetta Chimica Italiana*, the principal journal for publications of the Italian chemical society. [13] In these papers, Bonino gave support for a new hypothesis, differing from Henri's, that bands in the vicinity of

3.4 microns must be attributed more precisely to the variation of quantum oscillations set up by the C-H and N-H linkage of the molecule. Moreover, he showed how, in certain simple cases of organic liquid molecules, the intensity of the band in question can be expressed as a function of the hydrogen atom combined with carbon or nitrogen in the molecule. Bonino's interpretation was accepted soon afterward by the University of California physicist Joseph W. Ellis, who worked on the molecular absorption spectra of liquids. [14]

Bonino brought forward a further contribution to the theory of infrared spectra of organic liquids by incorporating the Bohr-Sommerfeld quantum conditions, including the correspondence principle of Bohr as well. This paved the way toward establishing a correlation between the physical and chemical image of molecules in the study of infrared spectra. From this series of papers on infrared spectroscopy, one can already observe the interdisciplinary character of Bonino's thought. In a lecture delivered some years later, Bonino offered these reflections on his chosen field of research:

> It is well known that the infrared spectrum represents the molecule and, in particular, the architecture of the molecule of organic chemistry as a physical reality. The structure formula of the organic chemists lacked this. In organic chemistry the structure formula made use of geometrical analogies to register in a concise way the chemical properties. The infrared spectrum, on the other hand, represents the molecule as a mechanical structure. It conveys the physical reality by showing the proper oscillations that all the atoms in the molecule can realize. Therefore, studying the infrared spectrum of a substance means finding *analogies* between the *physical* and *chemical reality* of the molecules. It is evident that these *analogies* establish concomitant relations that we often interpret as cause and effect between the *physical dynamism* of the molecule and its *chemical behavior.* [15]

Thus, according to Bonino, the correlations between the "physical reality" and the "chemical reality" of molecules were based on analogous structures. By finding these, one established a bridge between the dynamical molecular models of physicists and the statical models of chemists. This notion of analogy based on spectroscopic studies gave a new meaning to the word in chemical circles. For whereas the classical organic chemist used the concept of analogy to establish similarity relations between various members of a class of compounds, the spectroscopist was able to establish correlations between two very different models.

Bonino was the first chemist in Italy at the beginning of the 1920s to perceive and to exploit the importance of infrared spectroscopy for chemistry. The infrared technique offered considerable advantages for researchers working on the composition and structure of chemical compounds. While traditional methods of analysis, such as measurements of melting point or refractive index, yield information about one particular characteristic of the compound, an infrared spectrum offers several physical indicators. Despite their importance to infrared spectroscopy, however, Bonino's papers did not lead to a strong reception along the same lines within the Italian chemical community. Italian chemists continued to use classical methods of chemical analysis. Moreover, in the universities organic chemists were not interested in engaging in the delicate, labor intensive calibrations and adjustments that the infrared technique still required. Indeed, the high cost of instruments made it virtually impossible for academic laboratories in the post-war era to pursue such research.

4.4
The Scientific and Political Context

The 1920s were watershed years in both politics and physics, a period which witnessed the rise of the new quantum and wave mechanics as well as the ascent of Italian fascism. The years between 1925 and 1929 were highly significant ones for the construction and stabilization of the new regime. [16] Among the many important changes closely bound to the science of this period came the politicization of academic institutions. These were the years in which the academic world – its institutions and individual members, including Italian chemists – openly supported the new regime. [17] Important scientific centers for this support were Bologna and Pavia. This is not the place to go into details about the politicization of academic institutions in fascist Italy. Nevertheless, a brief reference to these events will help illuminate the influence of the political context and particularly Betti's political role in Bonino's academic career. Taking these earlier events into account makes it easier to understand both Bonino's involvement with fascism and his scientific politics during the late 1930s.

On 29 March 1925, the *Convegno Nazionale delle Istituzioni Fasciste di Cultura* was inaugurated in Bologna. On this occasion, scientists, artists, politicians, and journalists from all over Italy gathered and declared their support for the fascist movement. Under the presidency of the philosopher Giovanni Gentile, they decided to express this support in a document that was later called the "fascist manifesto." [18] Among the 44 natural scientists who were present, 11 were from the University of Bologna, including the chemists Mario Giacomo Levi, Ernesto Belloni, Gian Alberto Blanc, Raffaello Nosini, and Giuseppe Plancher. Two months later, the scientific society *Sips (Società Italiana per il progresso delle scienze)* held its 14th meeting from May 24–29, 1925 in Pavia. This organization brought together chemists, physicists, biologists, mathematicians, and physicians in a forum whose principal purpose was to unite these various disciplines, in order to orchestrate national scientific politics. In particular during the 1920s, *Sips* aimed to focus attention on relationships between theory and praxis. At the above mentioned meeting, several lectures were dedicated to possible applications of science for the national defense. Furthermore, a new section devoted to the military sciences was founded in order to facilitate collaboration between scientific research and military technology. In one particular lecture, the lieutenant colonel Natale Pentimalli emphasized the importance of chemical research for modern aviation warfare, thereby indicating how chemistry might play a pivotal role in supporting the new fascist regime. [19]

At the next meeting of *Sips*, held in Bologna from October 30 to November 5,

1926, Benito Mussolini chose not only to attend but even to intervene personally. In his talk during the opening ceremony, he attached great importance to the relationship between fundamental research at the universities and applied research in industry. In particular, he forcefully underlined the need to cultivate technological applications of chemistry for the national defense.[20] This effort, in fact, took priority over all others in the scientific politics of Italian fascism. Starting in 1926, Italy founded new university institutes, chairs, and chemical research laboratories, particularly in applied chemistry, electrochemistry, and physical chemistry. Parallel to these, several technical institutes were founded by private companies, including the technical institute of Ernesto Breda in the vicinity of Milan. The Breda institute was under the presidency of Nicola Parravano. As the Mussolini regime's preeminent chemist, Parravano worked for industry as well as for the national defense.[21]

In planning for the Bologna meeting of *Sips*, the executive committee submitted its program to Mussolini, asking him which lectures he would like to hear at the opening ceremony.[22] Mussolini chose, among others, Betti's lecture on "*Problemi ed aspetti della chimica della materia vivende*". In his presentation, Betti noted the role of glycerine as an initiating substance for the production of explosives. He then reported on the latest achievements in biochemistry and biology, emphasizing the importance of molecular asymmetry for organic matter.[23] Some years later Betti was elected as a Senator. Thus, both he and Bonino belonged to that group of chemists who enjoyed special favor with the government. In the early 1930s, all professors were obliged by governmental decree to swear their allegiance to fascism.[24] For Bonino and Betti, unlike many others, this represented a welcome opportunity to demonstrate their fidelity to the regime. Betti's good standing with Mussolini paved the way for the creation of a new institute of physical chemistry in Bologna. This was founded on Betti's initiative, with Bonino serving as its first director (Figure 4.1).[25]

The academic career of Bonino thus began in a context in which science, fascism, and national defense in Italy became tightly bound together. Bonino's strategy was to transform the institute in Bologna into a new kind of center for physical chemistry in Italy. He pioneered a new style of teaching, and promoted a new didactical approach to the various parts of physical chemistry with considerable success. In the traditional lecture courses in physical chemistry, the thermodynamical and kinetic-statistical methods were taught together in close connection with each other. However, Bonino found this approach didactically inferior and therefore introduced these methods independently so as to maintain a sharp distinction between thermodynamical and statistical thinking. Bonino taught these two approaches in two different academic years.[26] Furthermore, starting in 1927–1928, he completed his kinetic-statistic lectures by giving a short account of the new matrix and wave mechanics, including references to the classical paper of Walter Heitler and Fritz London.[27]

At this time, Bonino's research was concerned with physico-chemical problems and spectroscopical methods. In a parallel way, his laboratory activity turned to the electrolytic reduction of some natural organic substances. It was hoped that this

Figure 4.1 Giovanni Battista Bonino in his office in Bologna in the 1930s. (Courtesy of Andrea Concolato.)

research could contribute toward solving some important experimental questions with potential practical applications for industry and the national defense, such as the production of glycerine.[28] Between 1927 and 1929 Bonino supervized 35 graduate students, nearly all of whom took their degrees in three distinct areas of experimental research: six dealt with themes in molecular absorption spectra of organic liquids; a somewhat smaller group wrote on topics in the theory of strong electrolytes (known as Debye-Hückel theory), whereas the lion's share, around 25, took up topics in physical chemistry and electrochemistry.[29] Among this last group were Lazlo Brüll, Paolo Cella, and Reginaldo Manzoni-Ansidei, all of whom became active members of Bonino's scientific group during the 1930s.[30]

Bonino soon became well known outside Italy, owing to his research on infrared spectroscopy. The results from his papers were reported in the modern monograph by Jean Lecompte *Le Spectre Infrarouge*.[31] In recognition of his achievements, Bonino was invited by the secretary of the Faraday Society to participate at a meeting in Bristol in September 1929.[32] The theme of the meeting was "Molecular Spectra and Molecular Structure", and Bonino spoke "On the Infra-Red Bands of Hydrogen Combined with Carbon in Molecules of Organic Compounds."[33] What impressed Bonino most in Bristol were the talks of John Lennard-Jones and Chandrasekhara

Figure 4.2 From left to right Mario Betti, Chandrasekhara Raman, Alessandro Ghigi (Rector), and Giovanni Battista Bonino at the chemical institute of the University of Bologna in 1937. (Courtesy of Andrea Concolato.)

Raman. On this occasion, Lennard-Jones proposed a molecular orbital analysis for diatomic molecules, whereas Raman discussed his recent discovery, which later became known as the Raman effect.

In 1928 Bonino had opened the March 31 issue of *Nature* and read about the Raman effect for the first time. Soon afterward, he ended his long series of papers on infrared spectra by publishing two articles that dealt with the new Raman spectroscopy.[34] These were written in German and published in *Zeitschrift für Physik* in 1929. In the first of these two papers, dated May 1929, Bonino reported on the Raman spectra of dichloroethylene and tetrachloroethylene and discussed their relationship with the infrared spectra. The second paper was written with his co-worker Brüll and appeared in the fall of 1929. In it they studied the relationships between Raman spectra and geometrical isomerism.[35]

Bonino formed an acquaintance with Raman and, during the 1930s, his institute in Bologna was in regular contact with Raman's in Bankalore; Raman later visited Bonino's institute (Figure 4.2). This relationship led to a major reorientation of the methods utilized by Bonino's research team. Before meeting Raman, Bonino's team mainly relied on infrared spectroscopy. The infrared method, however, presented various technical difficulties. Bonino himself was quite explicit on this point, once remarking that "the process of mapping infrared spectra is, at best, a slow and tedious one."[36] Even in the 1920s, obtaining infrared spectra could become routine only if enough time and money were invested in the calibration and standardization of equipment. These conditions were usually only found at the largest chemical companies and thus were not available at typical research laboratories. Moreover, infrared spectroscopy did not make the lower frequencies accessible. With the discovery of the Raman effect, nearly the whole range of vibrational frequencies could be obtained relatively easily using simple and cheap apparatus. For these reasons, Bonino and his research group quickly turned to the application of Raman spectroscopy to organic compounds during the 1930s.

4.5 Scientific Contacts in Germany and Austria, 1931–1934

These new experimental methods in organic chemistry helped pave the way toward establishing important Italian-German scientific relations during the early thirties. Thus, from 1931 to 1934, Bonino began to visit various institutes and laboratories in Europe, in particular the German research institutes. One of his main aims was to learn more about how to set up a laboratory for investigating the structure of organic molecules. It should be emphasized that the Italian chemists had a long tradition of strong scientific relations with their colleagues in Germany. Thus Betti had earlier been in close touch with Hans Fischer at Munich. Moreover, Betti's predecessor in Bologna, Giacomo Ciamician, had been a member of the German Chemical Society. The chemical laboratory he founded in Bologna around 1900 was similar to those in Germany and Austria, where he had studied.[37] Thus Bonino's interest in cultivating contacts with German chemists was by no means new.

Through the mediation of Betti and Parravano, the "Reale Accademia d'Italia" helped finance Bonino's foreign scientific travels.[38] In a report in 1935 to the Academy, Bonino shed light on the motives that led him to undertake these trips:

> I think that in Italy it might be necessary to organize, as is happening in the United States of America, some laboratories where students might learn the importance of the new theoretical and experimental techniques of *physical organic chemistry*. I hope that in these laboratories it would be possible for the students to form a special way of thinking. It should be done in this way in order to understand chemical reactions, Raman spectra, dielectrical constant measurements, Eigenfunctions, secular equations, and exchange integrals.[39]

With these new methods, Bonino hoped to see the emergence of a new kind of Italian school of chemistry with strong ties to theoretical physics. In his report Bonino underlined the necessity of building a new institute in which the students could learn the basics of this new discipline: physical organic chemistry. The objective was to form a new interdisciplinary mentality in the Italian chemical environment. This mentality required a crossing or amalgamation of different scientific concepts and languages. All these concepts, as Bonino reported, belonged to different fields or subjects of study, which were usually called "disciplines".

Table 4.1 Bonino's visits to foreign laboratories, 1931–1934

Institution (Department Head)
Physical Institute in Graz (Kohlrausch)
Physical Institute in Breslau (Schäfer)
Physical Institute in Leipzig (Debye)
Chemical Institute in Leipzig (Weissberger and Wienhaus)
Zeiss company in Jena (Löwe and Hansen)
Institute of Organic Chemistry at the Technische Hochschule in Munich (Hans Fischer)
Institute of Physical Chemistry in Munich (Fajans)

Bonino did not give a clear definition of what he meant by "physical organic chemistry". We can make out, however, that he understood the new discipline as located between physical chemistry, organic chemistry, and quantum physics. Moreover, he underscored its interdisciplinary character.

Bonino was not able to visit any American institutes of physical organic chemistry, as he mentioned in his report: "Due to lack of funding I have visited, within the limitations of the Academy's financial possibilities, only some European institutes. The American institutes are at a very advanced level, and I think that a visit to the North American scientific institutes could be very interesting." [40] During these years Bonino visited seven institutes in Austria and Germany (Table 4.1).

Bonino's travels to Germany and Austria were undertaken in order to gather didactic, experimental, and theoretical ideas. His plan was to transfer know-how to Italy by organizing his own institute according to modern German models. In the fall of 1931 he visited the physical institute at the University of Graz whose director, Karl Wilhelm Fritz Kohlrausch, was well known internationally for his pioneering work on the Raman effect in organic compounds. Kohlrausch had just published his classic book *Der Smekal-Raman Effekt* when Bonino arrived. [41] In Graz, Bonino acquainted himself with the particular experimental device used by Kohlrausch's research team for the study of Raman spectra. Bonino's assistant, Cella, later spent time in Graz, and in April 1933 Kohlrausch visited Bonino's institute in Bologna. On this latter occasion, Kohlrausch held a seminar on applications of the Raman effect in organic chemistry and discussed strategies with Bonino for organizing the research program in Bologna. In the winter of 1932, Bonino visited Clement Schäfer at the physical institute in Breslau. This institute was well-known because two of its researchers, Bergmann and Frank Matossi, had worked out a special system of amplification with photoelectrical relays to confer a greater sensitivity in spectrochemical measurements. Bonino later sent his assistant, Manzoni-Ansidei, to Breslau to learn more about this.

Afterwards Bonino made a short stop in Leipzig in order to discuss electrolytic theory with Peter Debye. He also visited the chemical institute in Leipzig, where he met Weissberger. Bonino took a great interest in "his beautiful applications of modern physical chemistry to the field of organic chemistry." [42] Interestingly enough, Bonino's report did not refer to Werner Heisenberg, Friedrich Hund, or Erich Hückel, all of whom worked on quantum mechanics in Leipzig at this time. Here, however, it is necessary to point out that Bonino's principal interests centered

Figure 4.3 From left to right Hans Fischer, Arnold Eucken, and Giovanni Battista Bonino at a excursion to Toledo after the congress of Madrid in 1934. (Courtesy of Andrea Concolato.)

on the technical, experimental, and didactical aspects rather than theoretical ones. In a later report on his scientific and didactical activity to the Minister of Education, Bonino expressed his belief that in the new field of infrared spectroscopy experiment should take precedence over theory. On this occasion, he emphasized the need to obtain new experimental results which would establish the basis for a more coherent theoretical treatment. Still, he recognized the importance of recent theoretical work, noting that "both the matrix theory of Born-Heisenberg and the wave mechanics of Schrödinger allow a new reconstruction of the attempts made on the basis of the old quantum theory." [43]

In the winter of 1932 Bonino visited the scientific laboratories of the Zeiss firm in Jena, where he became acquainted with Löwe and with the director of the spectroscopic department, Hansen. He and Hansen discussed technical details concerning the modern spectrograph that Zeiss had built for the institute in Bologna for utilizing the Raman effect. At Zeiss, Bonino could thus learn further technical aspects of the Raman spectrometer and its potential applications. Soon thereafter, he and his collaborators developed an intensive research program with this spectrograph in order to study the constitution of organic compounds.

In 1932 Bonino spent considerable time at Hans Fischer's institute of organic chemistry at the Technische Hochschule in Munich (Figure 4.3). There, Bonino wished to become better informed about the organization of a major laboratory for organic chemistry with a section devoted to physico-chemical researches. At Fischer's institute this section was entrusted to Alfred Stern. This visit was very important to Bonino's future research on the constitution of aromatic compounds with five atoms (pyrrole, furane, and thiophene) and led to a regular cooperation between Bonino's institute and Fischer's. In 1933 Bonino worked at Fischer's institute on three separate occasions. Their joint research efforts centered on the utilization of Raman spectroscopy in studying the structure of heterocyclic compounds and their derivatives. Bonino and Fischer undertook this work together by way of a clear division of labor. The organic chemical work was carried out at Fischer's institute with the collaboration of Pietro Pratesi, who had studied earlier with Betti and Bonino in Bologna. The Raman spectroscopical research, on the other hand, was pursued at Bonino's institute with the help of his assistant Manzoni-Ansidei. For this research Bonino utilized the new micro-Raman method, which was developed by P. Grassmann at the Physical Institute of the University of Munich.[44] Bonino became personally acquainted with Grassmann, who gave him some useful advice on the micro-Raman method. In Munich he also became acquainted with other personalities in the local scientific community, including the physical chemist Kasimir Fajans and the organic chemist Richard Willstätter. Bonino had lengthy discussions with Willstätter about modern developments in organic chemistry as well as his own researches on the constitution of aromatic compounds.

4.6
Early Contributions to Quantum Chemistry

Bonino's first results in this direction were published in four papers, three in *Zeitschrift für Physikalische Chemie* in 1933 and 1934 and one in *Memorie della Reale Accademia delle scienze dell'istituto di Bologna*.[45] Through the study of the constitution and the aromatic character of the heterocyclic compounds, Bonino confronted the classical ideas of structural organic chemistry. In the case of heterocyclic compounds he and his collaborators emphasized that the classical structure formulae of organic chemistry could not account for the new Raman spectroscopic data. According to these, the existence of a double chemical bond for pyrrole, furane, and thiophene was improbable.

In Germany during the period 1931–32, Erich Hückel made an important start in this direction in a series of papers on the quantum mechanics of the aromatic and unsaturated molecules, which gave the first quantum theoretical description of benzene.[46] Two years later, in 1933, Pauling and his collaborators in America published a series of papers on the same subject. In his treatment of the benzene molecule, Pauling assumed that the wave function can be written as a superposition of the wave functions associated with the different valence bond structures according to the pictorial representation of classical organic chemistry. As Ana Simões and Kostas Gavroglu have analyzed in detail, Pauling followed a pragmatic approach and took an overall view close to established chemical traditions.[47] As I have described elsewhere, Hückel emphasized the insufficiency of the Kekulé structures in accounting for the chemical properties of benzene. There, thus, arose a dispute between Hückel and Pauling with deep methodological implications. [48] Hückel also discussed some similarities between the chemical behavior of benzene and other heterocyclic compounds such as pyrrole, furane, and thiophene. Furthermore, Hückel seized the occasion to underline the necessity for a further study of these compounds and their properties.

Following Hückel's indications, Bonino began to work on these problems using the new Raman spectroscopy. His original contribution consisted in a new formula for the heterocyclic compounds with alternate polarized bonding and in quantum resonance. In the 1920s and early 1930s, the idea of polarized bonding was by no means new. This notion had its roots in nineteenth-century Berzelian electrochemical dualism, which was revived by Bonino and other scientists by positive/negative carbon atoms in the aromatic compounds. [49]

Through his studies on the aromatic character of the heterocyclic compounds, Bonino inevitably confronted the classical problem of the structure of benzene. [50] He completed his work on the Raman spectrum of aromatic compounds which included benzene, presenting his results in April 1934 at the 9th International Congress of Pure and Applied Chemistry in Madrid. At the Madrid Congress, Bonino recommended a new formula for benzene; this formula, however, lacked a rigorous quantum mechanical grounding. Rather, it represented an attempt to summarize qualitatively some fundamental ideas in the wave-mechanical interpretation of benzene (Figure 4.4).

With this formula, Bonino brought together "the magisterial physical-mathematical work" of Hückel and Pauling's model, which showed "the necessity of admitting different structures in resonance." [51] In other words, Bonino's benzene formula was a bridge between the mathematical abstractions of quantum physics, on the one hand, and the intuitive graphical representations of organic chemistry, on the other. To define the valency of the carbon atoms in his formula, Bonino relied on the quantum mechanical reinterpretation of this notion. In 1928, Fritz London had published a paper in which he gave a quantum mechanical explanation of the classical notion of valency. There he showed a relationship between the valency numbers and the spectroscopical multiplicity, namely that valency = multiplicity −1. [52] This meant that the carbon atoms in Bonino's formulae for both the ^4S and ^4P configurations have valency three. In addition, according to Bonino, there were other interactions between the negative carbon atoms besides the proper bond due to this trivalency. Concerning these, however, Bonino could give no convincing explanation. He spoke about a vague "coordination" between the negative carbon atoms, referring to the book *La coordination des atomes dans la molécule* of the French chemist Georges Urbain. [53]

Bonino's principal goal with his formulae was to furnish the chemist with a new pictorial representation of chemical facts necessary for studying the chemical as well as the physical properties of the aromatic compounds. Bonino's contribution thus furthered the utility of the old structural formulae of organic chemistry by embracing certain physical properties. His considerations regarding the structure of benzene and heterocyclic aromatic molecules allowed him to foresee and to explain

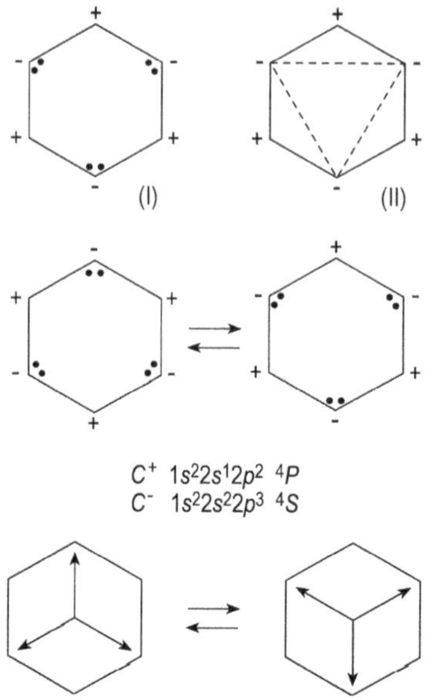

Figure 4.4. Bonino's 1934 formula for benzene. The internal arrows show the negative carbon atoms and the corresponding quantum resonance between forms I and II. The horizontal arrows indicate that the configurations of the carbon atoms have the same probability.

the phenomenon of orientation in an aromatic substitution, and particularly in monosubstituted benzene. This was possible through the measurement of dipole moments and Raman spectra.

Bonino's work can therefore be seen as an attempt to realize a synthesis between physical and organic chemistry by utilizing quantum physics. The power of this synthesis lay in its attempt to meet the conceptual demands involved in understanding the relationship between molecular physics and molecular constitution. Some aspects of the work of Bonino's research group contributed to framing this new theoretical synthesis, which the Bolognese called "physical organic chemistry." Among these were the concept of resonance, the schema of orientation in an aromatic substitution, and other aspects of physical chemistry such as measurement of dipole moments, coupled with experiments using infrared and Raman spectroscopy.

Bonino's new symbolic representation was utilized by other Italian chemists to investigate the chemical and physical properties of other classes of heterocyclic compounds. [54] Nevertheless, his formulae attracted little notice outside Italy. Why this was the case becomes clearer if we take into account the work of the Cambridge chemist Christopher Ingold. Ingold was comfortable working in the laboratory as well as behind his desk, and in the two domains of physical and organic chemistry. In 1934 he published a classic paper in *Chemical Reviews* dealing with the theory of organic reactions. [55] As Mary Jo Nye has analyzed in detail, Ingold presented a theory not only of aromatic chemistry but also of aliphatic chemistry, thereby unifying two major branches of organic chemistry while linking them with physical chemistry and with physics. He began with physical notions of electric polarization of molecules and submolecular groups and then went on to discuss the electron theory of valency.[56] Bonino's work was thus very fragmentary compared with Ingold's. Nevertheless, the Italian's contributions were without doubt original. Even if he made use of the most recent developments in quantum physics, Bonino's viewpoint and his symbolical representation were centered on chemical realities.

Raman and infrared spectroscopy proved valuable tools for establishing the symmetry of organic molecules. After the Madrid Congress, Bonino extended the agenda of his research group to include this new direction. The results they obtained were of great importance for the development of valency theory in the late 1930s. As we will see, they combined fruitfully the new experimental direction of Raman spectroscopy with group theory to reshape the boundaries of physical organic chemistry and opened the way for the new discipline of quantum chemistry.

4.7 Bonino's Place within Contemporary Research

The late 1930s proved to be crucial years for the genesis of quantum chemistry in Italy, which by 1935 had begun to take form as a new discipline, a process that consolidated by around 1940. Bonino's institute of physical chemistry in Bologna served as the focal point for this innovative research. A signal event for these developments came in 1935 when Bonino published a long paper in *Gazzetta Chimica Italiana* on the chemical constitution of benzene.[57] Here, in cogent language accessible to members of the chemical community, he described the development of the new quantum mechanics as well as its applications to chemical problems, in particular, the problem of benzene. Not only this, he also made use of the neologisms "quantistica chimica" and "meccanica-quantistica-chimica" to refer to an even newer discipline, one which combined "the precious sensibility of the classical organic chemist in the course of the construction of the new theories of organic chemistry with physical criteria."[58] Among these new theories, he discussed the "spin-valency" theory of Walter Heitler and Fritz London as well as its role in the theory of electron pairs, as set forth by Linus Pauling and John Clarke Slater.

Bonino also reported on the contributions of Friedrich Hund, Robert Mulliken, and Hückel on the molecular orbital method (MO-method), paying particular attention to the treatments of the benzene problem by Hückel as well as by Pauling. In discussing these, Bonino concluded in favor of Hückel's work, a verdict he based on new experimental findings, namely the Raman spectra obtained at his institute. By comparing these with results reported by other researchers, Bonino concluded that the existence of a double chemical bond for the benzene molecule was improbable. Beyond this critical result, however, Bonino gave several other reasons for preferring Hückel's approach over Pauling's. First, as Hückel himself had noted, since Pauling's research was based on the valence bond method (VB-method) it was not applicable to aromatic and unsaturated molecules. Second, Bonino indicated that Pauling's methodology was unsuitable for handling the coupling of the π-electrons since it led to results which were not in agreement with experiment. Third, by means of the MO-method, Hückel could explain the distinctive aromatic nature of rings with six π-electrons, including why such rings possess a remarkably stable configuration. Finally, Bonino underscored Hückel's success in accounting quantitatively for the problem of secondary orientation in monosubstituted benzene. Bonino's critical comparison of Hückel's and Pauling's approaches was thus based on chemical criteria. He concluded in favor of Hückel's approach because Hückel justified quantitatively with quantum mechanical methods the relevant qualitative chemical theories and empirical rules. [59]

Bonino expressed his admiration for Hückel's contribution to the benzene problem in the following terms: "The Hückel theory constitutes the best physical-mathematical effort which has been made to understand fully the difficult structure of benzene... Hückel's papers, however, are very difficult for the chemist." [60] Bonino took these difficulties into account, but proposed again the formula that he had recommended one year earlier at the Madrid Congress, now giving it a more explicitly quantum mechanical character (Figure 4.4). He calculated the resonance energy of the six aromatic electrons of his benzene formula with the molecular orbital method, viewing them in the self-consistent field of the six positive/negative carbon atoms and the other electrons. Then Bonino showed that the six aromatic electrons formed a closed group, whose interactions made the molecule stable. Furthermore, he took into consideration the central interactions between the negative carbon atoms. In connection with these, Bonino referred to the concept of resonance degeneracy (Resonanzentartung), a term introduced by Heitler. [61] Supported by Heitler's interpretation, Bonino deduced an attractive interaction between two negative carbon atoms having $2s^22p^3$ configuration when one of them is in the excited state $2s2p^4$. This showed that Bonino's benzene formula was capable of incorporating carbon atoms in an excited state, a feature Bonino was able to extend to heterocyclic compounds such as pyrrole, furane, and thiophene.

4.8
The Advent of Group Theory in Bonino's Work

In the fall of 1935, the scientific society *Sips* held its 24th meeting in Palermo. On this occasion, Betti gave a lecture on modern developments in organic chemistry, the second part of which was dedicated to relationships between organic and physical chemistry, especially to the quantum mechanical theory of valency. Betti emphasized that the VB-method justified the classical structural theory of organic chemistry, whereas the aim of the MO-method was not to understand the proper individual bonding between the atoms but rather the symmetry of the intramolecular field. This symmetry, according to Betti, constituted the "physical reality" of an organic molecule and could be deduced from spectroscopic findings. [62] One and a half years earlier in Madrid, Bonino had concluded his presentation with similar reflections on the approach of the classical organic chemist, thereby making his own methodological preferences clear:

> The modern attempts of the quantum mechanical study of molecules have shown that it is not possible to proceed in this field without making use of the mathematical concept of group. This makes me wonder whether perhaps the organic chemist had not always applied, more or less consciously though in coarse form, some of the concepts of group theory. Perhaps in this part of pure mathematical logic we find the primary essence of the structural argument. [63]

Starting around 1936, Bonino began to apply the molecular orbital theory to the study of aromatic organic molecules utilizing the methods of group theory and spectroscopy to deduce their molecular constitution and symmetry. His first works in this direction were two papers on the molecular structure of naphthalene and the symmetries of thiophene. In his first paper, Bonino used group theory to deduce the numbers and the types of proper oscillations for the naphthalene molecule as well as the symmetry of the configurations corresponding to the classical structural formulae of Emil Erlenmeyer and Richard Willstätter. He furthermore showed on the basis of Raman and infrared spectra that the fundamental level of the molecule can be considered in the symmetrical configuration of the type D_{2h}. [64] This constituted a strong confirmation of Hückel's earlier theoretical findings. [65]

Afterwards, Bonino deduced that the molecular configuration of naphthalene corresponded to Willstätter's formula representing not the fundamental level, but the first excited state of the molecule. A little later Bonino's second paper appeared in the *Atti della Reale Accademia Nazionale dei Lincei*. [66] Here he presented, in a comprehensible form for chemists, the principal ideas of group theory, and particularly Eugene Wigner's theorem on group representations. This result established a fundamental link between the latter notion and the number of vibrational states in a molecule. According to this theorem each type of proper oscillation corresponds to an irreducible representation of the symmetry group of the molecule. [67] Afterwards, Wigner's theorem, and group theory in general, played a pivotal role in Bonino's work for the study of problems pertinent to the relationship between chemical structure and Raman spectra. Bonino's methodology consisted in reanalysing the classical structural formulae of organic chemistry by means of their symmetry characteristics and then using Raman and infrared spectra to test critically the limits of this analysis. Substantially, Bonino followed in Hückel's footsteps regarding theoretical considerations, but he attempted to enrich these and to support them with new experimental evidence.

Between 1936 and 1937, Bonino's research group widened its investigations to include a number of other organic aromatic compounds. Led by his principal assistant, Manzoni-Ansidei, and other members (including E. Lucchi, L. Cavallaro, and P. Mascherpa), they examined the Raman and infrared spectra to determine the symmetries in the molecular structures of anthracene, phenanthrene, pyrrole, furane, and other organic compounds. In this work, Bonino and his collaborators outlined the limits of the formulae of classical organic chemistry for representing graphically the organic compounds with aromatic character. Bonino expounded his research approach very clearly in the fall of 1937, when he attended the "Réunion International de Physique-Chimie-Biologie" held in Paris.

This meeting took place from September 30 to October 9, 1937 in connection with the Exposition Internationale. Bonino presided over the session on organic chemistry, in which he gave the inaugural conference on "les spectres Raman en chimie organique." [68] In this lecture he discussed the principal lines of research in organic chemistry that had emerged with the advent of Raman spectroscopy. According to Bonino there were three such approaches. The first, so-called phenomenological research, established analogical relationships between Raman frequencies and chemical constitution. This line was exemplified by Bonino's first paper on Raman spectra and geometrical isomerism, written with his co-worker Brüll. [69] This approach, however, proved useless for determining the constitution of aromatic organic compounds. Therefore, Bonino argued that *"le chimiste est obligé sur ce point de conformer davantage sa pensée à celle du physicien, d'approfondir les théories, de chercher à mieux séparer ce qu'il a de symbolique dans ses formules de ce qui peut représenter, avec une probabilité suffisante, une réalité physique."* [70] This suggested a second, dynamical approach, in which the molecule is treated as an ensemble of masses oscillating around their equilibrium positions under the action of a particular system of forces. Bonino emphasized how this dynamical aspect guides the research sensibility of the organic chemist, but he also noted the difficulties in finding a mathematical expression for the equations that describe the molecule as a dynamical system. Finally, Bonino came to the third line of research, which took the symmetry properties of a molecule as the central empirical data. This is the approach pursued by Bonino after the congress in Madrid in April 1934. In Paris he had been offered some reflections on the current status of the theory and its methods:

> Dans le problème de la construction d'une molécule, selon la méthode de Hund-Mulliken-Hückel on écrit, en approximation d'ordre zéro les auto-fonctions moléculaires comme des combinaisons linéaires des auto-fonctions électroniques des atomes constituants. De toutes les combinaisons possibles, celles qui conviennent au problème doivent être conformes aux représentations irrèductibles du groupe de symétrie de la molécule. [71]

Bonino's lecture met with success and gained a nod of approval from various colleagues, including Walter Hückel, the brother of Erich. [72] Furthermore, Bonino renewed his friendship with Peter Debye and formed an acquaintance with Richard Kuhn, the president of the German Chemical Society. Nevertheless, in Paris he only touched upon the relationships between structure, valency, stereochemistry and the concept of symmetry. He took these arguments into deeper consideration the following year on a different occasion in Germany.

4.9
Bonino's Quantum Mechanical Concept of Coordination

In January 1938, Bonino received a letter from Kuhn. After mentioning the nice time they had passed together in Paris, the latter invited him in the name of the

German Chemical Society to participate in its annual meeting to be held in Berlin May 7, 1938.[73] Accepting this invitation, Bonino prepared a lecture on "Organische Chemie und Symmetrie" in which he addressed the implications of quantum mechanics for organic chemistry. On this occasion, Bonino indicated that the MO-method of Mulliken and Hund was consistent with Alfred Werner's (1866–1919) way of thinking about valency. He then showed how, using quantum mechanical considerations and group theory, it was possible to extend the coordination theory of Werner to organic compounds. As we have seen, since his student days, Bonino was well acquainted with Werner's theory of coordination. This is not the place to go into details about Werner's theory of valency,[74] who had been Professor of Chemistry at Zurich. In his famous paper of 1891, he attempted to replace the old valency theory with its rigid, directed valencies by a more flexible approach. He therefore differentiated between the concept of valency and that of affinity. According to Werner, affinity is an attractive force acting equally from the center of the atom toward all parts of its spherical surface. Furthermore he defined valency as an empirically determined numerical relationship independent of valency units.[75] Without assuming the existence of directed single valency forces, Werner was able to explain the structure of inorganic as well as of organic compounds. Werner's views on affinity and valence were further developed in the form of the well-known coordination theory with its notions of primary valency (Hauptvalenz) and secondary valency (Nebenvalenz).

What led Bonino in his lecture to emphasize the fundamental importance of Werner's theory of valency, and to extend Werner's coordination theory to organic compounds? Bonino's principal interest was to consider the doctrinal aspect of organic chemistry. In particular, he focused on the postulate of the tetravalency of the carbon atom and the pictorial representation of organic compounds. Benzene, naphthalene, and other aromatic compounds pointed to the contradictory character of this postulate. As I have emphasized, following Hückel's contributions, Bonino showed experimentally that the classical structural formulae were insufficient for representing these aromatic organic compounds graphically. Therefore, in his Berlin lecture Bonino distanced himself from August Kekulé's doctrine of rigid, directed valencies and the principle of the tetravalency of the carbon atom, adopting instead the more flexible alternative of Werner.

Bonino's theoretical approach required a relatively deep understanding of group theory, and ran parallel to work undertaken concurrently by John Van Vleck.[76] After Hückel's work, Van Vleck's exerted the strongest influence in shaping Bonino's research program during the late 1930s. Bonino followed these theoretical developments closely, and made one significant original contribution to them. He showed that the distribution of affinity forces around the carbon atom of the methane molecule depends on the number of electron eigenfunctions of the carbon atom as well as on the symmetry of the intermolecular field including the atomic orbitals of the attached atoms.[77] Thus, Bonino saw his work as completing and justifying Werner's concept of coordination by means of the molecular orbital method and group theory. Bonino continued to work in this new direction, publishing two papers in *Gazzetta Chimica Italiana*. Using group theory, he studied once again the methane molecule as well as the constitution of the ions CO_3^{2-} and ClO_3^- by appealing to the quantum mechanical coordination theory as a unifying conception for the valency actions. [78] Mario Rolla, a member of Bonino's research group, published further contributions on the constitution of inorganic ions in which he provided experimental support for Bonino's theoretical approach by means of Raman spectra. [79]

Bonino's contribution to the problem of chemical bonding, in which he treated the concept of coordination from a group-theoretic point of view, considered all possible configurations with coordination number four. His findings were not known far and wide, however, and aroused little interest in the international scientific community. Nevertheless, there were others who pursued a similar line of research. One year after Bonino's contribution, George Kimball from Columbia University published a paper in which he treated the problem of "directed valence" from a group-theoretic point of view. Kimball dealt with all possible configurations with coordination numbers ranging from two to eight. [80]

The experimental support provided by infrared and Raman spectra proved to be very important to Bonino's theoretical contribution. The Raman spectrometer served as the key instrument in Bonino's laboratory. With it he solidified chemical research in Italy during the late 1930s and gave quantum chemistry its identity. The technical capabilities of these instruments, reinforced by theoretical tools like group theory and the methods of quantum mechanics, reshaped the boundaries of the disciplinary activity located between physical organic chemistry and quantum physics, and gave rise to the new discipline of quantum chemistry. These processes gained momentum at the institutional level through Bonino's initiatives, one of which was the introduction of spectroscopy as an optional subject in the chemical curriculum at the Faculty of Science at the University of Bologna. Starting in February 1939, Bonino's assistant Manzoni–Ansidei held the first assistant professorship in spectroscopy. [81] Thus an essential part of the experimental basis for quantum chemistry began to consolidate within the Italian educational system. Indeed, those disciplinary topics of central importance for quantum chemistry were the subject of Bonino's lecture courses in Bologna starting around 1935. Among other physical-chemical issues Bonino taught his students the principles and techniques of the new quantum mechanics along with its applications to chemistry. [82]

4.10
Encroaching Political Developments

As I have emphasized, the new fascist regime in Italy promoted interest among scientists in solving problems of potential importance for industry and the national defense. Evidently these policies helped stimulate the research programs at several scientific institutes during the late 1920s and early 1930s. One of these, as we have seen, was Bonino's institute in Bologna. There, during the 1930s, a special center for military applications operated in cooperation with the institute of physical chemistry. Within this *Centro di Studi del Servizio Chimico Militare,* Betti and Bonino studied problems proposed by the Ministry of War. [83] Based on the present state of my research, I am not in a position to describe the nature of this enterprise more closely. In particular, the relationship between this military research and Bonino's experimental and theoretical researches during the 1930s remains unclear.

By the mid 1930s, Bonino had already begun to reap rewards and recognition from his peers and the Italian regime. In the spring of 1935, the Italian Association of Chemistry held its fifth national congress, at which Bonino was awarded the "Premio Morselli" for his "modern vision" of the benzene problem and for his work on the structure of the heterocyclic aromatic compounds using Raman spectroscopy. At this congress a number of chemists, among them Bonino, appeared in fascist uniform. At the opening ceremony, the regime's pre-eminent chemist, Parravano, outlined the importance of Bonino's "fascinating researches." Parravano went on to proclaim the conceptual insufficiency of the older mechanical models, emphasizing the necessity of abandoning them in order to follow the "fine, tenuous, courageous and fruitful new direction of theoretical physics." [84]

Two years later, Parravano informed Bonino that on his recommendation the "Reale Accademia d'Italia" had awarded him the "Premio Mussolini per le scienze." [85] At the award ceremony Parravano attached importance to Bonino's scientific contributions and, in particular, his "interpretation of the chemical bonds and reactivity of organic compounds based on the new views of quantum physics." [86] Thus, Parravano, the official spokesman for chemistry in the fascist regime, took a progressive line. This suggests that Bonino's new research field could proceed without fear of ideological interference, unlike what happened with regard to quantum physics and its applications under National Socialism in Germany.

Although such differences regarding scientific issues colored the respective atmospheres under Italian fascism and German National Socialism, in general cultural and political terms there was a strong convergence of interests. For example, the political relations between Italy and Germany grew closer with the Rome-Berlin axis of 1936. On October 26, 1936 the new foreign minister Galeazzo Ciano traveled to Germany. At a press-conference in Munich, Ciano proclaimed that both countries had decided to defend "the sacred cultural European patrimony", and therefore to intensify their cultural relations. [87] Thereafter the two countries developed more intensive contacts on all political, cultural, and scientific levels. [88] This common spirit of European fascist patrimony played an important part in Bonino's subsequent career, but it can already be seen in certain political discussions between him and other scientists at the time of the Paris international exhibition.

In Paris, besides the scientific exchanges, there were also sharp political discussions in which some democratic scientists protested against the political opinions of their colleagues from totalitarian states. After his return to Italy, Bonino replied with a reference to these tensions in the pages of the regime scientific review journal "La Chimica e l'Industria." There he characterized the protests as an "inopportune disharmony" without real purpose. Moreover, he emphasized that the direction modern scientific collaboration was taking at the international level, reflected the same forms of order, hierarchy, and conscious national responsibility that constituted the foundation of scientific research in the fascist countries. As Bonino stressed, "science and fascism will make an inseparable couple in the world", thereby essentially aligning himself with the new foreign and cultural politics of Mussolini's regime.[89]

This alignment was clearly a natural part of Bonino's political development, if we take into consideration that he had enjoyed long-standing favor with the regime. By the late 1930s, his career began to rise within the fascist academic world. After the death of Parravano, he became Italy's pre-eminent chemist, wielding great power, influence, and prestige. On June 16, 1939 Mussolini made him a member of the "Reale Accademia d'Italia", a choice made on the basis of a list submitted by its senior members.[90] He also succeeded Parravano as president of the Technical Institute at Breda. As the new director of this institute, Bonino participated in the October 1939 national conference dedicated to the applications of aluminum, magnesium, and their alloys. Here he spoke about applications connected with Italy's Air Force and national defense, proposing a new line of basic research for the Breda Institute.[91] Under the direction of his predecessor, Parravano, its research had been oriented toward thermodynamical chemistry as well as classical physical chemistry. Bonino hoped to shift the axis of basic research in the direction of microphysics, in particular towards new applications of quantum mechanics to chemistry, utilizing his own theoretical ideas with regard to the concept of coordination. Thus in his lecture he discussed the importance of quantum-symmetrical concepts for the study of electrons in crystalline structures. Referring to his earlier lecture in Berlin, he then emphasized how "through the modern chemical concept of coordination we have arrived at a suitable theoretical basis to deal with the problem of the interatomic and intermolecular forces in the field of the metallic alloys."[92] Bonino went on to emphasize the practical utility of ideas stemming from the new discipline of quantum chemistry in solving problems important for aviation and national defense.

The following year he addressed this theme again, when he took part in the fifth annual meeting of the German Academy for Aviation Research (Deutsche Akademie der Luftfahrtforschung) as a foreign guest (Figure 4.5). This took place on May 10–11, 1940 in Berlin. The academy had been founded in 1936, with Hermann Göring as its first president. It turned out to be a forum for discussions on atomic and nuclear research as well as other technological innovations in modern warfare. The members of the academy represented an elite from the worlds of science, economy, and the military. It was hoped that their exchanges would result in innovations in the production and operation of aircraft.[93] This particular meeting focused on the study of the physical and chemical processes during combustion. One of the fundamental problems in airplane motors at this time was the poor efficiency of the internal combustion in the engines. This phenomenon led to what German technicians called "Klopfen." Bonino's lecture described how the Raman spectra in paraffin could shed light on this problem. He pointed out that knowledge of the symmetry properties of paraffin's oscillations led to a better interpretation of the intensity of Raman-lines, which were closely connected with the phenomenon of

Figure 4.5 German Academy for Aviation Research. From left to right Prof. Schmidt, General-lieutenant Udet, Prof. Bonino, Vice-president of the Academy General-colonel Milch, Prof. Debye. (Courtesy of Andrea Concolato.)

"Klopfen." [94] In substance, Bonino presented a summary of researches he and his assistant Manzoni-Ansidei had made in Bologna. [95] These researches were very important at that time for warfare aviation.

On June 1, 1941, the "academic and scientist" Bonino held a discourse on the chemical aspects in the progress of the technology of flight at the general meeting of the "Reale Accademia d'Italia." Abounding in fascist rhetoric, he emphasized the contributions of chemistry for improvements in aviation, while proclaiming the necessity of the alliance between theory and praxis for the victory of the axis powers. [96] Thus, here once again, as on the other above-mentioned occasions, Bonino took the opportunity to speak clearly and with a sense of his duty as a "fascist scientist." Italian fascism held up a clear ideal of the scientist: he was not only a cultured person, but a technician whose work served the nation and the state. [97] Thus, Bonino's actions and words were fully in accord with this ideal of the fascist scientist, whose work went beyond theory and experiment to the realms of political and social engagement.

The political situation in Italy two years later precipitated the fall of the fascist regime. To dodge the consequences, Bonino transferred his operations to the University of Pavia. After the liberation of Italy, Bonino was able to pass the inquiries of the commission of clarification (comissione di epurazione) without a hitch. Returning in September 1946 to the University of Bologna, he began to work with new vitality. He and his research group, which gained some new collaborators, were able to make some important new contributions to quantum chemistry during the post-war period.

4.11 Conclusion

Within fascist Italy, Bonino was perhaps the only chemist who worked with ease as a theoretician and laboratory experimentalist. For Bonino, the interplay between quantum mechanical methods and the experimental interventions of Raman and infrared spectroscopy constituted the defining methodological core around which quantum chemistry took form as a new discipline in Italy during the late 1930s. Over and over again he discussed the relative merits and shortcomings of the molecular orbital and valence bond approaches. His principal interest in the conceptual issues was not confined to the discussion of the general aims and methods of quantum chemistry. As a chemist he wished to achieve a new, conceptually sound understanding and treatment of chemical bonding. To do this, Bonino embraced Werner's alternative viewpoint of valency and corroborated it through group theory. Thus group theory proved to be an abstract language capable of expressing the deeper meaning of valency for Bonino and his research team at Bologna. Perhaps the most striking and constant characteristic of Bonino's thought was his ability to cross boundaries between disciplines. Through mathematical notions, experimental measurements, and chemical concepts he was committed to the goal of articulating a unified conception of the "action of valency" in organic as well as in inorganic compounds. His career as the founder of quantum chemistry in Italy was as deeply influenced by the surrounding political events as was the new discipline by his scientific talents and personality.

Acknowledgments
I am grateful to Giancarlo Fini, the staff of the department of Biochemistry at the University of Bologna, Emma Scanavino-Bonino and her son Andrea Concolato for making the papers and documents of Bonino available to me. My special thanks to Dagniela Negrini at the Archivio storico dell'Università di Bologna. I wish to thank M. Epple, P. Mirone, P. Nastasi, and L. Paoloni for their stimulating contributions to this paper. I owe special gratitude to David Rowe, for his thoughtful and acute criticism, advice, and support of the English.

The following abbreviations are used:

ASUB – Archivio Storico dell'Università di Bologna

BPDB – Bonino Papers and documents, Department of Biochemistry, University of Bologna, Italy

BPDG – Bonino Papers and documents, in private possession of Emma Scanavino-Bonino and Andrea Concolato in Genoa, Italy

References and Notes

1 A. Karachalios, "Giovanni Battista Bonino e la scuola bolognese di chimica fisica, 1927–1944," in F. Abbri, M. Ciardi (eds.), *Atti dell'VIII Convegno Nazionale di Storia e Fondamenti della Chimica* (Arezzo, 28-30 ottobre 1999), Roma, Accademia Nazionale delle Scienze detta die XL, 285–302.

2 Y. M. Rabkin, "Technological innovation in science: The adoption of infrared spectroscopy to chemistry," *Isis* 78 (1987): 31–54; S. Nunziante Cesaro and E. Torracca, "Early applications of infra-red spectroscopy to chemistry," *Ambix* 35 (1988): 39–47; C. Meinel (ed.), *Instrument-Experiment, Historische Studien* (Berlin: GNT-Verlag, 2000), with further references.

3 BPDG, Cartella: Concorso Pisa, Documento 1B: Attestato di Laurea.

4 BPDG, Cartella: Tesina e Tesi di Laurea. The theme was "L'evaporazione dei liquidi secondo i principi della termodinamica."

5 BPDG, Cartella: Tesina e Tesi di Laurea. The theme was "Su una nuova proprietà dell'ione ferrocianidrico e di una sua applicazione alla ricerca ed alla determinazione del cloro-ione in presenza degli altri alogeno-ioni."

6 G. B. Kauffmann, *Alfred Werner – Founder of Coordination Chemistry* (Berlin: Springer, 1966); G. B. Kauffmann, "Arturo Miolati (1869–1956)," *Isis* 61 (1970): 39–47.

7 J. Stark, *Prinzipien der Atomdynamik:* III. Teil *Die Elektrizität im chemischen Atom* (Leipzig: S. Hirzel, 1915).

8 Bonino recalled his acquaintance with Betti on its commemoration at a solemn ceremony at the University of Bologna on January 9, 1943. BPDB, G. B. Bonino, "Mario Betti, Commemorazione letta nell'Aula Magna della Università di Bologna il giorno 9 gennaio 1943-XXI."

9 ASUB, Fascicolo personale di G. B. Bonino, Documento: Scheda personale.

10 BPDG, Cartella: Ditta A. Bevilacqua. Pratiche acquisto spettrometro ultrarosso, 1922. Documento: Bestellung von der Firma Andrea Bevilacqua an Firma Franz Schmidt & Haensch von 02/08/1921 über u. a. "Ultraspiegelspektrometer nach Rubens"; Rechnung der Firma Franz Schmidt & Haensch vom 12. Januar 1922 an Firma Andrea Bevilacqua, Genua über 6028 £.

11 A. Sommerfeld, *Atombau und Spektrallinien* (Braunschweig: Vieweg, 1919); Johannes Stark, *Prinzipien der Atomdynamik* (Leipzig: S. Hirzel 1915).

12 W. Coblentz, *Investigations of Infrared Spectra* (Carnegie Institution of Washington, n. 35, 1905); V. Henri, *Etudés de Photochimie* (Paris: Gauthier Villars, 1919).

13 G. B. Bonino, "Studi di spettrochimica nell'ultrarosso (I nota preliminare)," *Gazzetta Chimica Italiana* 53 (1923): 555–575; "II. Osservazioni sull'assorbimento dell'alcool," ibid. 53 (1923): 575–582; "III. Sull'assorbimento di alcune soluzioni," ibid. 53 (1923): 583–590 ; "IV. Sulla forma delle bande di assorbimento die liquidi nello spettro ultrarosso di vibrazione," ibid. 54 (1924): 457–465; "V. Ricerche statistiche sulla forma delle bande di assorbimento die liquidi," ibid. 54 (1924): 465–472; "VI. Sul confronto di alcuni precedenti risultati teorici con le curve sperimentali di Coblentz e su notevoli coincidenze numeriche con la composizione chimica delle molecole," ibid. 54 (1924): 472–485; "VII. Sull'assorbimento di alcuni chetoni omologhi," ibid. 55 (1925): 335–340; "VIII. Sull'assorbimento di alcuni idrocarburi alogeno sostituiti," ibid. 55 (1925): 341–348; "IX. Sull'origine delle bande di assorbimento impropriamente attribuite ai gruppi CH_3 e CH_2'" ibid. 55 (1925): 576–582; "X. Sull'influenza delle azioni intermolecolari sulla forma delle bande ultrarosse di assorbimento nei liquidi," ibid. 56 (1926): 278–286; "XI. Sulla banda doppia degli alcool a 3,0–3,4 μ," ibid. 56 (1926): 286–291; "XII. Sulla forma delle bande degli alcool in relazione al numero ed alla posizione degli ossidrili nella molecola," ibid. 56 (1926): 292–296; "XIII. Sull'andamento die coefficienti di assorbimento riferibili ai CH nelle bande degli alcool," ibid. 56 (1926): 296–300; "XIV. Su una banda da attribuirsi all'idrogeno aminico," *Atti Soc. Ligustica Scienze-Lettere, Genova* 5 (1926): 35–48.

14 In 1926 Ellis wrote: "There have just come to the writer's attention two papers by Bonino which contribute to the identification of the well-known band near 3.4 μ as a member of a C-H series ... Bonino's work contains slightly more convincing evidence

of the independent absorption by a C-H group than a similar earlier investigation by Henri"; J. W. Ellis, "Band series in infrared absorption spectra of organic compounds," *Physical Review* 28 (1926): 25–35, on 25.

15 BPDB, G. B. Bonino, "Lo spettro ultrarosso nell'indirizzo moderno della ricerca scientifica in chimica e biologia," Manuscript of the lecture delivered in Rome on December 29, 1954 at the High Institute of Health. (My emphasis)

16 On this argument there is an extensive literature. In my opinion the following books offer a lucid interpretation of this period with further references. G. Candeloro, *Storia dell'Italia moderna*, vol. 9 (Milano: Feltrinelli, 1988); A. J. De Grand, *Italian Fascism. Its Origin & Development* (Lincoln: University of Nebraska Press, 1982); B. Mantelli, *Kurze Geschichte des italienischen Faschismus* (Berlin: Verlag Klaus Wagenbach, 1998); G. Salotti, *Breve storia del fascismo* (Milano: Bombiani, 1998).

17 G. Marino, "La chimica e i chimici tra le due guerre mondiali," *Atti della Accademia Gioenia di Catania*, anno CLXV (1999): 7–23.

18 About this manifesto see H. Goetz, *Intellektuelle im faschistischen Italien: Denk- und Verhaltensweisen (1922–1931)* (Hamburg: Kovaç, 1997).

19 *Atti della società italiana per il progresso delle scienze*, Quattordicesima riunione, Pavia, 24–29 maggio 1925 (Roma, 1926), 30.

20 B. Mussolini, Discorso in *Atti della società italiana per il progresso delle scienze*, Quindicesima riunione, Bologna, 30 ottobre-5 novembre 1926, (Roma, 1927), 30.

21 V. Caglioti, "Nicola Parravano e la sua opera scientifica" in *Sviluppo della scienza dei metalli. Celebrazione del centenario della nascita del socio Nicola Parravano* (Roma: Accademia Nazionale delle Scienze detta die XL, 1984).

22 A. Ghigi, *Autobiografia*, edited by Mario Spagnesi (Ozzano Emilia, Bologna: Istituto Nazionale per la Fauna Selvatica, 1995), 179–181.

23 M. Betti, "Problemi ed aspetti della chimica della materia vivende," in *Atti della Società italiana per il progresso delle scienze*, Quindicesima riunione, Bologna, 30 ottobre – 5 novembre 1926, (Roma, 1927), 16–28.

24 H. Goetz, *Der freie Geist und seine Widersacher. Die Eidverweigerer an den italienischen Universitäten im Jahre 1931* (Frankfurt am Main: Haag und Herchen, 1993).

25 ASUB, Verbale delle adunanze della Facoltà di Scienze Matematiche, Fisiche e Naturali, seduta del 28/03/1928 e del 20/11/1929.

26 BPDG, Cartella: Concorso Pisa. Documento: Relazione sulla carriera e sull'attività didattica del Prof. Bonino. BPDB, Lezioni di Chimica-fisica impartite nella R. Università di Bologna. La Chimica-fisica sotto il punto di vista termodinamico vol. I, (1924–1925). La Chimica-fisica sotto il punto di vista cinetico statistico vol. II, (1925–1926).

27 BPDG, Cartella: Stabilità. Documento: Schema del corso di lezioni in Chimica fisica impartito dal Prof. Bonino agli studenti della Facoltà di Scienze. IIa Parte (Svolta negli anni 1927–28 e 1929–30), La chimica sotto il punto di vista cinetico statistico.

28 BPDG, Cartella: Stabilità. Documento: Relazione sull'attività didattica e scientifica per l'anno accademico 1928–29.

29 BPDG, Cartella: Stabilità. Documento: Laureandi diretti personalmente dal Bonino nell'elaborazione della Tesi sperimentale nel trienio 1927–28–29.

30 Ibid. The themes of these degrees were: L. Brüll "Attività ionica nella soluzione di solfato di zinco," P. Cella "Ricerche sperimentali sulla costante del campo interno nei liquidi e relazioni col potere rotatorio," R. Manzoni-Ansidei "Effetto Raman nei cloroderivati delle paraffine."

31 J. Lecompte, *Le Spectre Infrarouge* (Paris: Presses Universitaires de France, 1928).

32 BPDB, Scattola Bristol, lettera di Marlow a Bonino del 29 dicembre 1928.

33 G. B. Bonino, *Transactions of the Faraday Society*, XXV (1929): 876–888.

34 The first contribution to Raman spectroscopy from Italy was a paper by A. Carelli published in *Atti della Reale Accademia Nazionale*, Rome in August 1928. Then in the February 9, 1929 issue of *Nature* the physicist F. Rasetti reported the vibrational Raman spectra of carbon monoxide and carbon dioxide. For further informations see D. A. Long, "Raman Spectroscopy in Europe for the past fifty years," in *Proceedings*

of the Sixth International Conference on Raman Spectroscopy, Bangalore, India, 4–9 September 1978, vol. I, (London, Philadelphia, Rheine: Heyden, 1978).

35 G. B. Bonino, "Bemerkungen über das Ultrarotspektrum einiger Halogenverbindungen," *Zeitschrift für Physik* 54 (1929): 803–805; G. B. Bonino und L. Brüll "Ramanspektrum und geometrische Isomerie. Über das Ramanspektrum der beiden Formen von Dichloräthylen," *Zeitschrift für Physik* 58 (1929): 194–199.

36 Bonino,"Lo spettro ultra-rosso."

37 G. B. Bonino, "Giacomo Ciamician," *Dizionario Biografico degli Italiani* 25 (1960): 118–121; R. Nasini, "Giacomo Luigi Ciamician," *Journal of the Chemical Society of London* 129 (1926): 996–1004; R. Seligardi, "Il laboratorio di Giacomo Ciamician all'Università di Bologna (1890–1921)," in G. Marino (ed.), *Atti del V Convegno Nazionale di Storia e Fundamenti della Chimica* (Perugia, 27–30 ottobre 1993), Roma, *Accademia Nazionale delle Scienze detta die XL*, 417–432.

38 This institution was founded on initiative of Mussolini. It was established on January 7, 1926 with a Cabinet decret, and was inaugurated on October 28, 1929 by Mussolini. For the duration of the fascist period it served as a bastion of the conservative and institutional culture.

39 G. B. Bonino, "Visite a laboratori scientifici esteri" in *Reale Accademia d'Italia, Viaggi di studio promossi dalla Fondazione Volta*, vol. II (1938): 5–12, on 7. (My emphasis)

40 Ibid. About the making of the physical organic chemical community in America see: L. Gortler, "The physical organic community in the United States, 1925–50," *Journal of Chemical Education* 62 (1985): 753–757.

41 K. W. F. Kohlrausch, *Der Smekal-Raman Effekt* (Berlin: Springer, 1931).

42 Bonino, "Visite a laboratori scientifici esteri," 10.

43 BPDG, Cartella: Stabilità, Documento: Note sull'attività scientifica e didattica del Prof. Bonino nel triennio 1927–28–29.

44 P. Grassmann, "Zur Methodik des Ramaneffektes," *Zeitschrift für Physik* 72 (1931): 240–241.

45 G. B. Bonino, R. Manzoni-Ansidei und P. Pratesi, "Über das Raman Spektrum des Pyrrols und einiger Derivate," *Zeitschrift für Physikalische Chemie B* 22 (1933): 21–44; G. B. Bonino, R. Manzoni-Ansidei, "Über das Raman Spektrum des Thiophens und Furans, sowie einiger ihrer Derivate," ibid. 25 (1934): 327–347; G. B. Bonino, R. Manzoni-Ansidei und P. Pratesi, "Über das Raman Spektrum einiger Pyrrolderivate. II. Mitteilung," ibid. 25 (1934): 348–362; G. B. Bonino, R. Manzoni-Ansidei, "Ricerche sull'effetto Raman in alcune sostanze organiche. Nota I," *Memorie della Reale Accademia delle scienze dell'Istituto di Bologna* 1 (1934): 3–6.

46 E.. Hückel, "Quantentheoretische Beiträge zum Benzolproblem. I. Die Elektronenkonfiguration des Benzols und verwandter Verbindungen," *Zeitschrift für Physik* 70 (1931): 204–286; id., "Quantentheoretische Beiträge zum Benzolproblem. II. Quantentheorie der induzierten Polaritäten," ibid. 72 (1931): 310–337; id., "Quantentheoretische Beiträge zum Benzolproblem. III. Quantentheoretische Beiträge zum Problem der aromatischen und ungesättigten Verbindungen," ibid. 76 (1932): 628–648.

47 A. Simões, "Converging Trajectories, Diverging Traditions: Chemical Bond, Valence, Quantum Mechanics and Chemistry, 1927–1937," Ph. D. Thesis, University Microfilms Inc., Publication # 9327498 (University of Maryland, 1993); K. Gavroglu and A. Simões, "The Americans, the Germans and the beginnings of quantum chemistry: The confluence of diverging traditions," *Historical Studies in the Physical and Biological Sciences* 25 (1994): 47–110; A. Simões and K. Gavroglu, "Different legacies and common aims: Robert Mulliken, Linus Pauling and the origins of quantum chemistry," in J.-L. Calais and E. S. Kriachko (eds.), *Conceptual Perspectives in Quantum Chemistry* (Dordrecht: Kluwer Academic Publishers, 1997), 383–413. See also the contribution of the same authors in this volume.

48 A. Karachalios, "Die Entstehung und Entwicklung der Quantenchemie in Deutschland," *Mitteilungen der Gesellschaft Deutscher Chemiker Fachgruppe Geschichte der Chemie* 13 (1997): 163–179; A. Karachalios, "On the making of quantum chemistry in Germany," *Studies in History and Philosophy of Modern Physics* 31 (2000): 493–510.

49 C. A. Russell, *The History of Valency* (New York: Humanities Press, 1971), 261–269.

50 For details on the benzene problem the reader is referred to the paper of S.G. Brush, "Dynamics of theory change in chemistry: Part 1. The benzene problem 1865–1945," *Studies in History and Philosophy of Science* 30 (1999): 21–79.

51 G.B. Bonino, "Lo spettro Raman die nuclei aromatici e nuove idee sulla loro costituzione," *IX Congresso internacional de Quimica Pura y Aplicada*, Madrid, 5–11 Abril 1934, tomo IV, gruppo III, secciones A y B, 5–96.

52 F. London, "Zur Quantentheorie der homöopolaren Valenzzahlen," *Zeitschrift für Physik* 46 (1928): 455–477.

53 Bonino, "Lo spettro Raman die nuclei aromatici," 53–54; G. Urbain, *La coordination des atomes dans la molécule: la symbolique chimique* (Paris: Hermann, 1933).

54 M. Milone and Müller Geza, "Sulla costituzione dell'eterociclo (C_2N_2O)," *Gazzetta Chimica Italiana* 65 (1935): 241–248; C. Toffoli, "Colore e costituzione nei derivati dell'indolo," ibid. 65 (1935): 647–659; A. Mangini "Ricerche nella serie dell'1-cloro-3.4-dinitrobenzolo," ibid. 65 (1935): 1191–2003; A. Mangini "Reagibilità die sostituenti e sostituzione nucleare nei derivati benzenici," ibid. 66 (1936): 300–308; A. Mangini and B. Frenguelli, "Su alcuni derivati naftalinici," ibid. 67 (1937): 358–367.

55 C. Ingold, "Principles of an electronic theory of organic reactions," *Chemical Reviews* 15 (1934): 225–274.

56 M.J. Nye, *From Chemical Philosophy to Theoretical Chemistry: Dynamics of Matter and Dynamics of Disciplines 1800–1950* (Berkeley: University of California Press, 1993), 196–223.

57 G.B. Bonino, "Sulla costituzione molecolare del benzolo," *Gazzetta Chimica Italiana* 65 (1935): 371–423.

58 Ibid., 373

59 Nevertheless we must point out that in regard to the problem of orientation Hückel limited his calculations to the inductive effects. In 1935, Pauling with G. Wheland, and some months later after Bonino's paper, published a paper, in which turned his attention to the problem of orientation in aromatic molecules using the method of molecular orbitals. (See G.W. Wheland and L. Pauling, "A quantum mechanical discussion of orientation of substituents in aromatic molecules," *Journal of American Chemical Society* 57 (1935): 2086–2095.) In his paper Pauling took into consideration as well the inductive effect as the resonance effect, and the polarizing effect of the attakking group. Furthermore he aroused a criticism of Hückel's contribution. Therefore Bonino had not the possibility in regard the phenomenon of orientation in substitution reactions to compare Hückel's and Pauling's approaches.

60 Bonino, "Sulla costituzione molecolare del benzolo," 399.

61 In 1934 Heitler published a long contribution on "Quantentheorie und homöopolare chemische Bindung" in which he reported on the possibility of the existence of an attractive interaction between two identical atoms in different excited states. W. Heitler, "Quantentheorie und homöopolare chemische Bindung," in *Handbuch der Radiologie*, Band VI, Teil. II, *Quantenmechanik der Materie und Strahlung* (Leipzig: Akademische Verlagsgesellschaft, 1934), 485–586.

62 M. Betti, "Sviluppi ed orientamenti della odierna Chimica Organica," *La Chimica e L'Industria* 17 (1935): 637–646.

63 Bonino, "Lo spettro Raman die nuclei aromatici", 90.

64 G.B. Bonino, "Sulla costituzione molecolare della naftalina," *Gazzetta Chimica Italiana* 61 (1936): 827–843.

65 In 1932 Hückel showed the equivalency of the four positions α and β in the naphthalene molecule as well as its aromatic character. Furthermore with the MO-method he showed that all the bonds in the naphthalene molecule can be regarded as equivalent so that the conception of single and double bonds lost its meaning. Thus he pointed out the limit of the Erlenmeyer structure to represent graphically the molecule of naphthalene.(See Hückel, "Quantentheoretische Beiträge zum Benzolproblem. III.") Substantially, Hückel in this paper continued his critical revision of the classical structural system of valency that he began with his study of the ethylene and benzene molecule. For further information see Karachalios, "Die Entstehung und Entwicklung der Quantenchemie in Deutschland," and Karachalios, "On the making of quantum chemistry in Germany."

66 G. B. Bonino, "Sulla simmetria molecolare del tiofene," *Atti della Reale Accademia Nazionale dei Lincei Rendiconti Classe di Scienze fisiche, matematiche e naturali,* Classe di Scienze fisiche, matematiche e naturali, Vol. XXIV (1936): 288–294.

67 E. Wigner, "Über die elastischen Eigenschwingungen symmetrischer Systeme," *Nachrichten der Gesellschaft der Wissenschaften (Göttingen), Mathematisch-Physikalische Klasse* (1930): 133–146.

68 G. B. Bonino, "Les spectres Raman en chimie organique," Réunion Internationale de Physique-Chimie-Biologie, *Congrès du Palais de la Découverte Réunion Internationale de Physique-Chimie-Biologie, Congrès du Palais de la Découverte,* Paris, Octobre 1937 (Paris: Hermann, 1937), 275–295.

69 Long, "Raman Spectroscopy in Europe for the Past Fifty Years."

70 Bonino, "Les spectres Raman en chimie organique," 283.

71 Bonino, "Les spectres Raman en chimie organique," 285–286.

72 Bonino, "Les spectres Raman en chimie organique," 292.

73 BPDG: Lettera del presidente Richard Kuhn a Bonino del 14. 01. 1938.

74 Kauffman, *Alfred Werner-Founder of Coordination Chemistry.*

75 A. Werner, "Beiträge zur Theorie der Affinität und Valenz," *Vierteljahresschrift der Naturforschenden Gesellschaft in Zürich* 36 (1891): 129–169; id., "Contributions to the theory of affinity and valence," *Chymia* 12 (1966): 189–215. Translated by George B. Kauffman.

76 J. H. Van Vleck, "The group relation between the Mulliken and Slater-Pauling theories of valence," *Journal of Chemical Physics* 3 (1935): 803–806; J. H. Van Vleck, and A. Sherman, "The quantum theory of valence," *Reviews of Modern Physics* 7 (1935): 167–228.

77 G. B. Bonino, "Organische Chemie und Symmetrie," *Berichte der Deutschen Chemischen Gesellschaft* 71 (1938): 129–146.

78 G. B. Bonino, "Contributi ad uno sviluppo moderno del concetto di Coordinazione, 1) Nota introduttiva," *Gazzetta Chimica Italiana* 69 (1939): 763–779; 2) Sulla costituzione dell'ione ClO_3-, ibid. 70 (1940): 217–227.

79 M. Rolla, "Contributi sperimentali alla conoscenza della struttura degli ioni XO_3. I.Spettri Raman di clorati e bromati di cationi bivalenti," *Gazzetta Chimica Italiana* 69 (1939) 779–790; II. Spettri Raman di nitrati di ioni tri e tetravalenti, *Rendiconti della Reale Accademia d'Italia* 1 (1940): 563–571.

80 G. E. Kimball, "Directed valence," *Journal of Chemical Physics* 8 (1940): 188–198.

81 ASUB, Verbali delle adunanze della Facoltà delle Scienze. Seduta del 28 febbraio 1939.

82 ASUB, Registro delle Lezioni di Chimica Fisica ed Elletrochimica impartite dal Prof. G. B. Bonino negli anni accademici 1934–1940.

83 ASUB, Verbali delle adunanze della Facoltà delle scienze. Seduta del 29 maggio 1934.

84 Il discorso inaugurale di S. E. Parravano, *La Chimica e L'Industria,* 17 (1935): 307–310, on 309.

85 BPDB, Lettera di Parravano a Bonino del 12 aprile 1937.

86 Reale Accademia d'Italia, Premi Mussolini del "Corriere della Sera" assegnati nell'adunanza solenne in Campidoglio il 21 aprile 1937-XV, Roma 1937-XV: 7–8.

87 J. Petersen, *Hitler-Mussolini. Die Entstehung der Achse Berlin-Rom 1933–1936,* (Tübingen: Niemeyer, 1973); J. Petersen, "L'accordo culturale fra l'Italia e la Germania del 23 novembre 1938," in D. Bracher, L. Valiani (eds.), *Fascismo e nazionalismo,* (Bologna: Mulino, 1986), 337.

88 A. Hoffend, *Zwischen Kultur-Achse und Kulturkampf: Die Beziehungen zwischen "Dritten Reich" und faschistischem Italien in den Bereichen Medien, Kunst, Wissenschaft und Rassenfragen* (Frankfurt am Main: Peter Verlag: 1998).

89 G. B. Bonino, "Il Congresso del 'Palais de la Decouverte' Parigi 30 settembre-9 ottobre 1937-XV," *La Chimica e L'Industria* 19 (1937): 728–729.

90 Marino, "La chimica e i chimici tra le due guerre mondiali"

91 BPDB, G. B. Bonino, "Contributo dell'istituto tecnico Ernesto Breda alla conoscenza delle leghe leggere e ultraleggere ed alle loro applicazioni," in Convegno Nazionale per l'Applicazione dell'Alluminio, Magnesio e loro Leghe, Milano 20–22 ottobre 1939, Palazzo dell'Arte al Parco.

92 Ibid., 4.

93 J.-M. Hormann, *Elite in the Third Reich:*

The History of the German Academy for Aviation Research 1936–1945 (Garbsen: INFO-Verlag Jörg-M. Hormann, 1988).

94 G. B. Bonino, "Raman-Spektren und Klopffestigkeit einiger Paraffine," *Schriften der Deutschen Akademie der Luftfahrtforschung*, no. 9 (1939): 117–132.

95 G. B. Bonino, R. Manzoni-Ansidei, "Ricerche chimico fisiche sui carburanti," in *Memorie della Reale Accademia delle Scienze dell'Istituto di Bologna*, Serie IX, Tomo V, 1937–1938: 161–164; Tomo VI, 1938–39: 11–17.

96 G. B. Bonino "Aspetti chimici nel progresso della tecnica del volo," Discorso pronunziato nell'Adunanza Generale del 1 Giugno 1941-XIX per il solenne conferimento dei premi reali, *Annuario della Reale Accademia d'Italia*, vol. XIII, 1941-XIX: 11–17.

97 N. Parravano, "Il Fascismo e la Scienza," discourse delivered at the University of Padova in April 22, 1936, *La Chimica e L'Industria*, XVIII (1936): 222–226, p. 223.

5.
Between Disciplines: Jean Barriol and the Theoretical Chemistry Laboratory in Nancy

Marika Blondel-Mégrelis

My decision to write on Jean Barriol (1909–1989) and the Nancy Laboratory of Theoretical Chemistry in order to explore the emergence of new areas of chemical research between the classical disciplines was determined by a variety of reasons. The first chair in France officially named *Theoretical Chemistry* had been created at the *Faculté des Sciences de Nancy*. From the very start, Barriol has been considered to adopt this chair. This took place at the end of 1948, some months after the *International Symposium on the Chemical Bond* was organized in Paris, the first symposium to be devoted to quantum chemistry held after the war. It might be useful to compare this date with the first chair of theoretical chemistry ever created, in 1932 at Cambridge, England, with John Lennard Jones. Second, it is important because Barriol is not very well known and rarely mentioned in the historical literature, except for a history of physical chemistry in France. [1] Third, it is because Barriol was an unclassable man and an unclassable scientist. He founded an original laboratory of theoretical chemistry where experiment was regarded as first in the ranking and in the comparative weight of the two components of the scientific method – something unexpected, perhaps paradoxical.

Barriol studied a great variety of physical effects, mainly the interaction of matter with electromagnetic fields, and he did so in various modes as absorption, diffusion, permittivity, and polarizability – always in order to progress in the knowledge of molecular structure. Rumors were going round in the 1930s that physical chemistry consisted of whatever interested Jean Perrin and that chemical physics was anything that was found in the new *Journal of Chemical Physics*. [2] It seems, at first sight, that theoretical chemistry in Nancy was whatever interested Jean Barriol. As a matter of fact, the chair had been created for that purpose.

Foremost, I chose to present Barriol and theoretical chemistry in Nancy because it permits me to seize the opportunity of thinking about the borders inside chemical research and its separated partitions, as well as outside, with respect to other disciplines, bases, methods, techniques, and aims. Would the theoretical chemistry that had to be built in France constitute a new area of chemistry? Could a physical chemist be a theoretical chemist? Were the quantum chemist and the theoretical chemist one and the same person? Or, when did the transition occur?

The *Journal of Chemical Physics*, founded in 1933, would welcome papers "perhaps too mathematical for the *Journal of Physical Chemistry*, or too chemical for the *Physical Review*." [3] Barriol considered himself, while being student at the *Ecole Normale Supérieure*, not to be at the top level of mathematics – "the way of working of my mind was of another type" [4] – but he was also too mathematical to be a chemist; and he was too chemical to be a physicist. From the beginning, he was interested in entities generally called molecules by chemists, their movements and their properties. For Barriol, their dielectric properties were of the highest value when one wanted to have a look into the microstructure of matter.

5.1 Inspirations

Theoretical chemistry had yet to be created in France in the 1940s. There were prestigious models from abroad: Charles Coulson, Jan A. Ketelaar, Hugh C. Longuet-Higgins, Linus Pauling, and Robert Mulliken, who met in Paris in April 1948. In France, the *Centre de Chimie Théorique* had been created in 1943 by Raymond Daudel, under patronage of physicists like Louis de Broglie, Irène and Frédéric Joliot-Curie, and was located at the *Institut du Radium*. But Daudel was a physicist by formation, and as such he was considered by both the scientific community and the university hierarchy. Moreover, Daudel was interested in the refinement of calculation methods concerning isolated molecules, and in the applications to biological problems. These were to become the central part of Alberte and Bernard Pullman's preoccupations. [5] This was not the case with Barriol, who was interested in molecules in their complex and very concrete environment. Moreover, the second name given to Daudel's laboratory in 1957, *Centre de Mécanique Ondulatoire Appliquée*, and the description of the aims, by Daudel himself: "to discover new phenomena, to discover new theories and new concepts concerning the quantum molecular sciences" indicates the difference of the two projects. In contrast to Daudel, Barriol kept searching his whole life long, "to obtain, from electric actions, information concerning matter." [6] The quantum chemistry methods would constitute a means, if necessary, to attain this end.

Though theoretical chemistry in Nancy had to be created, it did not come from anywhere. Barriol was not a member of a school, since he had grown up in science by himself. But there existed some sources of inspiration, some strong personalities or work on which he had deeply thought, that marked his way of working and thinking, and for which he expressed deference and admiration. Barriol had been for a short time, in 1930, and while being at the *Ecole Normale Supérieure* (ENS), a pupil of Robert Lespieau who taught there the theory of chemistry. It marked the beginning of a life-long thinking about the complexity of reality, the fact that we can approach reality only through representations. We construct these rough macroscopic models in our mind and make them work, extrapolating afterwards on another scale. He acquired a great admiration for the teacher and thinker who taught him "how long and keen the fight had been to lay down the representation of the molecule in the chemist's way of thinking." [7] Actually, Barriol would possess the same critical mind, the independence and originality that mark Lespieau's work. [8]

It is certainly not by chance that Barriol's first research, in collaboration with Pierre Donzelot, dealt with the Raman effect. Lespieau had been one of the first who had studied, often with Maurice Bourguel, the Raman effect, [9] in relation to chemical constitution. [10] Donzelot did some research on the same subject with Charles Prevost, another student of Lespieau, when Prevost was a professor of chemistry at the Nancy *Faculté de Pharmacie*. At the very same time Barriol also was around. A note to the *Académie des Sciences*, concerning the relation between Raman frequencies and interatomic distances, by Donzelot and Barriol, was presented by Lespieau and concluded: "It seems that the interatomic distance constitutes an essential characteristic of a molecule, around which one can group the properties, not only those related to the Raman effect, but also manyfold physical characteristics." [11]

When Barriol begun with his laboratory, he inherited what had been left from the ancient laboratory of Donzelot, the laboratory of physical chemistry. [12] But because the laboratories were poorly equipped in this after-war time, the spectroscopic equipment had already been taken by other laboratories. There remained little at the laboratory of physical chemistry, only something such as a spiritual inheritance and a tradition that wanted to found theoretical speculations on experiment. Two topics were further pursued: Raman spectrography, or, more generally, the studies of vibrations inside the molecule, and dielectric problems. The first assistants had to build their own installation for the measurement of dipolar moments, and a part of an installation for the measurement of the refractive index in order to determine the atomic polarization.

The collaboration with Donzelot had deep consequences. Convinced of the potential and originality of the young scientist, Donzelot, in 1948 *Directeur de l'Enseignement Supérieur*, was very active to reorganize education and research in France after the tempest of the war. He managed to create a new chair and to install Barriol in it. Moreover, the new chair of theoretical chemistry gave him complete freedom to work in fundamental fields concerning physics and chemistry, physical chemistry, experiment and theory, and experimentation and calculation. If there was a master figure for Barriol, it was certainly Paul Dirac. For five years, Barriol kept meditating on *The Principles of Quantum Mechanics*. As a prisoner he had the book with him and he taught it the other prisoners, trying to make one of the most abstract and difficult treatises clear and simple. In spite of the difficulties, the lectures were a success. Back in France, he published *Mécanique quantique*, the result of this particular pedagogical experience and of the many discussions with his comrade-prisoners. [13]

As a matter of fact, Barriol introduced the new formalism by precisely following Dirac's method, keeping the quantum state as a central notion, but in a more "intuitive way of presenting." Profoundly attached to the awareness of the "physical sens hidden behind the mathematical formalism" and to "solve problems that really have a practical value," [14] he insisted on the prime importance of experience, "our starting point." The outstanding characteristic of the new quantum mechanics was, according to Barriol, the compatibility or incompatibility of two values, and this can be decided only by experimental studies.[15] Barriol would remain very cautious concerning the new notions: The incompatibility of two values such as the position and the impulse of an isolated particle for instance leads to the "philosophical notion" of indetermination. But "this way of reasoning is very dangerous as such a behavior corresponds to no experimental techniques, as localizing a particle is an operation deprived of any physical significance."[16]

Barriol also had a great admiration for Edmond Bauer who combined experimental and theoretical research, and worked on group theory. Bauer's "remarkable study on the hydrogen bond", in 1938 co-authored with Magat, remained "the most satisfactory theory, the electrostatic one." Two decades later, it is still presented in Barriol's *La Constitution des Molécules.*[17] Bauer, too, tried to reduce the difficulties of Dirac's rigourous method, to preserve the physical sense of coefficients and functions in quantum mechanics, and to draw attention to the fact that taking group transformations into account simplifies the problem in the case of degeneration: "It dispenses from completely working out the dynamical problem, what is generally very difficult."[18] Bauer was the only one in France who was engaged in studies concerning the theoretical aspects of physical chemistry.[19] As a professor of mathematical physics in Strasbourg, he became later the successor to Perrin in the chair of physical chemistry at Paris (1945). He was convinced that physical chemistry derived from physics not only its various methods, but also its leading principles.

But perhaps the most determining factor happened to be the war. As prisoner for five years, Barriol's only instruments were paper and pencil. "The war obliged me to a complete theoretical activity". During this period, besides teaching, he worked on the group theory and its applications to crystalline and molecular vibrations. This topic would eventually constitute his doctoral dissertation in 1946.

5.2 Mathematics

The training to enter the ENS and the education at the ENS itself was quasi exclusively mathematical. The circle of students was selected and the tendency was not to make one's way toward experimental sciences.[20] For him, mathematics was easy, a game with simple and evident rules. He was fond of calculations; generally analytical ones, and his mind, he confessed, was an abstract one. Barriol's calculations had to be simple, elegant, and rather short. He usually retreated when they were complicated and sinuous. But he was delighted when Donzelot charged him to search for analytical solutions of problems they dealt with; and he was delighted later on, when his student and future successor, Jean-Louis Rivail, expressed the wish to stay with Coulson after his doctoral dissertation, to perfect calculation methods on molecules. Barriol was not so fond of calculating wave functions. He did not trust the theoretical way used by quantum chemistry methods to obtain quantitative results, and suspicious of the approximations used. Nevertheless, he had experience in the determination of the structure of graphite with the help of the molecular orbital method (1950) and the variational calculation of molecular polarizabilities (1953). Moreover, he was not tempted by long numerical calculations, done by hand in these times, nor by the emerging computational tools. The first computers were introduced in France at the end of the 1950s, and opened, as is generally said, a new era in theoretical chemistry, in which Barriol did not personally play a role, letting his students Rivail and then Daniel Rinaldi take the way. Barriol was convinced that it is impossible to have an elaborate theory without any recourse to mathematics – and how could it be otherwise – but his ideal was not to make a "mathematical chemistry", as Maurice Letort used to call it. Barriol always rejected a chemistry which one could do at a desk only, and he regarded mathematics as a tool for the job. Mathematics was a wonderful tool, indeed, but a limited one. The value of the result is limited by the value of the axiomatic, and the formalism is the only means to get generality and rigor. But as they have to be looked at as the more refined and elaborate expression of our latent experience, they give an access to a relative truth only.[21] The use of group theory which gave so elegant, quick, and complete results when applied to the study of spectra and to wave equations is particularly telling for Barriol's manner of using pure mathematical theory to exhibit the reasons why molecules are what they are, as Coulson would have said.

But these mathematical tools have to be used in the service of fundamental physical ideas. This opinion had already been expressed by Dirac: "to maintain physics on the foreground and examinate, as often as possible, the physical sense hidden under the mathematical formalism."[22] In the 1950s, the wave equation was insoluble, except for the molecules of hydrogenic character. As a matter of fact, the chemist "introduces just those functions which correspond to the behavior to be expected chemically."[23] Mathematical operations have to be guided: "In practically the whole of theoretical chemistry, the form in which the mathematics is cast is suggested, almost inevitably, by experimental results."[24]

There was much affinity between Coulson and Barriol, not only because of the many subjects they shared, but also because of their similar way of proceeding and thinking. They both conceded a high value, in many respects, to the determination of dipole moments. Both worked on methane (CH_4) and more particularly on the dipole moment of the C–H group, for which Coulson gave a direction when Barriol's simple model could not.[25] It is highly interesting to compare the way how the two authors express themselves to show that "experience" or "physical and chemical evidence" had to correct "the false inferences" or "deductions that square in no way with reality": the description of the carbon electronic structure fails to account for four equivalent bonds. "We have to admit that the C-orbitals that are useful for constructing the bonds are neither 2s nor 2p orbitals .... We can only get four equivalent bonds by abandoning the clear-cut division into atomic s and p orbitals."[26]

Both authors are much interested in fundamental questions, both required much from theory, as to its demands, bases, efficacy, and coherence, in the same time that they point to its limits. In perfect harmony they thought that "contrary to what is sometimes supposed, the theoretical chemist is not a mathematician, thinking mathematically, but a chemist thinking chemically." [27]

5.3
Quantum Chemistry

Barriol being a chemist, of course, proved to be much interested in quantum mechanics, very early for the epoch and for France. As early as 1949, when he already was the director of the newly created Laboratory of Theoretical Chemistry, but left it rather desert because he had been called by Donzelot to reorganize the University of Sarre, he started pioneering work on graphite with Jacques Metzger who stayed at Sarre at that time. By the application of the molecular orbital method to the hexagonal lattice of graphite, considered as a planar indefinite molecule, he calculated the repartition of the electronic energy levels. This confirmed the metallic character of graphite. [28] From this first study on, which was presented at the First International Symposium on Graphite (Paris, 1949), he became to be considered as the theoretician of the subject. Let us note that, ten years later, he took on the question again and, after the 1958 Symposium on Carbon at Buffalo, he considered a limited planar lattice. This time, the chemist's intuition led him to correct the results of the calculations and to modify the hypothesis concerning the electron distribution.

In his numerous studies to determine the polarizability of a molecule, based on measurements of the influence of an electric field, Barriol considered the action of an electric field on the electrons (of an atom or a molecule). This was an eminent theoretical value, as it depends on the dimensions of the molecule, and Barriol, after having treated the problem with the help of a simple dynamic model (localized charges and draw-back forces), stated: "It is necessary to adopt the quantum point of view." [29] But there is a difficulty: Even if the energy and the function relative to the fundamental state are well known, this has not to be the case for the excited states. In 1953, he published with Jean Régnier a quick and simple calculation of the rigorous value of the energy of an excited state, on the basis of the variation method. The note was presented at the *Académie des Sciences*, in the section of theoretical physics by de Broglie. [30] This was one of the most fruitful papers of the laboratory. The method, effective to calculate the π-electron polarizability in ethane, would soon be extended to the evaluation of the transition energy, and applied to ethylene, butadiene, and benzene. A more elaborate approximation was proposed and justified at a 1957 CNRS meeting. [31] But what is most interesting in this example is the way how Barriol developed for the occasion "a simple and quick method to study a given excited state in the molecule." Such is Coulson's commentary after the presentation of the paper at the conference. And he comments that "M. Barriol's wave function is probably more precise than the usual LCAO one."

The essential requirements of Barriol's methods – to be simple, quick, and precise – were fulfilled thanks to something like intuition, something like a knack for passing around the difficulties. In the first memoir in which the calculation method was given, the authors introduced "an operational relation" that "suggested a certain form of the excited state function for which the evaluation would become much easier." The method revealed to be fruitful. The second approximation of the same method is obtained by "adopting the little more general function belonging to the same type of symmetry." In these researches, Barriol did not proceed according to a rational deductive method. He guided the calculation along a cautious, pragmatic way, the aim being to evaluate the transition energy and to obtain a wavelength value in agreement with the experimental data. If this kind of intuition – a word often used – that guided the author to make a choice according to the form of the function was rather of a mathematical type, it was not the same in the second study on graphite. The infinite model of constant width is more complicated than the indefinite one but closer to reality. The application of the M.O. calculation lead to nearly the same results as for the repartition of the π-electrons. But the consideration of many experimental effects and the chemist's intuition obliged to reexamine the results of the calculation. The electronic density inside the band had to be reduced in the interior and had to be balanced by an excess on the borders. Being in contact with matter for a long time had developed in the chemist's way of concluding something like a feeling that had more weight than the most sophisticated, rational, mathematical deduction.

5.4 Pragmatism

For the theoretician Barriol, chemistry was not "deductive, synthetic." [32] Progress in the demonstration did not proceed on a straight line. As reality is complex, there is not only one possibility to approach it. And you may go faster if you are pragmatic: if the simpler and quicker means work, do employ them. I will give two examples for that. First, the model of the harmonic oscillator is very simple, but it is very far from reality. However, it works and permits to explain the variation of polarizability with frequency (the dispersion phenomena), an achievement that could not be arrived for example with Mossoti's static molecular model. [33] The calculation method gives the C–H moment a value very close to that given by literature, and the method is easy to use. [34] Second, the hypothesis which is bound to the use of the last model that located charges remain constant during the deformation of a molecule placed under the influence of an electric field is disputable. That means that the distribution of ionic and covalent structures is not altered. However, the results for methane are very acceptable. In the case of hydrochloric acid they are unacceptable. And we understand why: The ionic character of the chemical bond (resulting of the hybridization of the purely ionic and purely covalent forms) changes with the internuclear distance. "One imagines the necessity of looking for something else." [35] However, on the condition that the dipolar moment (qr) is corrected by the induced one (due to the polarization of the chlorine ion by H^+) and, according to Peter Debye's method, it is possible to obtain an acceptable value for the polarizability α.

With this particular example of a located, invariable charge model, Barriol used a method that would be frequently used in his laboratory, particularly in the many studies on the Onsager model: to work on a very simple model and to adjust it punctually for one case or another. Other authors calculated the atomic polarizability of a molecule according to a dynamic model based on absorption and dispersion infrared measurements. But the problem is to determine the charge value participating effectively in polarization. [36] Barriol, for his part, did work on the simple model of located, invariable charges, with very disputable hypotheses: Things are certainly not like this, "but there are some difficulties to find a more elaborated model, with which it would be possible to do calculations." [37]

The question of electric charge distribution in atoms and molecules became a constant preoccupation of Barriol. The notion is at the center of his theoretical preoccupations. It permits "to simplify the theoretical approach of many problems: chemical reactivity, interactions between molecules and even intensities of vibration bands." In 1971, he would take on again the problem, by quantum chemistry techniques, with Rinaldi and Rivail. The method, founded on population analysis by LCAO calculations, would propose a new definition of point charges. The whole electronic structure is represented by a distribution of point charges and atomic moments. [38] The method presents the great advantage of being valid for multiatomic molecules of any geometry.

5.5
The Foundations

Being a pragmatic, adopting sometimes very rough approximations, required simplicity, efficacy, rapidity, but that does not mean that Barriol sacrificed the scientist's principles on the altar of the results. On the contrary, one finds here the occasion to point out that this theoretical chemist at work was theoretical in the most classical sense of the word: The one who thinks about the bases, who works on the foundations. Every time Barriol followed his intuition, every time he adopted a rough, simple, and far-from-reality model, he justified his choice and showed, a posteriori, why it worked. It has been shown above how the non-acceptable results for polarizability, when extended from methane to hydrochloric acid, constituted an occasion to reexamine the precedently admitted hypothesis; and more precisely, the theoretical significance of what was considered as a charge. Let us examine two other examples.

The metallic model of the electron distribution in organic substances is an approximation: One considers that, in a molecule where single and double bonds alternate, π-electrons are free to move along the chain. This simple model obtained incontestable success in the prevision of absorption spectra of many coloring matters. The advantage is that the geometry of the molecule alone is required for the calculation of the absorption frequencies, without any need of dynamic constants. This is not the result of a particular form of the wave function. Actually, it is possible to have the same advantage, starting from the LCAO approximation used for the calculation of the resonance energy. Calculations are a little longer but no arbitrary correction is needed. Most important for us is that one can see a link between the LCAO classical method and the metallic model. These calculations throw light on the reasons why this simple model gets so good results. The two methods give the expression of the transition energy without any reference to energy quantities, but these energy quantities implicitly intervene as they condition the structure and dimensions of the concerned molecules. [39]

The Onsager model, on which Barriol kept thinking and working, tries to render an account of the strong interactions between molecules. As a matter of fact, Barriol would continue generalizing and extending the model of Onsager to mixtures of liquids, associations of molecules (H-bond), and anisotropic molecules. The starting point of all was the theoretical study of the dielectric behavior of liquid mixtures and their possible associations. [40] Actually it seems as if Barriol had been doing his utmost, for half of his life, to pass around the difficulties of a statistical consideration in all these systems, giving preference to a refinement of models in which the molecular surroundings are considered like a continuous medium. The last efforts to extend the model to mixtures of polar species were made in 1971. The anisotropy of the molecule had been already taken into account by many authors, with the use of an ellipsoidal cavity. But Barriol was not satisfied by the fact that the ellipsoidal form of the cavity was related to the geometrical shape of the molecule. "It is reasonable to relate it directly to the anisotropy of polarizability." [41] He would later insist on the fact that in contrast to other authors, his extension of the model had been made by a physical and not a geometrical way.

At the end of the 1960s, Barriol gave up the mono-molecular cavity model in exchange for a multi-molecular one (but for few molecules only), and constructed a purely statistical theory of the dielectric constant, that could take into account fluctuations in time. [42] And what is interesting for our purpose is that, with his statistical theory, he found out the same results as those derived from the preceding model, thanks to hypotheses "that throw light on the conditions of validity of the relations deduced from these calculations". [43]

Theoreticians construct models, more and more sophisticated ones, but they must not forget that these are constructions of our faculty of thinking; they are not things that exist in reality. Barriol remained particularly careful about this gap and conscious of the danger of confusing the two orders. Constructed by the chemist, models are sometimes so fruitful, they permit so well to clutch at things that he may follow the force of concrete representation they generate. Many famous chemists fell in the trap. When Pauling treated the van der Waals radius, he argued as if the bond between two chlorine atoms were, in fact, one pair of electrons. Giving these conditions, the van der Waals bond length is greater than the covalence bond length, as there are two pairs of electrons between the two chlorine molecules. In the paper just mentioned, Barriol insisted on the conclusions: the model of Onsager with ellipsoidal cavities "gives an acceptable picture." But as this volume depends on the polarizability of the corresponding molecule and on the high frequency permittivity of the liquid, the picture is not in agreement with a real hole inside the liquid. "This cavity appears more useful as a model which allows us to calculate ..."

5.6
Experiment

Though Barriol was not an experimenter, neither by taste nor by training, he was fond of experiments made by others. His laboratory was mainly a place where measures were performed, during the first period at least, and all his students had to work first on experimental theses before doing, later on, whatever they liked. It was not only a matter of policy. At that time, in France, you could not climb the career ladder if you were suspected to be a theoretical chemist. It was a starting point. Precise measures of dielectric permittivities of associated binary and ternary systems for instance, were interpreted by means of theoretical results, within the frame of the extension of the Onsager model. Theory had to render an account of the experimental results, including the questions asked by experiment, for instance the disagreements between experimental results concerning associations in multifarious dilution conditions. [44] Conversely, experimental data were perhaps a pretext of developing a theoretical activity.

Barriol was interested in matter, in the most concrete and tangible matter, matter that you seize in a macroscopic way, matter that you have to describe in a dynamic way. [45] By experiment, you can approach, change, and study matter. That is why, although being not the experimenter, he was always by the side of the experimenter: suggesting the program, encouraging, eager for coming results, with his eternal and kind smile. And when an abnormal result happened to appear, for instance a break in a straight line, he jubilated, rushed home and came back the next morning, with a lot of sheets covered with calculations. But the ultimate question always was: what is, physically speaking, the meaning of this aberration?

All his life, Barriol kept taking care of maintaining a formation of theoreticians and experimenters of various specialities, often in the same person. Experimental and theoretical preoccupations conjugated closely. But times changed: At the end of the 1970s, the experimental studies concerning the non-linear dielectric effect (molecular liquids) had to be abandoned for want of staff. Studies of the effect of molecular anisotropy on dielectric properties were nevertheless carried on thanks to earlier laboratory data. At the beginning of the 1980s, when Barriol, although retired, kept on working in the "dielectric" research team, two of the experimental groups had disappeared: the *microonde* and the NMR groups. But, as can be read in a report of the mid-1980s, if the experimental techniques became less diverse, "theoretical chemistry that formerly occupied a relatively marginal position inside the formation, takes for the first time in the history of the laboratory a leading role." [46] In 1990, one year after Barriol had died, "the transition toward a complete theoretical chemistry formation was finished." Not only because of the departure of the last group of experimenters (X-ray spectroscopy) but also thanks to important new computational resources. [47]

5.7 Jean Barriol's Theoretical Chemistry

If one considers theory to be at the opposite side of experiment, then Barriol was not a theoretical chemist. Was he, in these conditions, a physical chemist? "It is within the area of physical chemistry that our laboratory took the occasion to perform the most significant advances" said Barriol, modestly. As a matter of fact, most papers were published in the *Journal de Chimie Physique* and the notes in the *Académie des Sciences* were mostly read in the physical chemistry section. Like Bauer, Barriol was interested in the theoretical aspects of physical chemistry. But, more generally, he can be considered as a kind of generalist, working at the center of the chemist's preoccupations – the structure of molecules and the nature of their interactions – who dominated general questions of fundamental chemistry, who could go ahead, rather far, in specialized directions of high interest, who was able to distribute problems to more specialized disciplines and, vice versa, to think about the responses.

His interest was focused, but the ways to approach it were multiple. The great diversity of his lectures, often pioneering ones (on quantum mechanics, group theory, solid state theory, statistic mechanics) testify this fact as well as the diversity of his students' careers does: Marc Grosjean, the first student who designed an apparatus to measure dipolar moments, became a specialist in electronics in the Roussel laboratories. Jean Régnier (measures of the atomic polarization of non-polar gases) was appointed as professor of physical chemistry at Montpellier. Rivail became Barriol's successor as head of the theoretical chemistry laboratory. Jean-Louis Greffe is professor of mathematics at the ENSIC. Rinaldi, one of the very few who did not pass through a preliminary experimental training in the lab, defended a "thèse de spécialité" in quantum chemistry,[48] then a *thèse de Doctorat de Sciences physiques*, Mrs Pullman being an invited member of the jury."[49] He is now head of the theoretical chemistry team, together with Rivail.

If Barriol could be judged, as well for the subjects he treated as for his mind, we could perhaps place him between John H. Van Vleck and Amyand D. Buckingham. Van Vleck, often quoted by Barriol, had studied the dispersion of CH_4 and HCl, had been professor of physics, of mathematics, and of natural philosophy, and had worked at the borderline of quantum chemistry and chemical physics.[50] Buckingham worked in physical chemistry and theoretical chemistry, and was profoundly interested to "learn more and more about the nature of matter through the measurement and interpretation of optical, electric, and magnetic properties of molecules."[51] He was the successor to John E. Lennard-Jones and Longuet-Higgins at Cambridge. The relations with the Cambridge laboratory became tighter in 1975–1978, when by a convention, Rivail, Rinaldi, and A. Cartier perfected a variational calculation of electronic multipole molecular polarizabilities.

More than just a scientist working at the borderlines of many disciplines, Barriol has to be seen as being in the classical tradition that makes theoretical chemistry a central discipline in which other disciplines have their roots: it takes into account all the various properties and unifies them inside the frame of a theoretical explanation. In order to approach the interaction phenomena in liquid phase, the way of proceeding has a well-marked physical character, but the aim is undeniable chemical. Even limited to the liquid state, molecular interactions include a multitude of phenomena that cover, only by themselves, a great part of chemistry.[52] In 1990, the laboratory, asserting itself officially as a "completely theoretical chemistry formation" would claim the status of a chemical science, with all its applications and interfaces with other disciplines.

If the activities of the laboratory in this field are said to be at the borders of quantum chemistry and statistical thermodynamics, these two disciplines are declared to be "techniques." The problems raised by molecular liquids and solvent effects can be solved, or at least simplified by these techniques. This is firmly stated everywhere: the method of calculation of molecular orbitals for the σ-bonds was developed in the laboratory (Rinaldi, 1969), for instance, by giving some indications about the configuration of a molecule. The value and direction of a dipolar moment constitutes a properly quantum chemistry method to be applied to the advancing of the essential problems in the laboratory. In the same way, statistical mechanics or statistical thermodynamics constitute methods that were elaborated to render an account of the systems studied by chemists and physicists. In *Eléments de Mécanique Statistique*, these methods are well said to constitute the "second step," the first step being taken by quantum chemistry that studies the stuctures and properties of the constitutive particles.[53]

But what is perhaps more interesting is that, at the time of his most important activity, Barriol enumerated the techniques used in the theoretical laboratory of Nancy to solve the problem of three-dimensional structural studies on isolated molecules, static studies on associated solutions, and dynamic studies of associations. They were five techniques explicitly mentioned by Barriol, four experimental and one theoretical. Theoretical calculations are situated on the same level as the four experimental techniques, which included dielectric polarization, nuclear magnetic resonance, microwave spectrography, and dielectric relaxation.[54]

More than a discipline on the borderline, theoretical chemistry, as it was practiced by Barriol, was a central discipline. A place where a lot of other activities were converging, and where they found their ultimate aim and unification. A place of foundations, of discussion, of control, a place where, on the contrary of what is generally admitted, theory and experiment intimately conjugate in order to enlighten our minds. The reexamination of the theory of the dielectric properties of mixtures of dipolar anisotropic molecules leads to an Onsager type model with ellipsoidal cavities that gives an acceptable picture of the electric permittivity of mixtures of polar compounds (1972). The critic of the Onsager cavity notion and the statistical treatment of polarization made it possible to give a more precise basis to Fröhlich's theories (1969). Models are increasingly discussed and rediscussed and the approximations are compared with those obtained by more rigorous methods: for instance, the cavity approximation can be considered as a first approximation of the statistical theory (1974). What reinforces Barriol's preference for models was the fact, that quantic and statistical calculations are often "arduous". In that way, it is possible to say that, for Barriol, and it has been seen above in the case of the metallic model, these calculations often founded the use of models. Not only the results are discussed, but also the methods are compared. At least, if theory throws light on experiment and has to answer its questions, experiment gives the impulse that has to control theory increasingly. As a theoretical chemist, Barriol kept working unceasingly at constructing the scientific method as a subtle and always active combination of theory and experiment, in order to construct a coherent and human picture of our world.

References and Notes

1 J. Gueron, M. Magat, "A History of Physical Chemistry in France," *Annual Review of Physical Chemistry* 22 (1971): 1–23.

2 Mary Jo Nye enumerates the topics treated in the *Journal* as "molecular spectroscopy and molecular structures, the quantum mechanical treatment of electronic structure of molecules and crystals and the problem of chemical binding, the kinetics of chemical reactions from the standpoint of basic physical principles, the thermodynamic properties of substances and calculation by statistical mechanical methods, the structure of crystals, and surface phenomena." M. J. Nye, *From Chemical Philosophy to Theoretical Chemistry. Dynamics of Matter and Dynamics of Disciplines* (Berkeley: University of California Press 1993), 254. Many of these were considered by Barriol, as we will see later in this chapter.

3 K. Compton, letter to G. N. Lewis, 6 August 1932, quoted by Nye, *From Chemical Philosophy to Theoretical Chemistry*, 252.

4 J. Barriol, *La Science, le Réel et l'Indicible*, unpublished manuscript, 2.

5 Alberte and Bernard Pullman kept exploring the electronic structures of molecular systems and are the founders of quantum biochemistry in France. Cf. M. Blondel-Mégrelis, *Theoretical Chemistry in France. The early years*, unpublished manuscript.

6 J. Barriol, "Le laboratoire de chimie théorique," *Bull. Ass. Amic. Anc. Él. Inst. Chim. et de l'ENSIC* 10 (1965).

7 J. Barriol, *La Science, le Réel et L'Indicible*, 26. J. Barriol had a copy of Lespieau's *La Molécule chimique* (1926).

8 G. Dupont, "Robert Lespieau, 1864–1947," *Bull. Soc. Chim. France*, 16 (1949): 1–7.

9 Maurice Bourguel had realized a wonderful apparatus that death did not permit him to utilize. R. Lespieau, "Notice sur les travaux de M. Bourguel," *Bull. Soc. chim. France*, 3 (1933): 1145.

10 Cf. the many papers on chemical constitution and the Raman effect in *Bull. Soc. chim.* (1930–32).

11 P. Donzelot, J. Barriol, "Relation entre les fréquences Raman et les distances interatomiques," *Comptes rendus*, 204 (1937): 1867.

12 Barriol initiated experimental research in 1935 with Donzelot, who created the laboratory of Physics at Nancy Faculté de Pharmacie, and devoted to Raman spectra studies. In 1937, Donzelot became Professor of Physical Chemistry at Nancy Faculté des Sciences. For more details, cf. M. Blondel-Mégrelis, "Jean Barriol et l'éclosion de la Chimie théorique à Nancy, *Philosophia Scientae*, to be published.

13 J. Barriol, *Mécanique quantique* (Paris 1952).

14 P. Dirac, *Les Principes de la Mécanique quantique*, Preface, (Paris 1931). This traduction had been used by Barriol.

15 J. Barriol, *Mécanique Quantique*, 20–21

16 J. Barriol, *Eléments de Mécanique quantique* (Paris 1966), 24–25

17 J. Barriol, *La Constitution des Molécules* (Paris 1958), 219–23.

18 E. Bauer, *Introduction à la Théorie des Groupes et à ses Applications à la Mécanique quantique* (Paris 1933), 60–61.

19 Gueron, Magat, "A History of Physical Chemistry," 12.

20 Dupont, "Robert Lespieau", 7.

21 Barriol, *La Science, le Réel et L'Indicible*, 33–6.

22 Dirac, *Les Principes de la mécanique quantique*, Preface.

23 C. A. Coulson, *Valence*, (Oxford 1952), 62.

24 *Ibid*, 108.

25 J. Barriol, J. Régnier, "Polarisation atomique et moment de la liaison C-H," *Journ. Chim. Phys.*, 49 (1952): 213. The authors refer to C. A. Coulson, *Trans. Farad. Soc.*, 38 (1942): 433.

26 J. Barriol, *La Constitution des Molécules*, (Paris 1958), 88–89; Coulson, *Valence*, 187–8.

27 Coulson, *Valence*, v.

28 J. Barriol, J. Metzger, "Application de la méthode des orbitales moléculaires au réseau du graphite," *Journ. Chim. Phys.*, 47 (1950): 432.

29 J. Barriol, *Les Moments dipolaires* (Paris 1957), 21.

30 J. Barriol, J. Régnier, "Contribution au calcul variationnel des polarisabilités moléculaires," *Comptes rendus* 237 (1953): 307.

31 J. Barriol, J. Régnier, "Contribution à l'étude des orbitales moléculaires de l'éthylène. Transitions N-V et N-T," *Calcul des Fonctions d'Onde moléculaires*, Colloque CNRS (Oct. 1957): 205–211.

32 According to Daudel, chemistry was "a purely analogical and essentially analytical science. It becomes, little by little, deductive, synthetic ... It has now reached a satisfactory state of abstraction ... One is no more content with ascertaining facts, one demonstrates." R. Daudel, *La Chimie théorique* (Paris 1943), introduction.

33 J. Barriol, *Les Moments dipolaires*, 12.

34 Barriol, Régnier, "Polarisation atomique et moment de la liaison C-H."

35 J. Barriol, J. Régnier, "Polarisation atomique et moment des liaisons OH et CH," *Journ. Chim. Phys.* 51 (1954): 9.

36 J. H. Van Vleck, *The Theory of Electric and Magnetic Susceptibilities* (Oxford 1932), 45.

37 J. Régnier, *Contribution à la détermination précise de la polarisation atomique à l'état gazeux*, Thèse présentée à la Faculté des Sciences de Nancy (4 Juill. 1957), 15.

38 D. Rinaldi, J. L. Rivail, J. Barriol, "Analyse de population dans les calculs LCAO: charges et moments atomiques," *Theor. chim. Acta* 22 (1971): 291.

39 J. Barriol, S. Régnier, "Relation entre le modèle métallique des électrons p et les orbitales moléculaires LCAO," *Journ. Chim. phys.* 53 (1956): 929; S. Régnier, J. Barriol, "Relation entre le modèle métallique des électrons et les orbitales moléculaires LCAO," *Cahier de Physique* 75–76 (1956): 63.

40 J. Barriol, A. Weisbecker, "Contribution à la théorie de la constante diélectrique des solutions," *Comptes rendus* 259 (1964): 2831.

41 J. M. Thiébaut, J. L. Rivail, J. Barriol, "Dielectric studies of non-electrolytic solutions," *Journ. Chem. Soc., Far. Trans. II*, 68 (1972): 1262.

42 J. L. Greffe, J. Barriol, "Méthode d'étude des interactions dipolaires en phase liquide. Théorie statistique de la constante diélectrique," *Journ. Chim. Phys.* 11–12 (1972): 1579.

43 *Rapport d'Activité du CNRS*, ERA 22 "Interactions moléculaires," June 1974.

44 J. Barriol, A. Weisbecker, "Contribution à l'étude de l'effet de solvant sur la liaison hydrogène," *Comptes rendus*, 265 (1967): 1372.

45 Cf. for instance, J. Barriol, J. L. Rivail, *Eléments de Mécanique statistique des Systèmes en Interaction forte*, 25 (1974).

46 Unité de Recherche Associée, "Interactions moléculaires", *Rapport d'Activité* 1 (1982–86).

47 Unité de Recherche Associée, *Rapport d'Activité* 19 (1986–90).

48 D. Rinaldi, *Contribution à l'étude des systèmes des électrons sigma par une méthode LCAO. Application à la détermination des propriétés électriques des molécules et associations moléculaires* (1969).

49 D. Rinaldi, *Application des méthodes semi-empiriques à l'étude des propriétés électriques moléculaires et des interactions entre molécules* (1975).

50 Cf. Nye, *From Chemical Philosophy to Theoretical Chemistry*, 260.

51 A. D. Buckingham, "Molecules in optical, electric, and magnetic fields: a personal perspective," *Ann. Rev. Phys. Chem.* 49 (1998): XIII.

52 Unité de Recherche Associée "Interactions moléculaires," *Rapport d'Activité* 1–3 (1980–82).

53 Barriol, Rivail, *Eléments de Mécanique statistique*, introduction.

54 Equipe de Recherche Associée au CNRS, "Interactions moléculaires," *Rapport d'Activité* (1967–69).

Part II
From Radiochemistry to Nuclear Chemistry and Cosmochemistry

6. From Radiochemistry to Nuclear Chemistry and Cosmochemistry

Xavier Roqué

Elements – their arrangement, abundance, production, and uses – continued to be a major concern to scientists, not just to chemists, through the twentieth century. In the decades up to World War II all gaps in the periodic table up to uranium (92) were filled – except for promethium (61), synthesized after the war together with a dozen unstable elements heavier than uranium – and the table itself was understood in terms of atomic structure. The new heavy elements, however, were radioactive, that is, unstable, and their decay deprived the hard-won nineteenth century concept of the chemical element of one of its defining traits, permanence. While the instability of an element was something of a contradiction in terms for many an orthodox chemist, it suggested new questions about the evolution and abundance of elements in heaven and earth. These questions claimed the attention of many scientists, as they linked the terrestrial study of elements, particularly of the new radioactive elements (radiochemistry), with the study of elements in the universe at large (cosmochemistry).

Naturally enough, these developments brought substantial changes in the practice and the disciplinary structure of chemistry, which the papers that follow seek to illuminate in different shades of detail. Radiochemistry developed alongside the physics of radioactive elements and nuclei, updating the nineteenth century practice of making substantial use of physical means of analysis. The papers by Ruth Sime and Brigitte Van Tiggelen, on element search and production in the interwar years, illustrate how delicate the balance of disciplines could be. Sime focuses on the search for elements heavier than uranium (the so-called transuranium elements) between 1934 and 1938, offering a detailed and thoroughly documented argument about the roles of chemistry and physics in the investigative process that led to fission. Van Tiggelen contrasts Walter and Ida Noddack's success in discovering rhenium with their failure in confirming the existence of what they called masurium (element 43, technetium), focussing on their analytic strategy and its significance for the boundaries of chemistry. Equally significant, if less well known, were the renewed links to the earth sciences and the astronomical sciences, discussed in Helge Kragh's overview of the rise of geo- and cosmochemistry. Although the chemistry of the universe only came of age as a scientific discipline in the 1950s, it had a rich background in nineteenth century geochemistry, meteoritics, and astrochemistry. Kragh's paper fits with those on radiochemistry because it is the concern with elements that makes cosmochemistry a chemical discipline, despite all its relations to other sciences.

Taken together, these papers provide insights on the structure and cognitive content of twentieth century chemistry. They all deserve careful reading and, rather than trying to summarize them, I have organized my presentation around a number of salient issues which relate them to one another: the value and status of physical evidence in chemical disciplines; the relationship between the identification of an element and its production; the contrast between natural and artificial elements; and the disciplinary dynamics that attended all of these changes.

6.1 Physical Evidence in Chemical Disciplines

The transmutation of elements now counts as the most revolutionary implication of the discovery of radioactivity, but initially the mere existence of radioactive elements was a serious enough challenge to established chemical practice. This had to do, first and foremost, with the use of physical means of analysis. Such use was anything but foreign to chemistry: the discovery, isolation, and production of new elements was often facilitated by new physical techniques in the nineteenth century, the voltaic battery and the spectroscope coming readily to mind. In the first half of the twentieth century, the search for elements involved electrometers and ionization chambers, X-ray and electron diffraction devices, and the mass spectrograph, among other tools. The authors in this part of the book document such uses. For example, Kragh shows that X-ray diffraction, famously pioneered by Henry G.-J. Moseley, was crucial to Victor Moritz Goldschmidt's collection of data on atomic and ionic radii, while Francis Aston's mass spectrograph, which got him the Nobel Prize in Chemistry for 1922, served him well in geochemical analysis and in establishing the relative abundance of isotopes. Spectroscopic measurements allowed Georges Urbain to confirm the existence of the rare earth elements he had been able to isolate and led the Noddacks in their successful search for rhenium (see van Tiggelen's paper).

The nature of these tools led to disputes concerning their analytic value and their evidential status. As had been the case with astrochemistry, radioactivity was deplored by some laboratory chemists on account of the kind of evidence it rested upon. Kragh points out, with regard to the elements discovered by the astrochemists, that "to most chemists, an element that could not be isolated and whose atomic weight could not be determined was not an element"; and the same applies to the new radioactive elements. As Henry Armstrong, professor of chemistry at the Central Technical College in London put it in 1906, "no one has yet handled 'radium' in such quantity or in such manner that we can say what it is precisely". In the summer of 1906, the elementary nature of radium was discussed in *The Times*, following a letter of 9 August in which Lord Kelvin questioned transmutation theory.

Oliver Lodge, Frederick Soddy, Robert J. Strutt (Lord Rayleigh), and Arthur S. Eve, among others, contributed letters to the debate. At issue was the evidence put forward by radioactive scientists: while some judged it sufficiently extensive and compelling, others felt that "a vast fabric of speculation" had been reared upon "a narrow basis of solid experimental facts". It is difficult to assess just how widespread the sceptical attitude was, yet a letter from Otto Hahn in Berlin to Ernest Rutherford in Manchester, written a few days before the radium controversy, suggests it was in no way limited to British chemists: "In the institute where I am going to settle down [Emil Fischer's Chemistry Institute], people know only very little on radioactivity. If they hear of something in connection with Ra, they always seem skeptical". [1] The radium debate resonates with disciplinary clashes dealt with in other parts of this book, such as those surrounding the uses of mathematics in chemistry. To Arthur Smithells addressing the British Association for the Advancement of Science in 1907, the "invasion of chemistry by mathematics" was particularly evident in the case of radioactivity, which he described as a "chemistry of phantoms". [2]

This much was recognized by the founding mother of radioactivity when she referred to the new science as "la chemie de l'invisible". With this denomination Marie Curie stressed that evidence for the existence of new radioactive elements did not rest on conventional gravimetric studies of atomic weight, performed on weighable samples; rather, radioactive elements made themselves apparent through radiations invisible to the naked eye. When the Curies and Gustave Bémont announced the discovery of radium, all they had was a tiny sample of barium chloride whose activity could not possibly be due to any of the elements that were known to form it, and which showed one unaccounted for spectral line. The discovery of both radium and polonium was based on the Curies' ability to compare, by means of an electrometer and a piezoelectric device, the activity of radioactive samples, thus finding their way through fractional precipitation. Electron physics was not just essential to guide the separation process: in the absence of pure samples that would allow a determination of the atomic weight of radium, spectral analysis provided the clearest evidence one could hope for that a new element had been found and allowed scientists to rely on atomic weight estimates from impure samples. Its importance was not lost on Eugène Demarçay, the prestigious spectroscopist who analyzed Curie's radium salts, and reassured Curie he was ready to stand by her: "I think like you that 225 must be very close to the actual [atomic weight of radium] and I still believe that the proportion of Ba in your radium is too small to affect the units. I would not hesitate to express this opinion publicly, you can attribute it to me in your note". [3]

These techniques helped to sort out elements and, as it turned out, also isotopes. The nature of radioactive *elements* was not initially clear. Did they all correspond to distinct chemical elements? To researchers in radioactivity, the word element long denoted a radioactive species, be it a proper chemical element or an isotope. Thus Lise Meitner described protactinium "not just as a long-lived radioactive element, but also as chemical one". The title of one of Otto Hahn's papers on the same element is also telling: "Das Protactinium als radioaktives und als chemisches

Element".[4] The use radiochemists made of physical instrumentation makes them into transition figures between traditional analytical chemistry, based on the treatment with known compounds and the observation of reactions, and today's intensely instrumentalized analytical chemistry, "which allow one to discriminate chemicals in terms of their physical properties", as Davis Baird has put it. Baird further observes that, prior to 1920, physical identification always followed on chemical separation and manufacture whereas from 1950, as instrumentation grew in importance, elements in substances could be identified and controlled without separating them. His claim that chemistry thereby underwent a "scientific instrumentation revolution" is not wholly substantiated by the papers that follow. Quite aside from the nineteenth century precedents referred to above, the techniques of the radioactive scientists clearly put them in an intermediate position between both traditions.[5]

6.2 Identification and Production

Ironically, by the time he decried it, Smithells's chemistry of phantoms was already at the basis of a new mid-size chemical industry, the radioelements industry. While some doubted its existence, radium fetched prohibitive prices. Widespread scepticism regarding the existence of radioelements was among the reasons that lead radioactive scientists to try and isolate them in substantial amounts. Marie Curie excelled in this practice, as witnessed by her quest for pure radium metal, successfully completed in 1911, which incidentally earned her a second Nobel Prize. The need for weighable samples was also felt by Walter and Ida Noddack, as Van Tiggelen points out. The Noddacks decided to produce rhenium in substantial amounts, because physical evidence would not do for them and many of their colleagues. However, I do not believe that attempts at purifying and producing ever greater amounts of radium and other substances were simply prompted by the disbelief of old-guard chemists, as has often been claimed (particularly by Curie's biographers). Rather, it seems to me that accumulation was driven by both scientific and industrial impulses. A pure and powerful source was essential to the conduct of radioactivity, and more often than not it could only be obtained with substantial industrial help. The issues of identification and production went hand in hand and cannot thus be easily separated.

This appears to have been the case with Curie, as I have argued elsewhere.[6] For Curie, knowing an element meant being able to produce it in fair amounts: the concentration of elements was an important goal in itself, but it was also essential in order to the in-depth analysis of radioactive substances. She applied this policy to radium, polonium and to elements in the actinium series, though short lifetimes often prevented Curie and her coworkers to reach their goal no matter the industrial resources at their disposal. While this was a characteristic feature of Curie's work on radioactivity, this does not mean she was alone. To give another example: In 1907 Otto Hahn had barely discovered mesothorium (MsTh, a mixture of MsTh1, an isotope of Ra, and MsTh2, an isotope of Ac) when he began work on the synthesis of the new element for the German chemical company Knöfler. In the same letter to Rutherford in which Hahn suggested a name for the new element, he talked about his work on concentration and asked for discretion on behalf of his industrial patrons. [7] Crucially, Hahn had not yet isolated MsTh, nor did he know its chemical properties in great detail; in fact, he needed Knöfler's facilities and, above all, the company's remains from the production of thorium, to go ahead with the laborious identification procedure. Much the same happened in 1917 with protactinium when Hahn and Meitner resorted to the radium-producing firm of Buchler & Co. in Brunswick, Germany in order to get raw material (100 g of pitchblende from which both uranium and radium had been extracted). Protactinium was later to be manufactured by the Auergesellschaft. [8] Besides Curie and Hahn, many radioactive scientists were engaged in consulting activities of just this kind, from Stefan Meyer in Vienna to Bertrand Boltwood in Yale.

These examples show that production was often undertaken *before* an element had been shown to exist in significant amounts. In the case of rhenium as discussed by Van Tiggelen, its properties were sufficiently promising to *Siemens und Halske* to help the Noddacks to manufacture it. The couple knew that the commercial availability of rhenium would elevate it from a *Privatelement* to a public element. Indeed, whereas by the end of 1926 the Noddacks had managed to isolate 2 mg of rhenium, five years later the annual production amounted to 120 kg and the price of rhenium, the rarest element on earth, had diminished so much that every chemical laboratory was able to purchase some. In due time, the barest inkling at an element's existence would suffice to build entire plants. As Glenn T. Seaborg said of the Hanford works, where plutonium was produced during World War II as part of the Manhattan Project: "The plants defy description with their massive structures and their intricate maze of equipment, piping, and remotely operated controls. The preliminary design of these plants was under way at a time when the world supply of plutonium was invisible to the naked eye". [9]

The significance of chemical isolation was such, that it rivalled with physical discovery. This tension was not apparent in the case of radium because Curie herself was involved both in its discovery and its preparation as pure metal. Had purification been the work of any other scientist, he or she might have claimed and perhaps got credit. Take protactinium, discovered in 1918 by Hahn and Meitner in Germany and independently by Soddy and John Carson in Great Britain. The new element was identified by the radiation it emitted, and its atomic weight and half-life determined by the laws of radioactive decay, but, even when Hahn and Meitner made a try with the help of the Auergesellschaft, a sample was only produced in 1927–1929 by a researcher of their institute, Aristid von Grosse. When von Grosse began being credited in some journal articles and textbooks as the discoverer of protactinium, Meitner and Hahn had to write a clarifying note. [10]

The connections between radioactive science and industry, prefiguring those between nuclear science and industry, were well in place long before the Manhattan Project, a conclusion that echoes Nicolas Rasmussen's remarks on molecular genetics not being the first life science to become biotechnology (see his paper in this volume). On the other hand, production has often been misunderstood as scientifically subordinate (conditional on proper knowledge of an element's properties), if not superfluous. However, we have seen that confidence in the existence and properties of new elements often followed their production, not the other way round. I have argued that this was also the case with the discovery of new particles in the 1930s, like the positron. At the time, physicists draw a contrast between the detection of natural events and its artificial production under controlled conditions in the laboratory – a contrast that has also attracted the attention of our authors, and to which I now turn.

6.3 Natural Versus Artificial Elements

The isolation of naturally occurring elements had its counterpart in the artificial synthesis of elements and isotopes, doubtless a feature of twentieth century chemistry. The transition was not smooth, as the case of masurium/technetium makes clear. Many scientists did not see the advantage of high-energy machines over natural processes. Gilbert N. Lewis's reference to "the great laboratories of the stars", like G. Hale's to "the vast laboratory of nature" (both quoted in Helge Kragh's paper), clearly posed the question of natural diversity versus artificial control. Lewis admitted that "we cannot plan the processes occurring in the stars", though these were infinitely richer than processes taking place in a terrestrial laboratory. However, a scientist's confidence was often based on his or her control over the experimental conditions, and laboratory scientists often strove to reproduce natural phenomena in the laboratory. The positron is a case in point. Carl Anderson's evidence for the existence of a positive electron, presented late in 1932, was based on photographs of cosmic ray tracks, and this made it difficult for such prominent physicists as Niels Bohr or Wolfgang Pauli to accept the existence of a new particle. Bohr wanted the new particle "to be produced ... under conditions which can be completely controlled experimentally." Positrons only carried the day when physicists were able to manufacture them at will in all terrestrial experiments.[11]

Sime describes a similar situation concerning the relative value of radioactive sources and the new machines for the production of high energy particles. She points out that, before 1934, "the primary sources of nuclear data were the natural radioactive species"; in this regard, the title of Rutherford, Chadwick, and Ellis's book, *Radiations from Radioactive Substances*, could not better match its contents. The transition from radioactivity to nuclear physics hinged on the use of radioactive sources, as opposed to artificially accelerated particles. The prominent role they accorded to natural radiations, distinguished the old generation of radioactivists from the new generation of nuclear physicists.

The same tensions are at play in the Noddacks' insistence on isolating naturally existing elements, as opposed to obtaining them by means of nuclear fusion. According to Van Tiggelen, this made them into chemists, possibly geochemists, but not radiochemists – perhaps nuclear chemists would be more appropriate. The

Noddacks' geochemical methods proved successful in the case of rhenium, yet they failed in the case of element 43, technetium, which they named masurium and claimed to be "naturally occurring". To the Noddacks, who lacked a pure sample of masurium, it was essential to possess a plate with the element's spectrum; the artificial production of an isotope of element 43 (technetium, Tc) by means of a cyclotron – the work of Carlo Perrier and Emilio Segrè in 1937 – did not amount to the same as locating the element in nature. The contrast between disciplines was here drawn not in connection with mathematical tools, but in connection with the use of the last nuclear toy: the cyclotron. It would take a new generation of chemists willing to manipulate high-energy machines to create nuclear chemistry.

6.4 Discipline Dynamics

In discussing the evidence for the existence of new elements, the papers of Van Tiggelen and Sime throw light on to the disciplinary relations of radiochemistry, for agreement on matters of evidence is a clear sign of disciplinary unity. Indeed, scientists pursuing radioactive research got over the hostility or the indifference of some colleagues by creating a new discipline, complete with textbooks, specific courses, conferences, journals, and institutes. Long before Wolfgang Pauli addressed a gathering of physicists in Tübingen with his famous "radioaktive Damen und Herren," radioactive researchers styled themselves as "the radioactivists" – the title of a superb thesis by Jeff Hughes.[12] In 1911, Franz S. Exner referred to "radioaktive Menschen" in a letter to Rutherford; for his part, ever since his return to Germany in 1906, Hahn was trying "to spread radioactive enthusiasm in our own still 'rayless' fatherland".[13]

Most scientists involved in radioactive research had a background in chemistry or physics, and up to World War I little distinction was made between the physical and chemical aspects of radioactive research. As Ruth Sime points out, radioactivity split after the war. In 1917, to give an example, the radioactive section at the *Kaiser-Wilhelm-Institut für Chemie* in Berlin-Dahlem split into a physical section (headed by Meitner) and a chemical section (headed by Hahn). In some sense, however, the field retained its unity: radiochemistry was kept much alive at the *Institut du Radium* in Paris, and this expertise helped in the discovery of artificial radioactivity, when phosphorus had to be isolated in three minutes. The subdisciplinary divide was informed by a common interest in radioactive substances. This division did not so much reflect the *independence* of radiophysics and radiochemistry, as the *mutual confidence* of their practitioners. As Sime puts it: Physicists and chemists "collaborated across a pronounced disciplinary divide... they trusted each other's expertise without always understanding each other's limitations".

However, Sime goes on to stress that "the pursuit of synthetic elements brought together nuclear physicists and radiochemists in what may be called a 'neo-classical' period of interdisciplinary research", after the relative independence of the 1920s. Collaboration was again essential to clarify the behavior of uranium under neutron bombardment, and both physics and chemistry contributed false premises, the confidence in small nuclear changes in the case of physics, and the belief that the transuranium elements would have the chemical behavior of transition elements. Hahn later tried to attribute the discovery of fission solely to chemistry, but Sime shows unambiguously, with a wealth of documentary evidence, that physics played a crucial role in the discovery, not to mention the political circumstances that prevented Meitner from getting full credit for her contribution. [14] It is intriguing to consider that the laboratories mostly engaged in the discovery of fission were precisely those where chemical and physical expertise sat easier together: the *Kaiser-Wilhelm-Institut für Chemie* in Berlin, to be sure, but also the *Institut du Radium* in Paris and Fermi's group in Rome. We can surmise that disciplinary cooperation gave a clear edge to these laboratories over those with looser connections between nuclear physics and chemistry – such as the Cavendish laboratory at Cambridge.

While radioactivists coalesced early, cosmochemistry was much longer in the making and long retained its pluridisciplinary outlook, as Kragh's rich and rewarding paper makes clear. It is impossible in a short introduction like this to give justice to this paper. What I find most interesting about it, and at the same time most distressing to some extent, is that he provides no tidy image of discipline building, but rather describes a complex pattern of relations. Some are rather suggestive: In the 1920s, for instance, prominent radiochemists such as George de Hevesy or Friedrich Paneth turned to cosmochemistry, while the cosmochemical work of Goldschmidt, the founding father of cosmochemistry, established links between nuclear physics and cosmology at a time these sciences were not at all related. Vís-a-vís this situation, instead of looking for a ready-made definition Kragh wisely clarifies the appearance and meaning of terms such as "cosmochemistry". Some of the research fields that contributed to cosmochemistry were relatively short-lived fields which barely managed to cohere and show some unity before being absorbed into a bigger or simply a different discipline. One is reminded of the emergence of cosmic ray physics, which for a while competed with radioactivity. Cosmic ray physics hardly amounted to a discipline, yet it was an important focal point for some nuclear scientists in the 1920s and 1930s, during which period it shared some of cosmochemistry's most ambitious objectives, such as providing a unified account of the origin and evolution of the elements. Robert A. Millikan exemplifies this trend, together with the scientists referred to in Kragh's paper: John Nicholson, Walther Nernst, and William Harkins.

We may conclude making with Kragh the simple but nonetheless sound point that the study of interdisciplinary areas of research such as these demands interdisciplinary historians. In much the same way as chemists and geologists ignored each other in the early days of geochemistry, historians of chemistry and historians of other sciences have largely neglected the fertile areas of contact between disciplines. Cross-disciplinarity is not a late twentieth century phenomenon, but has been going on ever since disciplines were formed. Scientists have often found it advantageous to share a disciplinary identity, yet the progress of science seems to demand disciplines constantly to reorganize and to rethink themselves, resulting in short-lived subdisciplines or branches, like cosmic physics, cosmic chemistry or radioactivity in the first decades of the twentieth century. Radioactivity was rather successful as a probe into the structure of atoms and nuclei yet it was also a relatively short-lived discipline, for already in the 1930s young radioactivists preferred to be called nuclear physicists. Of course, radioactivity continues to be an active field of research, but no scientist today would define himself or herself as a *radioactivist* in the way that Rutherford, Meitner or Curie did. The fields of cosmochemistry and nuclear chemistry offer, as witnessed by the papers that follow, fascinating insights into the complex disciplinary dynamics of twentieth century chemistry.

Acknowledgments
This work has partly been supported by the Spanish DGICYT under research program no. PS96–1169. I acknowledge permission of the syndics of Cambridge University Library to quote from the Rutherford Papers.

References and Notes

1 O. Hahn to E. Rutherford, 4 August 1906 (Rutherford Papers, Cambridge University Library, Add 7653/PA/H 4).
2 Smithells's 1907 presidential address to the chemical section of the BAAS is quoted in Chapter 3 in this book. Also: H. Armstrong, *The Times*, 10 August 1906, and "The radium controversy," anonymous leading article, *The Times*, 18 August 1906.
3 E. Demarçay to M. Curie, 11 July 1902 (Papiers Pierre Curie, Bibliothèque Nationale, n. a. f. 18434).
4 L. Meitner, "Über das Protactinium," *Die Naturwissenschaften* 6 (1918): 324–326, on 324; O. Hahn, "Das Protactinium als radioaktives und als chemisches Element," *Die Naturwissenschaften* 16 (1928): 453–457.
5 D. Baird, "Analytical Chemistry and the 'Big' Scientific Instrumentation Revolution," *Annals of Science* 50 (1993): 267–290.
6 X. Roqué, "Marie Curie and the Radium Industry: A Preliminary Sketch," in X. Roqué and Soraya Boudia (eds.), *Science, Medicine and Industry: The Curie and Joliot-Curie Laboratories*, special issue of *History and Technology* 13, number 4 (1997): 267–291.
7 O. Hahn to E. Rutherford, 10 February 1907 (Rutherford Papers, Add 7653/PA/H 11); O. Hahn, "Einige Erinnerungen an das Radiothor und das Mesothor," *Physikalische Blätter* 17 (1961): 570–576. Knöfler had decided not to submit a patent but keep the chemical concentration procedure secret. All the same, Frederick Soddy, one of its first customers, submitted a German patent on "The production of the radioactive element Mesothorium" and forced Knöfler to pay 28,000 marks in order to keep the right to produce the substance (to no avail, for another German company, the Auergesellschaft, contested the patent on the grounds that the chemistry of mesothorium could be gathered from existing publications, and won the case).
8 R. L. Sime, *Lise Meitner. A Life in Physics* (Berkeley: University of California Press, 1996), 67.
9 G. T. Seaborg, quoted in Samuel Glasstone, *Sourcebook on Atomic Energy* (Princeton, NJ: van Nostrand, 1950), 3rd ed., 1967, 635.
10 O. Hahn, L. Meitner, "Notiz über die Entdeckung des Protactiniums," *Die Naturwissenschaften* 19 (1931): 738.
11 X. Roqué, "The manufacture of the positron," *Studies in History and Philosophy of Modern Physics* 28 (1997), 73–129.
12 W. Pauli to L. Meitner and H. Geiger, 4 December 1930, in K. von Meyenn (ed.) (with the cooperation of A. Hermann and V. Weisskopf), *Wolfgang Pauli. Wissenschaftlicher Briefwechsel*. Band II: *1930–1939* (Berlin: Springer, 1985); Jeff A. Hughes, *The Radioactivists: Community, Controversy,*

and the Rise of Nuclear Physics (PhD dissertation: University of Cambridge, 1993).

13 F. S. Exner to E. Rutherford, 21 February 1911, and O. Hahn to E. Rutherford, 4 August 1906 (Rutherford Papers, Add. 7653, E83 and H4, respectively).

14 Besides Sime's chapter, see her *Lise Meitner. A Life in Physics.*

7. The Discovery of New Elements and the Boundary Between Physics and Chemistry in the 1920s and 1930s. The Case of Elements 43 and 75

Brigitte Van Tiggelen

The search for new elements must undoubtedly have been an appealing enterprise for young chemists around the turn of the century. Eventually, successful research could lead to the most praised reward: a Nobel Prize. After the acceptance of the periodic law and system of Dimitrii Mendeleev, the search for yet undiscovered elements became a more organized and rationally based investigation, but nevertheless the identification and "manufacture" of new elements was made possible only by a set of techniques and instruments developed in the realm of both physics and chemistry.

After Niels Bohr's proposal of the atomic model and the reinterpretation of the periodic system by Moseley's rule, the search for the missing elements received a new impetus and numerous were the attempts to fill the last gaps in the periodic table. Several rare earths were isolated and, by 1924, only five elements were still to be discovered: numbers 43, 61, 75, 85, and 87. Four of those elements are the natural radio-elements 43, 61, 85, and 87 and, for this reason, they escaped the searches led by chemists who were missing the right tool to isolate or even produce them in observable quantities. This was not the case with rhenium (element 75), the last stable element to be discovered and also one of the less abundant in the earth's crust.

A claim was made in 1925 for the discovery of the elements 43 and 75, by Walter Noddack, Ida Tacke, and Otto Berg.[1] Whereas the existence of element 75, named rhenium by the discoverers, was finally acknowledged by the scientific community during the 1930s, the existence of masurium (43) was not secured until Carlo Perrier and Emilio Segrè produced an artificial isotope of 43 in 1937 as a product of a nuclear fusion.[2] The ultimate naming of this element as technetium took place in 1947 and was not challenged at that time neither by the German chemical community nor by the Noddacks themselves.[3] More interesting is the belief in which the Noddacks stayed for the rest of their lives, that element 43 was a "natural" element, present in the earth's crust but in such a small proportion that it was hardly observable with contemporary chemical means.

All this suggests that the comparative history of the discovery of elements 43 and 75 constitutes a suitable case study to investigate the boundaries of chemistry and physics, and is expected to shed some light on these boundaries between 1920 and 1940, especially since both discoveries were claimed by the very same scientists. In the one case (rhenium), the usual "traditional" means of investigation proved to be successful, whilst in the other (masurium/technetium) it led to failure and discredit. Meaningful is the reaction of the actors, the Noddacks, who were unable to understand that "times had changed". The fact that Ida Noddack suggested the possibility of nuclear fission in 1934, in an attempt to interpret and correct Enrico Fermi's claim to have "produced" elements 93 and 94 by bombarding uranium [4] and that, at the very same time, she proved unable to draw the conclusion from this for making the light unstable isotopes of element 43 she was looking for, is also very symptomatic.

7.1 Rhenium: A Success

As soon as Mendeleev set up his table of the chemical elements, he noticed that some elements close to manganese most probably did exist, but were still to be isolated and identified. He named them "eka-manganese" elements, more precisely eka- and dvi-manganese. [5] But after half a century of search, claims, and withdrawals, these elements were still missing and some even questioned their existence. [6] Moseley successfully explained other anomalies in the state of the periodic table in 1913: he established a simple relationship between the X-ray spectra of an element and its atomic number. Switching from the atomic weight to the atomic number, the system lost some of its anomalies, but at the same time some vacancies remained to be filled in. Moseley investigated all X-ray spectra of known elements from aluminum [atomic number (Z) = 13] to gold (Z = 79) and reached the conclusion that four elements were still missing: those with the atomic numbers 43, 61, 72, and 75. The method of X-ray spectroscopy soon became the "signature of the element" and was used for instance by Georges Urbain to confirm the existence of the rare earth elements he had isolated. From then on, since the X-ray spectra could be calculated, the missing elements could be detected and distinguished from other known and established elements. For various reasons, the program proved to be very difficult to achieve. The missing elements were extremely rare or even not naturally occurring, an exception being hafnium, which is widely spread in nature but had escaped earlier detection and, because of its close resemblance to zirconium, was so to speak hidden behind this element. [7]

Out of the three missing elements left, two were the eka-manganeses. Ida Tacke [8] and Walter Noddack [9] decided to look for these two unknown metals. It remains unclear how Ida got into contact with Walter and how they decided this joint search for 43 and 75. In some accounts, Walter claims that he had already begun before World War I, but had to interrupt his work for military duties. [10] Ida on the other hand puts the starting point for this investigation about 1922. According to the laboratory notebooks, the work began in 1921–22. [11] Although we do not know precisely when and why the research on this topic was initiated, the

Tabelle 1.

Voraussage einiger chemischer Eigenschaften der Ekamangane.

43	75

Niedrige Oxyde

dunkle, in Säuren schwerlösliche Körper, leicht weiter oxydierbar. *)

Höhere Oxyde.

43	75
XO_3 hellgefärbt (gelb) in Wasser lösl. bildet wasserlösliche Salze: $Me^I_2XO_4$, die XO_3 anlagern.	XO_3 hell (gelb oder weiß) in Wasser lösl. bildet wasserlösliche Salze: $Me^I_2XO_4$.
X_2O_7 rosa oder gelb, Nadeln lösl. in Wasser sublimiert bei 350—400° leicht zersetzlich gibt lösl. Salze vom Typus Me^IXO_4	X_2O_7 weiße Nadeln schwer löslich in Wasser, lösl. in HII sublimiert bei 400—500° ziemlich beständig gibt lösl. Salze Me^IXO_4

Sulfide.

Die höheren Oxydationstufen hydrolysieren leicht, die niedrigen geben graue oder braune Sulfide.

Halogenverbindungen.

Leichtflüchtige Fluor- und Chlorverbindungen: $XHal_6$ u. $XHal_7$.

Figure 7.1 Predictions of the chemical properties of the eka-manganeses. [13]

documentation provides more information on how the discovery was organized, what kind of evidence was regarded as decisive and how the discovery was made a convincing one for their colleagues.

The first step was to document the search for other elements. They, especially Ida, managed to collect an incredible amount of literature, abstracts and summaries about the elements and the methods used to identify them. [12] As a second step came the description of the missing elements the team – not yet a couple – was looking for, based on an understanding of the periodic table that was both analytical and geochemical. They deduced or tried to foresee the physical and chemical properties of the elements 43 and 75 by comparison to their neighbors (e.g. manganese). The aim was two-fold: to establish precise means of identification of the missing elements and to build analytical procedures to isolate them from others that bear close resemblance in their chemical and physical behavior (Figure 7.1).

Their geochemical interpretation of the periodic table led the Noddacks to the conclusion that the missing elements had not been identified earlier because of their extreme rarity, and they were even more rare than any known element. They rejected the fact that the missing elements would form specific ores, since in that case they would have accordingly been discovered by then. The expectation to find both eka-manganeses elements at once (i. e. in the same ores) was also linked to a similar deduction, since this had already happened to be the case for zirconium and hafnium, and for molybdenum and tungsten. This is the reason why the Noddacks were convinced that the search for one element would also lead to the discovery of its “brother element”. The geochemical approach also influenced them in the choice of the ores to be inspected. They first selected manganese ores but soon realized

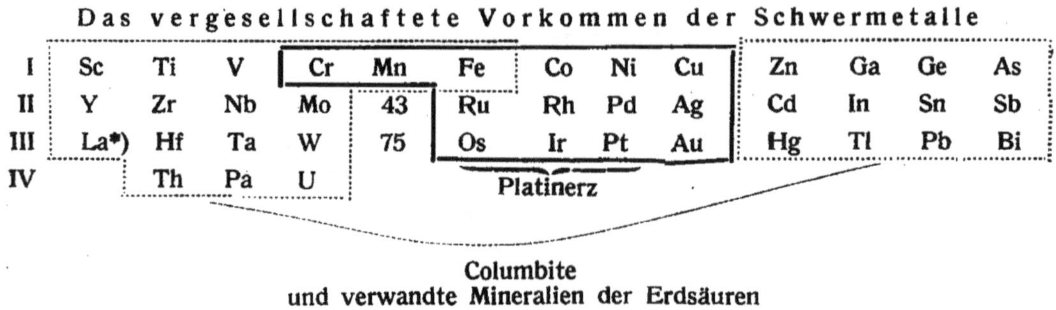

Figure 7.2 Part of the periodic table showing the connections of elements found in manganese ores and platinum ores.[1]

Häufigkeit der Elemente in der Erdrinde.

Sc	Ti	V	Cr	Mn		Fe	Co	Ni	Cu	Zn	Ga	Ge	As
	$2 \cdot 10^{-3}$	$3 \cdot 10^{-5}$	$3 \cdot 10^{-5}$	$7 \cdot 10^{-2}$		10^{-2}	$3 \cdot 10^{-6}$	$3 \cdot 10^{-5}$	10^{-7}	10^{-6}	10^{-9}		10^{-7}
Y	Zr	Nb	Mo	43		Ru	Rh	Pd	Ag	Cd	In	Sn	Sb
$2 \cdot 10^{-6}$	$6 \cdot 10^{-5}$	10^{-7}	10^{-7}	$\sim 10^{-13}$		$2 \cdot 10^{-12}$	10^{-11}	10^{-11}	10^{-9}	10^{-8}	10^{-8}	$7 \cdot 10^{-6}$	$7 \cdot 10^{-8}$
La	Hf	Ta	W	75		Os	Ir	Pt	Au	Hg	Tl	Pb	Bi
$6 \cdot 10^{-7}$	$6 \cdot 10^{-6}$	$5 \cdot 10^{-7}$	$5 \cdot 10^{-7}$	$\sim 10^{-12}$		$2 \cdot 10^{-11}$	$2 \cdot 10^{-11}$	10^{-9}	10^{-9}	10^{-9}	$4 \cdot 10^{-8}$	$4 \cdot 10^{-7}$	10^{-9}
	Th		U										
	$7 \cdot 10^{-5}$		$7 \cdot 10^{-6}$										

Figure 7.3 Table of the proportions of elements in the earth's crust.[1]

that they would have to broaden the search to platinum ores and beyond (see Figures 7.2 and 7.3).

Therefore, at the end of these preliminary reflections, the Noddacks were able to know exactly what to look for (chemico-physical properties and behavior), where to look (ores and enriched fractions), and how to look (characteristic chemical reactions).

At the beginning of their quest, the Noddacks were using only chemical procedures to enrich the selected ores and to test the presence of one or both of the eka-manganeses. An early success proved to be irreproducible.[14] After two years, they turned to another means of identification, optical spectroscopy, but soon realized how difficult and uncertain it would be to identify the faint lines of an unknown element amidst thousands of lines, sharp or not, of well known elements. X-ray spectroscopy, in contrast, seemed to offer both precision and certainty: the spectral signature of yet unknown elements could unequivocally be detected by this method. Noddack and Tacke collaborated for a while with the department of the *Institut für physikalische Chemie* in Berlin but, when in 1924 Karl W. Hausser of *Siemens und Halske* contacted them, a new kind of collaboration took place. Scientists at *Siemens und Halske*, a major manufacturer of electrical supplies, were also looking for manganese homologues from an industrial point of view. The foreseeable properties of element 75 were most interesting. The element to be discovered could eventually replace tungsten for the manufacture of light bulbs and thus perhaps allow a circumvention of patents held by one of the competitors of *Siemens und Halske*, General Electric. From the scientific point of view, *Siemens und Halske* would provide the team with the X-ray spectroscopic analysis. [15] Berg was to take care of this part of the investigation and was joined by Ida in the summer of 1924, who learned the technique and performed some measurements on her own, leaving most of the chemical work to Walter. [16]

In May 1925, at last, the X-ray measurements seemed to indicate the presence of both elements 43 and 75. The first mention of the discovery was made at a very high level, at a meeting of the *Preußische Akademie der Wissenschaften*, thanks to Walther Nernst who had been Walter Noddack's mentor. [17] The results were published in the prestigious journal *Die Naturwissenschaften* soon after, in a two-fold contribution. A first section of the paper is devoted to the analytical and geochemical part of the investigation, whereas the second section deals exclusively with the X-ray spectroscopy, [18] and Ida co-authors both contributions. In Walter's and Ida's mind however, the quest was not yet finished. Walter Noddack publicly declared that the crucial point to assess the existence of these two new elements was to produce samples and to hand them over to colleagues. [19]

Walter and Ida (just married) were very much aware of the fact that the only possibility to establish their credibility was to produce weighable samples of both rhenium and masurium, and decided to explore new ways of enrichment and production. One obvious way was to find ores that contained more rhenium and masurium than those already used. To achieve this, the couple undertook several trips to Scandinavia and Russia in 1926–1930. [20] The Noddacks convinced *Siemens und Halske* that rhenium could become an industrial product very soon; thanks to *Siemens'* sponsoring (40 000 Marks) and material help (a small laboratory was set up at their disposal for several months), the first gram of a rhenium salt was obtained in 1929. [21] At this point, the investigation of the chemical properties of rhenium could begin and they showed that most of the predictions made on the basis of the spectra alone were not far from reality.

Rhenium manufacturing soon became a business. Having looked vainly for natural ores, the couple was approached by W. Feit of *Vereinigte Chemische Fabriken* in Leopoldshalle. [22] Feit asked them to check an industrial residue, molybdenum glass, and Ida and Walter reached the conclusion that this industrial by-product contained more rhenium than usual, since some kind of enrichment procedure had already taken place. [23] Soon thereafter, rhenium was found in incredible amounts in the molybdenum glass of Mansfeld and, by the 1930s, about 120 kg of the least abundant element on earth's crust was produced. Mostly (and still now) needed for thermocouples, rhenium was also used as a catalyst. [24]

There are many similarities to the case of Marie and Pierre Curie, and their experience with the manufacturing of radium, to draw from the successful discovery of rhenium: the emphasis on the analytical enrichment procedure, the determination to produce weighable samples, the art of taking patents and mastering the production of the new element and, last but not least, the writing of a monograph on it.[25] Becoming a public element did not prevent rhenium remaining Noddacks' child.

In the hunt for yet undiscovered elements, the Noddacks were far from being alone and the hunt for manganese homologues had begun much earlier. It is thus not surprising that their claim of victory was quickly challenged and questioned.[26] Rhenium was by far the most controversial, and claims of the priority on masurium did not occur. The controversies were of two kinds: some questioned the results and argued that the evidence for having found elements 43 and 75 was non-existent; and others raised the priority debate, convinced they had seen element 75 before the German team. In both cases, the Noddacks' work was under suspicion and much of the Noddacks' early work on rhenium was a matter of closing the controversy successfully.

One year after the Noddacks had claimed victory, Wilhelm Prandtl gave a talk at the *Münchener Chemische Gesellschaft*, stating that back in 1913 he and his young colleague A. Grimm had prepared columbite in very much the same way as the Noddacks did. But these ores did not confirm the presence of element 43 or 75 when submitted to a spectrographic test. Even an early sample provided by the Noddacks did not show the main spectral lines as illustrated in the *Zeitschrift für angewandte Chemie*. Prandtl's interpretation was the following: since the main spectral lines of element 75, La and Lb, are very close to the main spectral lines of zinc and a spectral line of tungsten, Noddack and Tacke had most probably got confused and misinterpreted the spectrogram.[27] This point was of course vigorously countered by Walter and Ida[28] and, since only the manufacturing of rhenium could end the debate, they devoted most of their energy and time to it.

At least two different teams of chemists claimed, soon after the victorious announcement of the discovery of rhenium and masurium, that they had observed them previously. F.H. Loring and John G.F. Druce were actually looking for element 93 (viewed as tri-manganese) and pretended to have discovered not only element 93, but also eka- and dvi-manganese in pyrolusite and crude salts of manganese.[29] The Noddacks were able to prove less than two years later that these manganese ores were very poor in rhenium, about the poorest of all ores containing element 75.[30] Thus, the claim of Druce and Loring could be countered. V. Dolejsek and Jaroslav Heyrovsky were also working on manganese salts and made use of a new technique, polarography. They thought they had detected element 75 in an anomaly of the polarographic curve[31] and also questioned the chemical purity of the samples used by the Noddacks and hence the authority of the spectrograms of these samples. It was soon proved that this anomaly was not related with the presence of element 75 whatsoever.[32] The last opponents, O. Swjaginzew, M. Korsunski, and N. Seljakow, entered the scene in 1927, pretending to have found no dvi-manganese in platinum ores. They even tested the ores from the Gorablogodatski region (Ural) used for the 1925 claim and reached a negative conclusion.[33]

In their struggle to make their discovery acknowledged, the Noddacks were very much supported by the skepticism of their national competitors. Prandtl and his coworkers, A. Grimm and W. Francke, demonstrated on several occasions that the measures of Loring and Druce, or Dolejsek and Heyrovsky were not reproducible.[34] As in most controversies, the value of the chemical work achieved by both parts was subjected to heavy and sometimes impolite criticism. However, the Noddacks mastered the debate for two reasons.[35] First of all, the methodical preparation in their search of eka- and dvi-manganese had prepared them for the controversy. Testing systematically all kinds of ores, they were able to provide their opponents with sound answers. Also, their chronological priority was clear, as long as they could document their discovery. Less easy to defend were their arguments about the methods of identification and their correct use: the spectrographic measurements were harshly discussed. The team was mostly saved by their use of chemical – and more precisely analytical and mineralogical – methods and knowledge, which allowed them to eventually produce pure rhenium salts in weighable samples.

In 1929, rhenium was acknowledged by the *Deutsche Atomgewichtskommission*, and in 1930 the element was officially given the atomic weight of 188.71, suggested by Ida and Walter Noddack from analysis of the disulfide.[36] Otto Hönigschmid and R. Sachtleben gave a correction soon after,[37] based upon the conversion of silver perrhenate into bromide. A year later, in 1931, Francis W. Aston, provided with a rhenium sample by Walter Noddack, was able to identify the two main isotopes of rhenium (185 and 187) with the help of mass spectrometry; and he gave a value of 186.22 in good agreement with Hönigschmid's last value of 186.31.[38] At this stage, the discovery and priority were left unquestioned and, after ten years of continuous labour, the Noddacks enjoyed the recognition of the national chemical community by receiving the Liebig Medal of the German Chemical Society in 1931. Both Ida and Walter were elected members of the *Deutsche Akademie der Naturforscher Leopoldina* in 1934. The Scheele Medal of the Swedish Chemical Society was awarded in 1934 to Ida and other signs of international recognition can be found in nominations for the Nobel Prize. Ida was nominated in 1933, 1935, and 1937, jointly with Walter, who was also nominated in 1932 and 1934.[39]

7.2
A Failure: Masurium

Actually, the name masurium was never widespread in the German or the international literature. The Noddacks themselves remained silent about this element after their claim of discovery. The first reason is that by the end of the 1920s, they mainly concentrated on the production of weighable quantities of rhenium in order to prove its existence to the chemical community. As the Noddacks argue, masurium was left aside for a while since the controversy concentrated on rhenium.[40] Once they had won that battle, it should have been the turn of element 43 to be manufactured. Since the awaited occurrence of masurium was even less than that of rhenium, they perfectly knew they would have enormous difficulties to isolate it in usable quantity.[41] Progressively, although the discovery of rhenium proved to be a success, the failure to document the discovery of element 43 undermined the credibility of the Noddacks. Had they offered a public withdrawal, things might have been different for them in the following 40 years. [42]

The making of element 43 was eventually achieved by two nuclear means. In 1937, Perrier and Segrè obtained several isotopes of element 43 by bombarding a molybdenum plate with a strong deuteron beam, thus as a result of fusion. [43] In 1939, Otto Hahn and Fritz Straßmann identified element 43 as a product of the nuclear fission of uranium. [44] By 1940, Segrè and C. S. Wu were able to provide confirmation of this second way of production of element 43. [45] The making of this element was thus achieved following a very different approach, compared to the manufacturing of rhenium through analytical and enrichment procedures.

This new kind of tool to find the missing elements was celebrated by Segrè at the beginning of one of his papers. [46] He says that every time physics has provided chemistry with a new tool, it allowed the discovery of new elements, and he lists radioactivity, spectroscopy, and now the cyclotron. Nevertheless, the proof of the existence of and the actual detection of the isotopes of element 43 is first of all chemical, very much in the same way as Ida had criticized Fermi's conclusion in 1934. Segrè even published the "radioactive" part in another journal! [47]

By 1933, the price of rhenium, the rarest element on earth, had lowered so much that every chemical laboratory was able to purchase some. [48] At this time, when the Noddacks looked towards public and private funding, they were not able to find anybody ready to provide them with the necessary amount of money to isolate masurium. The industrial partnership which had been crucial in the manufacturing of rhenium was not considered: the properties of rhenium had attracted *Siemens und Halske* and then *Leopoldshalle*, but this was not the case for the properties of 43. By the time the Noddacks were to cope with Masurium, Walter was appointed in Freiburg as *ordentlicher Professor der physikalischen Chemie* in 1935 and afterwards, in 1941, in Strasbourg where he was heading the *Institut für Photochemie* and lecturing on physical chemistry. Ida kept working as a research associate but they were unable to reproduce the spectra or to find the right ores to isolate masurium. When in 1937 Segrè found an isotope of 43 by bombarding molybdenum with deuterons, the Noddacks did not even react, though Segrè and Perrier in one of their first publications on element 43 mentioned their 1925 paper on the discovery of the eka-manganese elements. [49] In the same way, nothing was said by the Noddacks after 1947, when element 43 eventually received his name, technetium. The name masurium appeared in some publications but disappeared quickly and, by the 1950s, the name and with it the recall of the Noddacks' claim was forgotten or countered. [50]

On several occasions, letters were sent to the couple asking whether the discovery had received confirmation or not. [51] Walter Noddack just kept silent, and even more and more silent as the years went by. The Noddacks felt they were unable to provide any kind of proof. Not only were they unable to show a sample of masurium which would have, as had been the case for rhenium, closed the controversy on the existence of the element, but even the original photographic plate of typical spectral lines was never shown as a piece of evidence. [52]

7.3
A Comparison: From Hunting to Breeding

For a while, the man-made elements produced at the Berkeley cyclotron remained numbers. But in 1947, Friedrich Paneth suggested several rules for the naming of newly discovered elements. [53] It remained the right of the discoverers to suggest a name for the element(s) they discovered. But the advent of the nuclear production of new elements had somewhat affected the game: one did not discover an element anymore, but one or several of its isotopes. To Paneth, the discovery of any of the isotopes meant the discovery of the element and, as long as this event was firmly documented, the discovery should be acknowledged as such. To avoid any discussion, Paneth made it clear that there should not be any distinction whatsoever between natural and artificial isotopes. In his contribution, "The making of the missing chemical elements," Paneth also mentioned that his concern about rules in naming elements stemmed from a mispractice related to the misuse of masurium and illinium to designate isotopes that were the first representatives of hitherto missing chemical elements. And he developed the case of the Noddacks who failed to withdraw their statement, providing two anecdotes aiming to show how arrogant and undocumented the claim of discovery was. In the same issue of *Nature*, the name technetium was suggested for element 43, to recall the fact that it had been artificially produced. [54]

These personal attacks were left unanswered by Walter Noddack. By that time, the Noddacks had more urgent matters to deal with: after the war, Walter encountered many difficulties in finding an academic position again. Having been appointed to the German University of Strasbourg during the German occupation of France heavily influenced his post-war career, even though he passed the denazification. [55] He was appointed in 1947 as professor of chemistry at the *Philosophisch-Theologische Hochschule* in Bamberg, but privately expressed bitter feelings about not being admitted to higher academic positions. [56] However, the papers of the couple show continuous interest in rhenium and element 43. Ida kept up with the bibliography and in 1954–1955, they even offered a lecture on the "*Vorkommen und Anreicherung des natürlichen Elements 43.*" [57] During the lecture, they made use of the same old way: analytical procedures to isolate element 43 from rhenium and systematic enrichment of selected ores. Firmly convinced that 43 must be available naturally, Ida, who did not master English, ordered a translation of a report on the occurrence of technetium on the earth's crust. [58] It must have been quite a deception to uncover the negative conclusion despite a title that was so appealing.

Commemorating the 100th anniversary of Mendeleev's birth in Leningrad, Lise Meitner in 1934 provided an understanding of the periodic system of the elements based on properties of the nucleus. [59] She reviewed the last results of radiochemistry and nuclear physics to reach the conclusion that the table set by Mendeleev would be enlarged in the near future. Meitner did not illustrate her contribution with the usual periodic system; she made use of a chart drawn by Max Delbrück mapping all possible atomic changes observed from hydrogen to calcium. To her, this is the beginning of a "nuclear chemistry." Because of the existence of elementary particles, elements could be built or destroyed, changed into one another, just as molecules.

On the same occasion, Ida Noddack delivered a talk to the *Bezirksverein Groß-Berlin und Mark*, focusing on the gaps in the periodic system of the elements.[60] Rhenium and masurium are of course not considered as missing elements. Having reviewed the efforts made to discover elements 61, 85, and 87, Ida comes to the description of elementary particles. She is very aware of the fact that chemists might have to deal with them one day, but at this stage the neutron had no chemical properties: "*Für den Chemiker ist das Neutron z. Z. noch etwas Wesenloses.*" She passes then to the ending of the periodic system: are there elements to be found after uranium? Ida thinks that elements 94 to 96 might be obtained, although they would be incredibly rare and short-lived. But these discoveries would not complete the periodic table, Ida argues, since chemists were yet able to separate the isotopes of hydrogen and this achievement opens the way to a similar chemical separation of all isotopes, raising the number of elements from 92 to 260. Another natural system would by then be necessary and replace Mendeleev's table. To summarize this point, one could roughly say that Ida Noddack seems not to have admitted the crucial transition from atomic weight to atomic number in the definition of elements and that, according to her opinion, weight remained the distinguishing character of the essence of an element. Thus if there exists a variety of isotopes, they are, in the ultimate form of chemistry, by essence different kinds of elements.

7.4
The End of a Research Tradition

The comparative study of the discovery of elements 43 and 75 clearly describes a major change in research traditions. Missing elements are no longer manufactured by chemical means (analytical procedures of enrichment) but have to be produced with the help of physical instruments to come to existence. Since in this contribution the point of view of the actors was deliberately chosen, it is possible to gain a deeper understanding of the gap dividing chemistry and physics in the discovery of new elements. Whereas rhenium, though very rare, is naturally occurring, technetium had to be man-made to be observed. To the Noddacks who were looking for the natural element 43, the isotopes detected by Segrè and Perrier did not end their quest. The X-ray spectra were a means to identify the presence of elements, but the definitive proof, in their mind, was the isolation of the element or of pure salts of it in a weighable amount, and the opportunity to verify the foreseen chemical properties.[61] These properties were as much viewed as a "signature" of the elements, as were the X-ray spectrograms. Parallel to a different understanding of the laws underlying the periodic table, there is also a shift in the place of publication: whereas most of the rhenium story can be read through *Angewandte Chemie* and *Zeitschrift für anorganische Chemie*, technetium belongs to *Zeitschrift für Physik*, *The Physical Review*, and *The Journal of Chemical Physics*.

This shows very clearly the shift in the investigative tradition in chemistry in the search for the missing elements. To fill the gap of number 43, the Noddacks chose not to turn towards physics but towards a new emerging subdiscipline instead: geochemistry. Their hope was that these studies of the natural occurrence of the elements would show the way to the right ores to search for masurium. [62] With the development of the concept of *Allgegenwartskonzentration*, came along a "law" that every element was to be found in every mineral ore on the earth crust, if only in incredibly small amounts for some of them, which explains the fact that they could not be detected by ordinary and present means. [63] Since element 43 was to be found somewhere on earth's crust in very small quantities, the main point was to find out which ore would be the best carrier of this element. They also looked for 43 and 75 in meteorites but without much success. [64]

In their research, physics remained instrumental, through the use of spectroscopy for instance. In several survey articles on the geochemistry of rhenium, rare earths, and other elements, spectroscopy was presented as a tool but not as the ultimate proof of the existence of these elements. The ultimate proof was left to chemical behavior. [65] It was only because the chemical means are unavailable in the case of rare earths that the use of X-ray spectrography was developed. Other physical techniques, especially cyclotrons, were developed rapidly at the very same time. Segrè, and after him Paneth, did acknowledge the existence of this new tool as a means to find new elements just as radioactivity or spectroscopy did before. Nothing was said by the Noddacks on this matter. This is quite surprising, since they both were aware of the existence of these developments and they were also able to understand their conception. Walter Noddack was trained as a physical chemist under Nernst and published several papers. Ida knew perfectly well about the past developments in experimental nuclear physics, for she commented on the alleged discovery of elements 93 and 94 by Fermi in 1934. In their minds, it seems that the cyclotron was not the relevant tool to produce and identify what had to be discovered in the realm of nature and isolated by chemical means.

This point is emphasized by the general survey Ida Noddack gave of the science of chemistry, its aim and methods, in a popular book published during World War II. [66] Chemistry was presented as *Lehre der Stoffe*, science of matter, dealing only with weighable quantities (which is actually the crucial notion for the existence of elements of the International Atomic Weight Commission). [67] The chemist thus has to investigate the properties of the elements, their appearance in nature or in the universe, the reactions they experience with one another, and even the structure of matter, while the physicist has to deal with all other changes occurring in nature that do not affect the essence of matter (motion, heat, energy exchange, and others). At the boundaries of physics and chemistry, of course, lies physical chemistry, but the role of physics is instrumental both from the theoretical and experimental point of view. Therefore the different aspects of physical chemistry are related to the physical means of chemical changes. For instance, heat or light lie in the domain of thermochemistry or photochemistry, each of these specialities using everything that physics can tell and teach about heat or light.

Until the end of their lives, Ida and Walter Noddack kept an eye on the chemistry of rhenium and especially on that of technetium. This is especially remarkable when one has a look at the papers of Ida. It is true that she outlived her husband and also that she never found a position again. As a benevolent worker in her husband's *Geochemical Institute*, she had plenty of time to devote herself to old memories. In that respect, the Noddacks should no longer be called radiochemists, but rather chemists, geochemists, or photochemists, if subdisciplinary specifications were needed. What led them to success in the case of the rhenium appeared to be a complete failure in the case of element 43. They had in mind the example of the Curies and forgot, or were unable, to look at the contemporary developments by Hahn, Ernest O. Lawrence, and Glenn T. Seaborg.

References and Notes

1 W. Noddack, I. Tacke, and O. Berg, "Die Ekamangane," *Die Naturwissenschaften* 13 (1925): 567–574. This paper was translated into English: "Eka- and dvi-manganese," *Chemical News* 131 (1925): 84–85.

2 C. Perrier and E. Segrè, "Radioactive isotopes of element 43," *Nature* 140 (1937): 193–194; see also C. Perrier and E. Segrè, *Journal of Chemical Physics* 5 (1937): 712.

3 It is only in recent times that a polemic arose about the naming of element 43. See P. H. M. Van Assche, "The ignored discovery of element Z=43," *Nuclear Physics* A480 (1988): 205–214 and G. Herrmann, "Technetium or masurium – a comment on the history of element 43," *Nuclear Physics* A505 (1989): 352–360.

4 Ida Noddack's proposal of the fission process has awakened interest in the history of science since Gerald Holton unearthed the case in 1973. See T. Hopper, *She was ignored. Ida Noddack and the Discovery of Nuclear Fission*, Master's thesis, Stanford University, 1990, and F. Habashi, "Ida Noddack: proposes of nuclear fission," in M. F. Rayner-Canham and G. W. Rayner-Canham (eds.), *A Devotion to their Science: Pioneer Women in Radioactivity* (Philadelphia: Chemical Heritage Foundation – Mc Gill's University Press, 1997), 217–225.

5 D. I. Mendeleev, "Die periodische Gesetzmässigkeit der Elemente" *Liebigs Annalen der Chemie* (suppl) 8 (1871): 205.

6 J. G. F. Druce, *Rhenium. Dvi-Manganese, the Element of Atomic Number 75* (Cambridge: Cambridge University Press, 1948), on 1–4, lists several alleged discoveries: Davyum, ilmenium, lucium, nipponium for element 43 and uralium, pluranium, ruthenium, pelopium. For reports on unsuccessful search see e. g. C. H. Bosanquet and T. C. Keeley, "Note on a search for the missing element n. 43," *Philosophical Magazine* 48 (1924): 145; F. H. Loring, *Chemical News* 225 (1922): 309 and 386; W. Prandtl and A. Grimm, *Zeitschrift für anorganische und allgemeine Chemie* 136 (1923): 283.

7 On hafnium see H. Kragh, "Anatomy of a priority conflict: The case of element 72," *Centaurus*, 23 (1980): 275–301 and, on rare earths, C. H. Evans (ed.), *Episodes from the History of Rare Earth Elements* (Dordrecht: Kluwer, 1996).

8 Ida Eva Tacke (1896–1978) was the daughter of a lacquer and varnish manufacturer from Lackhausen, north of Köln. She was trained as an engineer at the Technische Hochschule in Charlottenburg and received the title of Dr. Ing. in 1921. She worked at AEG from 1921 until 1923, when she resigned to devote all her time to the search for the eka-manganeses. Married to Walter Noddack in 1926, she held research positions usually unpaid and informal, except for the period from Freiburg (1935) to Strasbourg (1944). See F. Habashi, "Ida Noddack," and J. A. Johnson, "German women in chemistry, 1895–1925 (Part II)," *N. T.M.* 6 (1998): 78–79.

9 Walter Noddack (1893–1960) began studying chemistry, physics and mathematics at the University of Berlin in 1912. Having volunteered during World War I, he received his doctorate in 1920 only, under the direction of Nernst on Einstein's law of photochemical equivalence. He became di-

rector of the chemical laboratory of the Physikalisch-Technische Reichsanstalt in 1922, and by 1927 was also director of a newly founded photochemistry laboratory in Berlin. In Freiburg, he was appointed as chairman of the physical chemistry department between 1935 and 1941, before he moved to Strasbourg. Geochemistry and Photochemistry were his major fields of research, see F. Szabadvary, *Noddack, Walter,* in *Dictionary of Scientific Biography,* vol. 10 (New York, 1974), 136 and references therein.

10 He states this early concern in a talk to the Deutsche Chemische Gesellschaft: W. Noddack, *Zeitschrift für angewandte Chemie* 38 (1925): 876.

11 The Noddack papers are kept at the Katholieke Universiteit te Leuven, Belgium.

12 See for example the notebook with publications lists on palladium, ruthenium, and rhodium, ca. 1922, K. U. Leuven Archives, Noddack-Tacke Papers, 63; and the notebook with an overview of the properties of the elements nearby elements 43 and 75, K. U. Leuven Archives, Noddack-Tacke Papers,108.

13 See fig.1 from I. Tacke, "Über die Auffindung der Elemente Masurium und Rhenium," *Japanisch-Deutsche Zeitschrift für Wissenschaft und Technik* 3 (1925): 365–370.

14 Draft of a letter of 6. September 1936 to Walter's nephew, K. U. Leuven Archives, Noddack-Tacke Papers, 1187. This has to be interpreted carefully, because on several occasions her memory is biased and the story reconstructed.

15 This collaboration was actually established by an industrial research contract, signed on 18 June 1925, K. U. Leuven Archives, Noddack-Tacke Papers, 107. The opportunity to patent their discovery had not escaped the team since they wrote patents on their own. See the list of thermocouple patents, ca. 1923, K. U. Leuven Archives, Noddack-Tacke Papers,108. As a consequence of their contract, the rights of the Noddacks' patent for the extraction of rhenium from manganese ores is held by Siemens und Halske. See J. F. G. Druce, *Rhenium,* 73.

16 She also chose at that time to drop her job at AEG (a competitor of Siemens und Halske) and to perform benevolent research at a plant of Siemens und Halske, Wernerwerk M. Draft of a letter of 6. September 1936 to Walter's nephew, K. U. Leuven Archives, Noddack-Tacke Papers, 1187.

17 W. Noddack and I. Tacke, "Zwei neue Elemente der Mangangruppe. Chemischer Teil," and O. Berg and I. Tacke, "Zwei neue Elemente der Mangangruppe. Röntgenspektroskopischer Teil," *Sitzungsberichte der Preussischen Akademie der Wissenschaften, Physikalisch-Mathematische Klasse* 19 (1925): 400. The public press was also alerted very early.

18 See W. Noddack, I. Tacke, and O. Berg, "Die Ekamangane."

19 See for instance the report of a conference given by W. Noddack at the *Deutsche Chemische Gesellschaft* on 13 July 1925, *Zeitschrift für angewandte Chemie* 38 (1925): 876–877; and draft of a letter of 6 September 1936 to Walter's nephew, K. U. Leuven Archives, Noddack-Tacke Papers, 1187.

20 About their travels, see I. and W. Noddack, *Das Rhenium* (Leipzig: Leopold Voss, 1933). These scientific travels were financed by the *Notgemeinschaft der deutschen Wissenschaft* with a total amount of 40 000 Marks.

21 I. and W. Noddack, "Die Herstellung von einen Gramm Rhenium," *Zeitschrift für anorganische und allgemeine Chemie* 183 (1929): 353.

22 Correspondence of W. Noddack with W. Feit, 1929–1930, K. U. Leuven Archives, Noddack-Tacke Papers, 326.

23 W. Feit, "Über die technische Herstellung des Rheniums," *Zeitschrift für angewandte Chemie* 43 (1930): 459. In her reminiscences, Ida ignores the fact that they were contacted by Feit (and not the other way around).

24 J. F. G. Druce, *Rhenium,* 72–74

25 I. and W. Noddack, *Das Rhenium* (Leipzig: Leopold Voss, 1933).

26 W. Prandtl, "Auf der Suche nach den Manganhomologen nr. 43 und 75," *Zeitschrift für angewandte Chemie* 39 (1926): 1049–1051

27 W. Prandtl, "Auf der Suche nach den Manganhomologen nr. 43 und 75."

28 W. and I. Noddack, "Über den Nachweis der Ekamangane," *Zeitschrift für angewandte Chemie* 40 (1927): 250–254.

29 J. G. F. Druce and F. H. Loring, *Chemical News* 131 (1925): 273, 289, 321, and 337.

30 W. and I. Noddack, "Über den Nachweis der Ekamangane," *Zeitschrift für angewandte Chemie* 40 (1927): 254. See also id., *Das Rhenium*, on 3 and 16.

31 V. Dolejsek and J. Heyrovsky, "The occurrence of dwi-manganese (At. N. 75) in manganese salts," *Nature* 116 (1925): 782–783.

32 A. N. Campbell, "The occurrence of dwi-manganese (At. N. 75) in manganese salts," *Nature* 116 (1925): 866

33 *Dwimangan in Platinerzen. Chemischer Teil*, O. Swjaginzew and *Röntgenspektroskopischer Teil*, M. Korsunski and N. Seljakow, in *Zeitschrift für angewandte Chemie* 40 (1927): 256–259. See also O. Svjaginstsev, "Dwi-manganese in native platinum," *Nature* 118 (1926): 263–264.

34 See also W. Prandtl, "Auf der Suche nach den Manganhomologen nr. 43 und 75;" and "Zur Frage nach den Vorkommen der Manganhomologen," *Berichte der Deutschen Chemischen Gesellschaft* 60 (1927): 621–623.

35 One can easily detect the aftermath of personal attacks in I. and W. Noddack, *Das Rhenium* and J. G. F. Druce, *Rhenium* without careful reading.

36 "Bericht der Deutschen Atomgewichtskommission," *Berichte der Deutschen Chemischen Gesellschaft* 63 (1930): 16; "Über einige physikalische Konstanten des Rheniums," *Zeitschrift für Electrochemie* 34 (1928): 629.

37 O. Hönigschmid and R. Sachtleben, "Revision des Atomgewichtes des Rheniums. Analyse des Silberperrhenats," *Zeitschrift für anorganische und allgemeine Chemie* 191 (1930): 309. See also *Nature* 126 (1930): 896.

38 F. W. Aston, "Constitution of rhenium," *Nature* 127 (1931): 591 and id., *Proceedings of the Royal Society* (A) 132 (1931): 487.

39 E. Crawford, J. L. Heilbron, and R. Ullrich, *The Nobel Population 1901–1937: A Census of the Nominators and Nominees for the Prizes in Physics and Chemistry* (Berkeley: Office for the History of Science and Technology, 1989). The nominations come from the German community and W. Nernst is one of their main supporters.

40 I. and W. Noddack, "Darstellung und einige chemische Eigenschaften des Rheniums," *Zeitschrift für physikalische Chemie* 125 (1927): 264.

41 Draft of a letter to Walter Noddack's nephew, 6 September 1939, K. U. Leuven Archives, Noddack-Tacke Papers, 1187.

42 The undocumented claim for the discovery of masurium seems indeed to have played a major role in the disregard of Ida's fission proposal: see F. Krafft, *Im Schatten der Sensation: Leben und Wirken von Fritz Straßmann* (Weinheim: Verlag Chemie, 1981), 314–320 and R. L. Sime, *Lise Meitner: A Life in Physics* (Berkeley: University of California Press, 1996): 168–171 and 271–274.

43 C. Perrier and E. Segrè, "Some chemical properties of element 43," *Journal of Chemical Physics* 5 (1937): 712.

44 O. Hahn and F. Straßmann, *Die Naturwissenschaften* 27 (1939): 15. A Japanese team had also produced an isotope of element 43, R. Sagane, S. Kojima, G. Miyamoto, and M. Ikawa, *Physical Review* 57 (1940): 750. See also O. Hahn and F. Straßmann, "Über die bei der Uranspaltung auftretenden Molybdän-Isotope," *Zeitschrift für Physik* 117 (1941): 789.

45 E. Segrè and C. S. Wu, "Some fission products of uranium," *Physical Review* 57 (1940): 552.

46 E. Segrè, "Element 43," *Nature* 143 (1939): 460–461.

47 C. Perrier and E. Segrè, "Some chemical properties of element 43," *Journal of Chemical Physics* 5 (1937): 712. "We will give more details on the radioactive side of this investigation in a later paper to appear in the *Physical Review*." See also C. Perrier and E. Segrè, "Some chemical properties of element 43," *Journal of Chemical Physics* 7 (1939): 155.

48 This was not previously the case, see for instance the letter from H. V. A. Briscoe to W. Noddack, 19 June 1930, and letter from the same to W. Feit, 15 March 1932. K. U. Leuven Archives, Noddack-Tacke Papers, 273.

49 C. Perrier and E. Segrè, *Journal of Chemical Physics* 5 (1937): 712.

50 O. Hahn and F. Straßmann use the name and the symbol in most of their war-period papers. See also *Angewandte Chemie* 62 (1950): 148. "Da sie [Noddack and Tacke] den Nachweis nicht sicherstellen konnten, ging der Name wieder verloren."

51 See for instance a letter from F. W. Aston

to W. Noddack, 6 March 1931, asking for masurium after having found the rhenium isotopes. K. U. Leuven Archives, Noddack-Tacke Papers, 724.

52 In M. Nicolini, G. Bandolini, and U. Mazzi (eds.), *Technetium in chemistry and nuclear medicine* (New York: Raven Press, 1986), Segrè recalls the following anecdote: Having asked in 1937 to see the X-ray plate showing Masurium, he was answered it had been broken.

53 F.A. Paneth, "The making of the missing chemical elements," *Nature* 159 (1947): 8–10.

54 C. Perrier and E. Segrè, "Technetium: the element of atomic number 43," *Nature* 159 (1947): 24.

55 File on W. Noddack's denazification, 1945–1947, K. U. Leuven Archives, Noddack-Tacke Papers, 764–769. On the value of the "denazification process" see R. L. Sime, *Lise Meitner*, on 356–357 and references therein. Ida was not affected by denazification, though her behavior during the nazi period was under investigation in 1963–1964. Official document, K. U. Leuven Archives, Noddack-Tacke Papers, 1231, and file on the legal investigation of Ida Noddack in 1963–1964, K. U. Leuven Archives, Noddack-Tacke Papers, 1232.

56 Letter from W. Noddack to I. N. Stranski, 14 November 1954, K. U. Leuven Archives, Noddack-Tacke Papers, 634. Having applied for a position in Berlin, he writes a letter to I. N. Stranski that his age is not a real obstacle but rather the fact that he had been appointed in Strasbourg during the war.

57 Correspondence with Fritz Straßmann, 1955, K. U. Leuven Archives, Noddack-Tacke Papers, 635; and draft of a paper on the natural element 43, K. U. Leuven Archives, Noddack-Tacke Papers, 917.

58 G. E. Boyd and W. V. Larson, "Report on the occurrence of technetium on the earth's crust," *Journal of Physical Chemistry* 60 (1956): 707–715. Translation in K. U. Leuven Archives, Noddack-Tacke Papers, 1316. See also the lists of publications on rhenium and technetium, 1954 and 1956, K. U. Leuven Archives, Noddack-Tacke Papers, 87.

59 L. Meitner, "Atomkern und periodisches System der Elemente," *Die Naturwissenschaften* 22 (1934): 733–739.

60 I. Noddack, "Das Periodische System der Elemente und seine Lücken," *Angewandte Chemie* 47 (1934): 301–305.

61 This interpretation is based on the correspondence of Ida with Otto Hönigschmid, 1934–1935. K. U. Leuven Archives, Noddack-Tacke Papers, 1032.

62 The link between both paths of investigation in the Noddack enterprise is clearly made by Ida in her "memory". Draft of a letter to Walter Noddack's nephew, 6 September 1936, K. U. Leuven Archives, Noddack-Tacke Papers, 1187.

63 This investigation of the relative abundance of all elements in the earth's crust and in meteorites was already well under way thanks to Victor Moritz Goldschmidt, see the contribution by Helge Kragh in this volume. It seems that the "Allgegenwartskonzentration" concept first appeared in Ida's work but was later developed by both Walter and Ida. I. and W. Noddack, "Die Häufigkeit der chemischen Elemente," *Die Naturwissenschaften* 18 (1930): 757; see also W. and I. Noddack, *Aufgaben und Ziele der Geochemie* (Freiburg: H. F. Schulz, 1937). The couple exchanged correspondence with V. M. Goldschmidt at an early stage of their investigation and were undoubtedly influenced by the founder of geochemistry. Correspondence with V. M. Goldschmidt, 1926–1933. In his first letter (3 October 1926), Goldschmidt sends W. Noddack some geochemical publications and offers to help the Noddacks in their search for eka-manganese elements. K. U. Leuven Archives, Noddack-Tacke Papers, 363.

64 See for instance I. and W. Noddack, *Das Rhenium* and references therein.

65 See I. Noddack, "Röntgenspektroskopische Untersuchungen an seltenen Erden," *Angewandte Chemie* 50 (1937): 28.

66 I. Noddack, *Entwicklung und Aufbau der chemischen Wissenschaft* (Freiburg: H. F. Schulz, 1942).

67 *Ibid.*: "Für den Chemiker ist alles das Stoff was er wägen kann."

8.
The Search for Artificial Elements and the Discovery of Nuclear Fission

Ruth Lewin Sime

The effort to synthesize artificial elements beyond uranium began in 1934, went on for several years with a number of apparent successes, and then came to an abrupt halt in 1938 when nuclear fission was discovered and scientists realized that they had not found a single new element in all that time – the entire four-year search for "transuranium" elements had in fact been the study of fission fragments.

The discovery of fission was a complete surprise and also a great shock, because it shattered fundamental ideas of nuclear behavior that had guided the investigation. The surprise was evident in the events of December 1938. On December 10, Enrico Fermi was awarded the Nobel Prize in physics. He and his group in Rome had been the first to irradiate uranium with neutrons and to propose that transuranium elements had been formed in the process. In his Nobel lecture, Fermi was so confident of the first two, elements 93 and 94, that he referred to them by name: ausonium and hesperium. But at that very moment, the Berlin team of Otto Hahn, Lise Meitner, and Fritz Straßmann was on the verge of identifying barium among the uranium products. By the end of the year, they understood that uranium had split, explained the fission process, and concluded that the "transuranium" elements were false. When Fermi published his Nobel lecture, he added a footnote to that effect, but by then ausonium and hesperium were themselves footnotes (if that) in the history of science.[1]

In 1963 the nuclear physicist Lise Meitner, who had been one of the leading participants in the work, would describe it in an article with the title "Wege und Irrwege zur Kernenergie", (somewhat awkwardly translated as "Right and Wrong Roads to the Discovery of Nuclear Energy").[2] The title (although not the article itself) may leave the impression that investigators somehow knew that they were on the path to nuclear fission, but that was not the case, of course. The misguided search for transuranium elements is a good example of what Percy Bridgman called the "illogical progress of science", in which the unknown is not a mere extrapolation from the known. In that sense, students and teachers may find it a useful illustration of science as a very human activity, complete with "errors and retraced steps". Certainly the search is of interest to historians of science, not least because it so clearly displays the prevailing scientific assumptions of that period and the process by which these assumptions were shown to be false.[3]

In 1934, nuclear physics was young and the neutron had only just been discovered, yet the transuranium project was approached with a remarkable degree of confidence. The concepts from chemistry and nuclear physics that framed and guided the investigation were never seriously questioned, even though the synthesis and identification of new elements was, by definition, a leap into the unknown. Similarly, researchers were relatively unconcerned about the limitations of their small-scale experiments, even though the experiments themselves were notoriously difficult due to the tiny quantities of radioactive material.

Their confidence may have arisen in part from the simplicity of the central scientific idea: get a uranium nucleus to capture a neutron and it will decay to the next higher element. In practice the work called for collaboration that was interdisciplinary by necessity, yet flawed for just that reason. Here, nuclear physics and chemistry and radiochemistry came together in nearly every experiment and in the conceptual framework overall, yet in many ways they were never truly integrated: chemists were still thinking chemically and physicists, physically. [4] They collaborated across a pronounced disciplinary divide: they came to the investigation with different questions; they compartmentalized responsibility for the experimental design and the data; and they trusted each other's expertise without always understanding each other's limitations. In some respects, their relationship was more codependency than true collaboration. In any event, the disciplinary divisions masked the complexity of the work and created false confidence in the results, especially at first, and particularly among chemists.

External factors also played a role. One such factor was the very prominence of the leading investigators, whose premises and conclusions were not challenged by younger scientists, even those with more data and better equipment. Another factor, it seems, was the lure of the artificial elements themselves: each was a prize, with attention and rewards beyond its inherent scientific value. For years, this narrowed the focus to the search for transuranium elements only, keeping investigators from alertly examining the entire range of phenomena before them. Together, these factors served to delay the discovery of fission but in the end they did not prevent it. In this sense, Lise Meitner had her title right, for the experiments and theories of nuclear physics and chemistry that were used to pursue the false transuranium elements did lead to the recognition of fission, and the scientists who worked most assiduously on the transuranium project were those who did, finally, succeed in making the discovery.

The immediate precursors to the search were the discovery of the neutron by James Chadwick in 1932 and the discovery of artificial radioactivity by Irène and Frédéric Joliot-Curie in early 1934. In the spring of 1934, Fermi and his group put the two together for the first time. Fermi was a theoretical physicist who had recently started an experimental program with a group of young physicists, including Edoardo Amaldi, Franco Rasetti, Emilio Segrè, and one chemist, Oscar D'Agostino. Realizing that the neutral particle was an ideal projectile for producing new artificial radioactive species and new nuclear reactions, they systematically went through the periodic table bombarding every available element with neutrons. Later, Lise Meitner would recall her fascination with these first experiments – she quickly duplicated them in her laboratory – and that seems to have been true of physicists everywhere. Ernest Rutherford, known for his distaste for theorizing, jovially congratulated Fermi on his "escape from the sphere of theoretical physics".[5] In Copenhagen Meitner's nephew, Otto Robert Frisch, was working as an experimentalist in Niels Bohr's Institute for Theoretical Physics. Frisch had a talent for languages and when *Ricerca Scientifica* arrived each week, physicists crowded around him for an "instant translation of Fermi's latest discoveries. And what an exciting time that was!"[6]

For physicists, the excitement lay in the rapid expansion of their field. Before 1934, the primary sources of nuclear data were the existing natural radioactive species and a handful of artificial nuclear reactions. Nearly all the natural activities had been found during the first two decades of radioactivity research, a period marked by the collaboration of physicists and chemists. Around 1918, when that "classical" era was essentially over, radioactivity research separated into the specialized subfield of radiochemistry and the new domain of what came to be known as nuclear physics. The natural activities remained the workhorses of nuclear physics, however: their transformations and patterns of instability provided insight into nuclear behavior; beta-gamma spectra were studied for evidence of nuclear structure and energetics; and alpha, beta, and gamma radiation were used, as they had been from the beginnings of radioactivity, for studying the interactions of energetic particles and radiation with matter. Although some artificial nuclear reactions were known since Rutherford first reacted alpha particles with nitrogen in 1919, the natural activities remained so essential to nuclear research that Rutherford, Chadwick and Charles D. Ellis gave the title *Radiations from Radioactive Substances* to a comprehensive treatise that covered virtually all that was known of the nucleus in 1930.[7] In just a few months in 1934, Fermi's neutron experiments dramatically opened the field with many new nuclear reactions and radioactive nuclei available for study.

When Fermi and his co-workers reached uranium, they found several new activities, all beta emitters. After chemical tests indicated that these were neither uranium itself nor the elements just below it, Fermi proposed that the uranium nucleus had captured a neutron and begun a chain of beta decays, producing elements 93, 94, and perhaps more: the first artificial elements.

This attracted the attention of everyone, even the popular press. One Italian newspaper hailed it as proof of Italy's rekindled grandeur under the Fascists; another reported that Fermi presented a small vial of element 93 to the queen of Italy.[8] Although the former claim was suspect and the latter a complete fabrication, the prospect of synthetic elements enthralled scientists as well, including chemists and radiochemists. In the 1920's, radiochemistry had been something of a scientific backwater, confined mainly to the refinement of existing techniques – a victim, it has been written, of its own "suicidal success" in having characterized essentially all the known radioactive species.[9] In 1934 radiochemistry came alive again. Glenn T. Seaborg, then a chemistry graduate student at Berkeley, has described how he "devoured" Fermi's early papers and then avidly followed the work of Hahn, Meitner, and Straßmann.[10] Also in Berkeley, Ernest O. Lawrence recruited Philip Abelson, one of the few graduate students in the cyclotron group with a background in chemistry, to look into the transuranium elements. [11]

The pursuit of synthetic elements brought together nuclear physicists and radiochemists in what may be called a "neo-classical" period of interdisciplinary research. The renewed collaboration of Lise Meitner and Otto Hahn is an example. As young scientists in Berlin, they had made names for themselves as leaders in the early years of radioactivity, but in the 1920s their research interests diverged and they had worked independently, Hahn in his section for radiochemistry in the Kaiser Wilhelm Institut für Chemie in Berlin-Dahlem, Meitner in her own section for physics in the same institute. In 1934, the impetus for the first experiments with neutrons was entirely physical, beginning with Fermi and spreading quickly to other physicists. Meitner repeated and verified Fermi's results on her own, but for transuranium elements she realized that she "could not get ahead in this field with physics alone." She needed an "outstanding chemist" and persuaded Hahn, probably the most prominent radiochemist of all, to work with her again. They were joined by Fritz Straßmann, whose specialty was inorganic and analytical chemistry, a few months later. [12]

The atmosphere was competitive. In Paris Irène Joliot-Curie and her coworkers were as knowledgeable about radioactivity and almost as expert in radiochemistry as the team in Berlin; some thought the ideas in Paris were more original and the outlook more creative. [13] For the members of the Berlin team there was also a political edge. Each was a target under the Third Reich: Meitner, "non-Aryan;" Hahn, anti-Nazi; and Straßmann, the principled younger person who refused to join Nazi labor associations and was thus unemployable outside the Institute. They hoped that high-profile research with international recognition would give them a measure of protection. (In retrospect, a belief in the protective power of internationalism in science may seem naive, but once fission was discovered, Hahn and his institute were indeed politically sheltered by their connection to weapons research.) The competition led to a pronounced rivalry between the groups in Paris and Berlin.

From the start, the work was framed by two guiding principles, one from physics and the other from chemistry, that turned out to be untrue. Physicists had always observed that nuclei, even radioactive nuclei, were quite stable: when radioactive decay or other nuclear reactions took place, the changes were small. Fermi's early neutron results were consistent with this. He and his coworkers found that neutron irradiation of light elements might knock out a proton or two or perhaps an alpha particle, but nothing bigger; with heavier elements, the reaction was always neutron capture. If the new artificial nucleus was radioactive, it always decayed by beta emission, forming the next higher element. Thus it seemed reasonable to propose that the new beta activities that resulted from the neutron irradiation of uranium would be higher elements beyond uranium.

The principle of small nuclear changes was given a theoretical basis by George Gamow. In 1928 he derived a successful theory of alpha decay, in which the nucleus is quantized and only small particles, such as protons or alpha particles, have a finite probability of tunneling through the nuclear barrier and escaping the nucleus. That

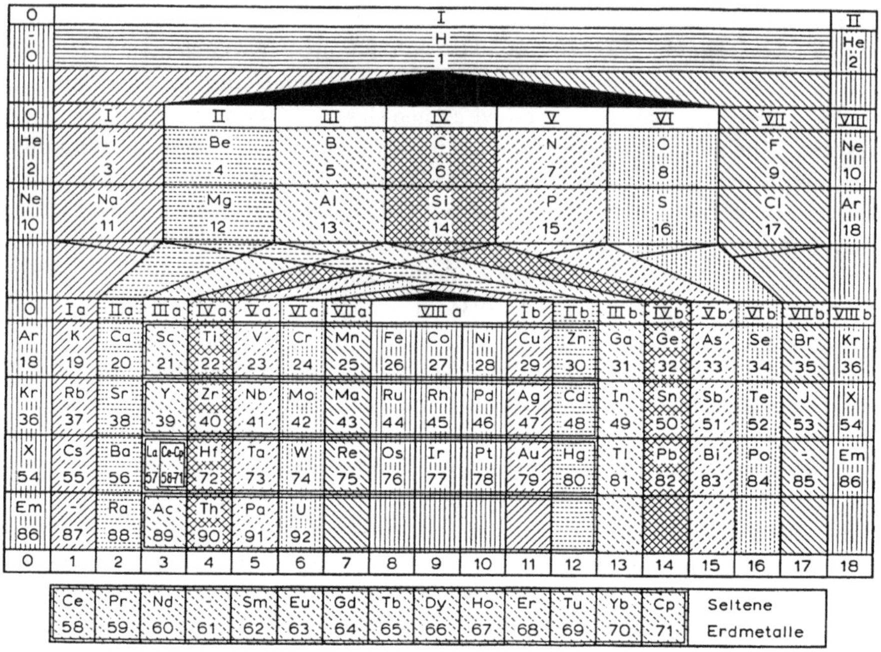

Figure 8.1 *Periodic system, 1920s and 1930s.* Prior to the 1940s, the lanthanides (rare-earths) were grouped separately as shown, but Th, Pa, and U (now classified as actinides), were considered to be transition elements, as shown in this table according to A. von Antropoff. From J. W. van Spronsen, *The Periodic System of the Chemical Elements: A History of the First Hundred Years* (Amsterdam, 1969), fig. 59, p. 160.

year, Gamow also devised a more classical theory, in which the sub-nuclear particles are bound together by surface tension, much like molecules in a drop of water. The liquid-drop theory successfully accounted for nuclear stability, and when mass spectroscopy provided precise isotopic mass data, the liquid-drop theory was quantitatively consistent with known nuclear mass defects. In the mid-1930's, Niels Bohr and Fritz Kalckar developed a theory of the compound nucleus, also based on a liquid drop, which accounted for nuclear reactions. [14] No theory predicted, and no physicist imagined, anything as disruptive as nuclear fission.

Chemists brought their own incorrect principle to the project: the prediction that transuranium elements would have the chemical behavior of transition elements. The actinides, including uranium, are now placed in a separate grouping below the rare-earth elements, but in the 1920s and 1930s thorium, protactinium, and uranium were classified as transition elements (Figure 8.1) because they each chemically resemble the transition element just above them far more than they resemble each other. By extension, chemists predicted that the transuranium elements would be transition elements too, taking their places below the third-row transition elements Re, Os, Ir, etc. This assumption was very strongly held. It was not shaken even after fission was discovered, and it delayed the discovery of element 93.

It is interesting to note, however, that the placement of the heaviest elements in the periodic table was subject to active discussion throughout the 1930s. Although periodicity itself provided little guidance at the outer limits of the periodic table, no one doubted that it would prevail, as the periodic system had proved itself adaptable to the discovery of many new elements. Since Mendeleev, new groupings had been added when the noble gases and the rare-earth elements were discovered, and the many natural radioactive species that first seemed to violate periodicity eventually vindicated it when isotopy was understood. Then, in the early 1920s, Bohr demonstrated that the periodic system was far more than a practical organizational scheme for chemists, but was instead a fundamental expression of the underlying relationship between chemical behavior and electronic structure. One of his great successes was his proposal of a *4f* sublevel for the rare-earth elements, which incorporated and correctly placed them in the periodic table; his prediction that the elusive element 72 would not be a rare-earth but a transition element very similar to zirconium led almost instantly to the discovery of hafnium, named for Copenhagen and his institute. [15] Bohr also predicted the existence of a second rare-earth series in a *5f* sublevel, but he was unable to predict just where the *5f* would begin. [16] In his scheme (Figure 8.2) he placed the second rare-earth series somewhere beyond uranium. He did so because the spectroscopic data for the electronic structure of uranium was inconclusive while the chemical evidence for uranium as a transition element was very strong.

Nevertheless, the position of the second rare-earth series remained an open question. From measurements of ionic volumes, the Norwegian geochemist Victor Moritz Goldschmidt proposed that the *5f* begins with actinium; from spectroscopic data, others suggested that it begins with thorium; still others made quantum mechanical calculations and proposed that the *5f* begins with element 93, or perhaps 95. [17] But the chemists engaged in the search for transuranic elements seem not to have paid much attention to the discussion: they regarded uranium as a transition element and extrapolated from there. Hahn and Meitner may have been especially inclined to do so because their discovery of protactinium in 1918 was based on its chemical similarity to tantalum. [18]

Uranium was doubly deceiving: its chemistry is that of a transition element although it has *5f* electrons, and its nucleus appears quite stable even though it is on the verge of explosive disintegration. This was bad luck, the more so because the two false principles from nuclear physics and chemistry dovetailed, preventing one from providing a check on the other for many years. The only published challenge came from Ida Noddack, who had discovered rhenium (element 75) in 1925, together with her husband Walter Noddack and Otto Berg. In 1934 she sharply questioned Fermi's chemical separations and his assumption of small nuclear changes. "It is conceivable," she wrote, "that when heavy nuclei are bombarded by neutrons, these nuclei break up into several *larger* fragments, which would of course be isotopes of known elements but not neighbors of the irradiated elements." [19] Noddack's reputation was mixed, in part because the claims she and her husband made for the discovery

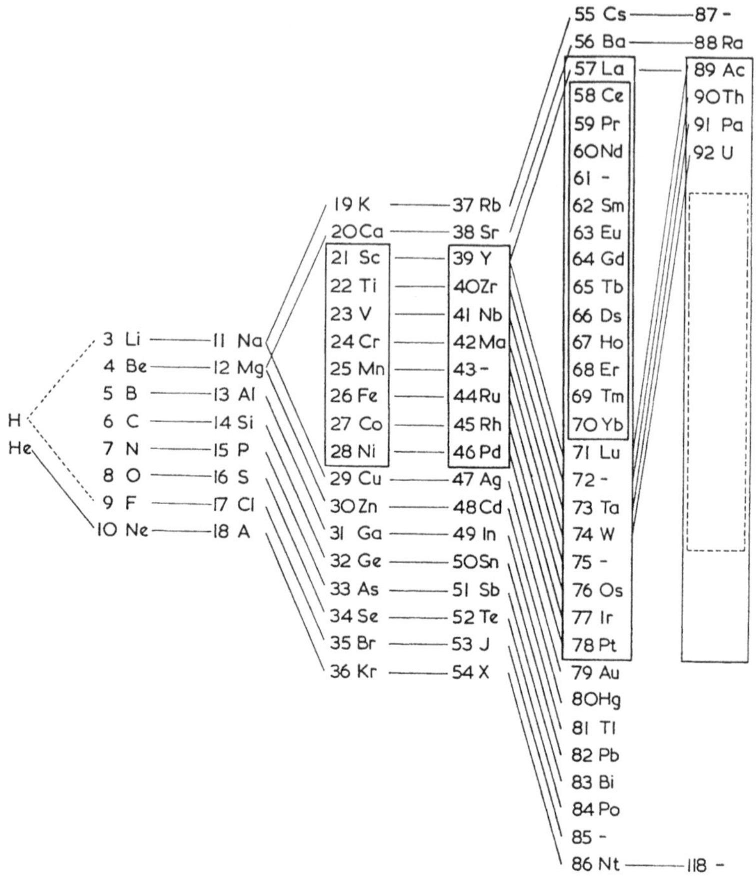

Figure 8.2 *Bohr's periodic system, 1922.* With the dotted lines, Bohr indicated that a second rare-earth group would begin somewhere after uranium. From Spronsen, *The Periodic System*, Fig. 55, p. 156.

of element 43 could not be verified. Moreover, she was neither a nuclear physicist nor a radiochemist. Her comments, in retrospect amazingly prescient, were mostly ignored. They were regarded as speculation, Noddack herself did not pursue them, and they did not influence the course of the investigation.[20]

Although physics and chemistry were responsible for the conceptual framework overall, radiochemistry defined the experimental approach and provided much of the initial data. The neutron sources then in use (usually radium or radon [a source of alpha particles] mixed with powdered beryllium) were weak, with the result that the new beta activities were not much stronger than the natural radioactivity of uranium and its decay products. In 1934, the Rome group chemically separated the new activities from uranium by co-precipitating them with manganese and rhenium compounds (both transition metals), which supported the notion that these were transuranium elements. In early 1935, Hahn, Meitner, and Straßmann devised a more quantitative precipitation using platinum and rhenium sulfides. This separation method fixed the direction of the investigation for the next several years. With few exceptions, their attention was focused on disentangling the activities in the precipitate while the filtrate, which contained uranium, its decay products, and (as it happened) quite a lot more, was ignored.

Thus circularity was built into the project from the start. Once fission was discovered and it was recognized that fission products, which are all beta emitters, can be found in any group across the periodic table, it was obvious that the "transuranium" precipitate consisted of isotopes of lighter nuclei (Figure 8.3). In Rome, Berlin, and elsewhere, scientists were separating out just those species with the chemical properties they expected to find. By 1935, Fermi and his group left the field and the Berlin team dominated the investigation. By 1937, they had found an impressive number of new radioactive species which they assigned to three distinct nuclear processes (Figure 8.4). Two processes (subsequently shown to be fission processes) showed an extended sequence of beta emitters that were ascribed to transuranium elements, while a third process was attributed (correctly) to the resonance capture of slow neutrons by uranium.[21]

Although the experiments themselves were interdisciplinary, the results were quite segregated. Chemical separations and radiochemical analysis were responsible for characterizing the radioactive species in the two long series. In these, the genetic sequences (parent ? daughter ? . . .) fit the chemistry expected for the transuranium sequence (ekaRe ? ekaOs ? . . .) so closely that it seemed they just *had* to be true. Moreover, as each new species fell into place, it added credence to the correctness of the entire sequence – and the appearance of not one but two long series seemed to verify each other. The fact that these two long series were parallel isomeric chains – all species had mass 239 – did not apparently seem strange, at least not at first, perhaps because Hahn had been the first to observe a genuine isomer in 1921.[22]

Physics was responsible for integrating the data from chemistry, radiochemistry, and physical measurements into nuclear reaction mechanisms that made sense. From reaction yields and cross-sections, measured under varying irradiation conditions and neutron energies, it was found that fast neutrons were effective in processes 1 and 2, an indication that the target nucleus was ^{238}U, but that the yield was greater when thermal neutrons were used, which is typical of neutron capture. For process 3, only slow neutrons of defined energy (25 eV) were effective, making it a typical resonance neutron capture process. Thus it appeared that all three processes were neutron capture, with the effective isotope in all three cases being ^{238}U.

Taken together, these conclusions created serious problems in physical interpretation. How could neutron capture by a single isotope initiate three such different reaction processes? How could the capture of just one neutron create such great instability that multiple beta decays were needed to alleviate it? Nuclear isomerism was known, but how to explain the *triple* isomerism of ^{239}U? Worst of all, how could one account for the *inherited* isomerism – for several generations – in the

$$
{}_{0}n + {}_{92}U \nearrow {}_{56}Ba \rightarrow {}_{57}La \rightarrow {}_{58}Ce \rightarrow {}_{59}Pr \rightarrow {}_{60}Nd \rightarrow \ldots
$$
$$
\searrow {}_{36}Kr \rightarrow {}_{37}Rb \rightarrow {}_{38}Sr \rightarrow {}_{39}Y \rightarrow {}_{40}Zr \rightarrow {}_{41}Nb \rightarrow \ldots
$$

$$
{}_{0}n + {}_{92}U \nearrow {}_{54}Xe \rightarrow {}_{55}Cs \rightarrow {}_{56}Ba \rightarrow {}_{57}La \rightarrow {}_{58}Ce \rightarrow {}_{59}Pr \rightarrow \ldots
$$
$$
\searrow {}_{38}Sr \rightarrow {}_{39}Y \rightarrow {}_{40}Zr \rightarrow {}_{41}Nb \rightarrow {}_{42}Mo \rightarrow {}_{43}Tc \rightarrow \ldots
$$

Figure 8.3 *Fission processes.* The uranium nucleus can split in many ways, of which two are shown here. Since fission fragments tend to be heavier with neutrons than stable isotopes of the same element, they each begin a sequence of beta decays, forming elements from virtually every group across the periodic table, including transition elements.

1. $U + n\ \{\text{fast, thermal}\} \rightarrow {}_{92}U\ (10'') \rightarrow {}_{93}EkaRe\ (2.2') \rightarrow {}_{94}EkaOs\ (59') \rightarrow {}_{95}EkaIr\ (66h) \rightarrow {}_{96}EkaPt\ (2.5h) \rightarrow {}_{97}EkaAu?$

2. $U + n\ \{\text{fast, thermal}\} \rightarrow {}_{92}U\ (40'') \rightarrow {}_{93}EkaRe\ (16') \rightarrow {}_{94}EkaOs\ (5.7h) \rightarrow {}_{95}EkaIr?$

3. $U + n\ \{\text{slow}\} \rightarrow {}_{92}U\ (23') \rightarrow {}_{93}EkaRe?$

Figure 8.4 *The "transuranium" elements, 1937.* In 1937 Meitner, Hahn, and Straßmann assigned the radioactive species they had found to three different reaction processes. The extended sequence of beta decays in processes 1 and 2 was assigned to elements 93, 94, etc., all with mass 239. (Half-lives are in parentheses; EkaRe denotes element 93 as the higher homologue of Re.) Processes 1 and 2 are in fact fission processes, but process 3 was correctly interpreted at the time as resonance capture of slow neutrons by ^{238}U to form ^{239}U, whose daughter, ^{239}Np, was identified in 1940.

parallel chains of processes 1 and 2? Despite the efforts of theoreticians, there were no satisfactory answers to any of these questions, and the problems grew worse as the data accumulated. Thus, while chemists were cheering the appearance of every new species, physicists were growing more troubled.

It was obvious that process 3 was the most normal – a resonance capture of slow neutrons to form ^{239}U, a beta emitter which necessarily decays to element 93. If the Berlin team had been able to detect this element 93 and determine its chemical properties, they would have realized that processes 1 and 2 were incorrect. But they did not do it. In Berlin, as Hahn later wrote, they were not very interested because they believed they had already discovered several transuranium elements in processes 1 and 2.[23] One of the difficulties was that their neutron sources were weak. But even if they had had a stronger source – Meitner was constructing an accelerator in her laboratory that was operational by the summer of 1938 [24] – it is questionable whether they would have used it to look for the element 93 of process 3.

At the time no one challenged this scheme, although everyone was aware of the difficulties. Irène Joliot-Curie and her co-workers were the Berlin team's chief competitors, and they verified the new species as they were discovered. A number of physicists tried to determine reaction mechanisms by physical means, but they covered their ionization chambers to screen out the natural decay of uranium and missed the huge ionization bursts from fission fragments. Straßmann thought he found barium among the uranium products in 1937, but when Meitner shrugged it off, he did not pursue it.[25] Using the Berkeley cyclotron as a neutron source, Philip Abelson had a neutron flux that was orders of magnitude more intense than the ones in Europe and produced far more radioactive species, yet he too verified the Berlin results. Later he attributed it to the "high reputation and prestige of the Fermi group." [26] For Seaborg, Hahn's 1933 monograph, *Applied Radiochemistry*, was his "bible" and he too accepted the Berlin results.[27] And in 1938, just before the fission discovery, chemist Lawrence L. Quill discussed the many difficulties in a 70-page review article, but he did not question the assignments in processes 1 and 2. Instead he accepted that the elements up to 97 were transition elements and proposed that the 5*f* sublevel could not, therefore, begin before element 98.[28]

Part of the problem was the extraordinary confidence level of the Berlin chemists. As the senior author for the chemical publications, Hahn repeatedly stressed in *Chemische Berichte* that there was "no doubt" about the existence of the transuranium elements, that they were unquestionably new elements, distinct from all previously known elements. Later, after the fission discovery, Hahn would blame physics for misleading the investigation with its assumption of small nuclear changes, without mentioning the mistaken predictions of chemistry. Hahn had non-scientific motives for separating himself from physics and Meitner and claiming the discovery of fission for chemistry only,[29] but in blaming physics he was tacitly admitting that physics had guided the investigation in ways that he had been unable to challenge.

This disciplinary divide affected Meitner as well. As the team's physicist, she was the scientific leader of the project, responsible for interpreting all the results, more aware of its difficulties, and more troubled by what she did not understand. Yet she was dependent on the chemists for data and she could not dispute their radio-analytical methods or argue for a different chemical approach. As it happened, the Berlin team's most crucial error was that it focused on the "transuranium" precipitate only and, with few exceptions, did not search the filtrate for new activities. Later Meitner recalled,

> [o]f course what we did was wrong, Hahn and I. I really think our misfortune is that we didn't search the filtrate. We couldn't search it because uranium was in it and ... our neutron sources were too weak ... The chemists absolutely didn't

> want to. I begged them to do it while I was there because I was so disturbed by it. Because I understand too little chemistry I naturally was always uneasy about what wasn't done ..." [30]

Clearly physics and chemistry both contributed errors to the investigation, but there was a difference: chemistry had no means for detecting the error, while physicists at least recognized that the overall pattern did not make sense – and was getting worse. Physics did not predict fission, but it defined the investigation and sustained it. In her publications, Meitner would often end on a dissatisfied note. "Perhaps one should look elsewhere for an explanation," she would write, or, "[t]his result is very difficult to reconcile with current concepts of the nucleus." [31] The breakthrough came from Paris, where Irène Curie had devised a method using paired counters and copper filters for measuring the uranium activities without separation. Early in 1938, she and her coworker Pavel Savitch reported a strong new beta activity, but they were uncertain of the chemistry. By the time Hahn and Straßmann looked into it, it was October 1938 and Meitner, who was of Jewish origin, was no longer in Berlin. [32] She had fled Germany in the summer of 1938 and taken a position in Stockholm. She and Hahn corresponded constantly. When Hahn and Straßmann separated the Curie activity and noted that it followed a barium carrier, they assumed it was an isotope of radium. As in processes 1 and 2, thermal neutrons enhanced the yield and there were multiple isomers.

It is at this point, when the supposed transuranium elements were out of the picture, that the problem was finally solved. It happened because the chemists were now in familiar territory, with elements of known chemistry and tested radiochemistry. When their findings contradicted the physics assumption of small nuclear changes, the discrepancy was apparent and had to be resolved. It can be argued that their interdisciplinary collaboration functioned best just when Meitner was physically separated from her Berlin colleagues, and it is ironic that their separation produced a correspondence that demonstrates how effective this collaboration was.

In her letters to Hahn, Meitner repeatedly questioned the radium result. The reaction ${}_{92}U \rightarrow {}_{88}Ra$ required the emission of two alpha particles. In an earlier study, Meitner had used Gamow's theory of alpha decay to calculate the energetics of a similar process, and she was convinced that thermal neutrons were not sufficiently energetic to cause uranium to emit even one alpha particle – and certainly not two. In November 1938, Meitner met Hahn at Bohr's institute in Copenhagen and, according to Straßmann, she "urgently requested that these [radium] experiments be scrutinized very carefully and intensively one more time." For Straßmann, Meitner had been the "intellectual leader of our team ... Fortunately [her] opinion and judgement carried so much weight with us in Berlin that the necessary control experiments were immediately undertaken." [33] These were the experiments that led directly to the finding of barium a few weeks later.

The radiochemical procedures required skill and delicacy but they were not conceptually new. To verify the presence of radium, Hahn and Straßmann tried to separate it from its barium carrier, using Marie Curie's method of fractional crystallization. When no separation occurred, they knew, with a certainty that reflected all of Hahn's experience in radiochemistry, that their "radium" had to be barium. The interpretation was up to physics, as it had been throughout the project. When Hahn informed Meitner of the barium, he was sure of the result but mystified by what it meant, and he turned to her "for some sort of fantastic explanation." Meitner instantly responded. "A major breakup seems very difficult to me, but in nuclear physics we have experienced so many surprises that one cannot unconditionally say: it is impossible." This was Meitner's moment of discovery. Before then she had rejected all but the smallest nuclear changes; now she was suddenly willing to consider complete nuclear disintegration. Her discovery is perhaps best described as an intuitive insight into the limits of existing nuclear theory. Within a week she and her physicist-nephew Otto Frisch, who was visiting her for Christmas, devised the first theoretical interpretation of the fission process, calculated the energy released, understood that the apparent transuranium elements were fission fragments, and predicted that only process 3 would lead to element 93.

The barium finding was published by the chemists in *Naturwissenschaften* [34] and the fission theory by the physicists in *Nature* [35] a few weeks later. This emphasized the separations – chemistry from physics, experiment from theory, in different journals, in different languages. The separations were an artifact of Meitner's forced emigration and the politics of the time. From that time forward, the interdisciplinary nature of the discovery was obscured and its history distorted, not least by the award of the 1944 Nobel Prize in chemistry to Otto Hahn alone. [36]

In any event, the collaboration between physicists and chemists that led to fission ended abruptly with its discovery. Their false assumptions shattered, physicists took the new result and ran with it. Fission was in their domain: they had the instruments, they did the experiments, they developed the theory. Within weeks fission would become a recognizable new area of research whose importance was magnified by the threat of war. Within a year, over 100 articles were cited in a comprehensive review of fission research. [37] And the physical measurements of the pre-fission years were still valid and useful. Shortly after the fission discovery, Bohr used the reaction cross-sections that Meitner had measured in 1937 for the three processes to deduce that the highly fissile isotope of uranium must be ^{235}U, and not ^{238}U.

In contrast, the chemists were left with little but fission debris – the earliest instance of radioactive fallout. The chemical data from the uranium investigation was essentially meaningless – the "transuranium" elements that had inspired such confidence turned out to be a messy cocktail of light elements from across the periodic table. [38] Moreover, chemists had not broken new ground with the fission discovery, since they were still saddled with the assumption that transuranium elements were transition elements.

This continued to block the discovery of element 93, even though everyone now knew just where to look for it. In the spring of 1939 Edwin McMillan and Emilio Segrè irradiated uranium in the Berkeley cyclotron to form the 23-minute uranium of process 3. They detected a new 2-day beta activity, but because the activity had rare-earth chemistry, they decided that it was just a fission fragment. Ironically, this mirrored the earlier error: where fission fragments were once thought to be transuranium elements, now the true element 93 was mistaken for a fission fragment. It took another year before McMillan, together with Philip Abelson, used physical and radioanalytical methods to identify the 2-day activity as a transuranium element, and then showed that its chemistry was indeed that of a rare-earth. [39]

This opened the road to the transuranium elements. A new generation of chemists who knew physics and appreciated big machines defined the discipline of nuclear chemistry. Glenn Seaborg and his associates discovered and characterized plutonium (element 94) in 1941 – making nuclear chemistry as important to weapons research as nuclear physics – and went on to discover nine more transuranium elements (his last, element 106, being named *Seaborgium* for him in 1997). From the known properties of the first several transuranium elements, Seaborg proposed in 1944 that the actinides, a second rare-earth series with electrons in the 5*f* sublevel, would begin with element 90, a proposal that was confirmed by the discovery of subsequent elements, giving the periodic table its current form. [40]

With this, the goals of those who first sought artificial elements beyond uranium were realized. The understanding of nuclear behavior was deepened by the discovery of nuclear fission, and the periodic system was extended and clarified by the synthesis of transuranium elements.

Acknowledgment

I am grateful to Professor Glenn T. Seaborg (1912–1999) for his warmth, encouragement, and generous interest in the science history of this period.

References and Notes

1 E. Segrè, *Enrico Fermi: Physicist* (Chicago: 1970), 216–17.
2 L. Meitner, "Wege und Irrwege zur Kernenergie," *Naturwissenschaftliche Rundschau* 16 (1963): 167–169; translation: "Right and wrong roads to the discovery of nuclear energy," *Advancement of Science* 19 (1963): 363–65.
3 P. W. Bridgman, "Impertinent reflections on history of science," quoted in R. Stuewer, *History as Myth and Muse* (Amsterdam, 1998), 5–7.
4 Quote from Coulson, cited by A. Simoes and K. Gavroglu in the introduction of their contribution to this volume.
5 Segrè, *Enrico Fermi*, 74–75.
6 O. Frisch, *What Little I Remember* (Cambridge, 1979), 88.
7 E. Rutherford, J. Chadwick, C. D. Ellis, *Radiations from Radioactive Substances* (Cambridge, 1930; reissued 1951).
8 L. Fermi, *Atoms in the Family* (Chicago, 1954), 91.
9 L. Badash, *Radioactivity in America: Growth and Decay of a Science* (Baltimore, 1979).
10 G. T. Seaborg, *Nuclear Milestones* (San Francisco, 1972), 5; Seaborg, introduction to Otto Hahn, *A Scientific Autobiography* (London, 1967), x. Translated and edited by W. Ley.
11 P. H. Abelson, "Discovery of Neptunium," in L. R. Morss and J. Fuger (eds.), *Transuranium Elements: A Half Century* (Washington DC, 1992), 52.
12 For the work of the Berlin team, see R. L. Sime, *Lise Meitner: A Life in Physics* (Berkeley, 1996), chapter 7.

13 Emilio Segrè, interview, 21 May 1985, Lafayette, California

14 R. H. Stuewer, "The origin of the liquid-drop model and the interpretation of nuclear fission," *Perspectives on Science* 2 (1994): 76–129.

15 A. Pais, *Niels Bohr's Times, in Physics, Philosophy, and Polity* (Oxford, 1991), 204–10; H. Levi, *George de Hevesy: Life and Work* (Bristol, 1985), 51–56.

16 Pais, *Niels Bohr's Times*, 207, states that Bohr began the second rare-earth group with element 90 (Th), but Bohr in his 1922 paper, reproduced in L. Rosenfeld (ed.), *Niels Bohr: Collected Works* (Amsterdam, 1972), 4:405, states that "no family similar to the rare earths occurs in the known part of the seventh period," a statement consistent with known chemistry and incorporated into his published table (Fig. 8.2).

17 J. W. van Spronsen, *The Periodic System of Chemical Elements: A History of the First Hundred Years* (Amsterdam, 1969), 317–20.

18 O. Hahn and L. Meitner, "Die Muttersubstanz des Actiniums," *Physikalische Zeitschrift* 19 (1918) 208–18; R. L. Sime, "The Discovery of Protactinium," *Journal of Chemical Education* 63 (1986): 653–57.

19 I. Noddack, "Über das Element 93," *Zeitschrift für Angewandte Chemie* 47 (1934): 653–55.

20 For the Noddack controversy, see T. Hopper, *She was Ignored*, master's thesis, Stanford University, 1990; P. van Assche, "Ignored priorities: First fission fragment (1925) and first mention of fission (1934)," *Nuclear Europe* 6–7 (1988): 24–25; Segrè, *Enrico Fermi*, 76; F. Krafft, *Im Schatten der Sensation: Leben und Wirken von Fritz Straßmann* (Weinheim, 1981), 314–17.

21 L. Meitner,O. Hahn, F. Straßmann, "Über die Umwandlungsreihen des Urans, die durch Neutronenbestrahlung erzeugt werden," *Zeitschrift für Physik* 106 (1937): 249–270.

22 Hahn, *A Scientific Autobiography*, 95–103.

23 Hahn, *A Scientific Autobiography*, 175.

24 B. Weiss, "Lise Meitners Maschine," *Kultur & Technik* (March 1992): 22–27.

25 Krafft, *Im Schatten der Sensation*, 220–21.

26 Abelson, "Discovery of Neptunium," 51–53.

27 Seaborg in Hahn, *A Scientific Autobiography*, ix.

28 L. L. Quill, "The transuranium elements," *Chemical Reviews* 23 (1938): 87, 138ff.

29 Sime, *Lise Meitner*, chapters 10, 11, 12.

30 Lise Meitner, interview by T. S. Kuhn, 12 May 1963: American Institute of Physics Oral History Project, transcript of tape 65a: 19–20.

31 Meitner, Hahn, and Straßmann, "Über die Umwandlungsreihen des Urans," 269.

32 R. L. Sime, "Lise Meitner's escape from Germany," *American Journal of Physics* 58 (1990): 262–67; also Sime, *Lise Meitner*, chapter 8.

33 Krafft, *Im Schatten der Sensation*, 208, 210.

34 O. Hahn and F. Straßmann, "Über den Nachweis und das Verhalten der bei der Bestrahlung des Urans mittels Neutronen entstehenden Erdalkalimetalle," *Naturwissenschaften* 27 (1939): 11–15.

35 L. Meitner and O. R. Frisch, "Disintegration of uranium by neutrons: A new type of nuclear reaction," *Nature* 143 (1939): 239–240.

36 E. Crawford, R. L. Sime, M. Walker, "A noble tale of wartime injustice," *Nature* 382 (1996): 393–6; E. Crawford, R. L. Sime, M. Walker, "A noble tale of postwar injustice," *Physics Today* 50:9 (September 1997): 26–33.

37 L. A. Turner, "Nuclear fission," *Reviews of Modern Physics* 12 (1940): 1–29.

38 H. Menke and G. Herrmann, "Was waren die 'Transurane' der dreißiger Jahre in Wirklichkeit?" *Radiochemica Acta* 16 (1971): 119–23.

39 E. Segrè, *A Mind Always in Motion: The Autobiography of Emilio Segrè* (Berkeley, 1993), 152–153; Abelson, "Discovery of Neptunium," in L. R. Morss and J. Fuger (eds.), *Transuranium Elements*, 53–55.

40 G. T. Seaborg, *The Transuranium Elements* (New Haven, 1958), 89, 140–4. G. T. Seaborg and W. D. Loveland, *The Elements Beyond Uranium* (New York, 1990), 65ff.

9.
From Geochemistry to Cosmochemistry: The Origin of a Scientific Discipline, 1915–1955

Helge Kragh

9.1
Introduction

Twentieth-century chemistry has evolved into numerous subdisciplines closely connected with subdisciplines from the other sciences. Physical chemistry, chemical physics, biochemistry, clinical chemistry, and nuclear chemistry are well known examples. [1] The connections to the earth sciences and the astronomical sciences are much less known and have received relatively little historical attention. Yet these interdisciplinary fields are extremely rich and have histories that are rewarding subjects for the historian of science. The purpose of the present paper is to provide a historical introduction to what is currently known as cosmochemistry, or the chemistry of the universe. In one sense, this is a modern branch of science, established only in the 1950s, but it can be plausibly traced back to the late nineteenth century, if not earlier. It can even be argued that it dates back to about 1800, when meteorites were first subjected to chemical analysis.

Among the fascinating aspects of cosmochemistry is its completely interdisciplinary nature. It is a thoroughly integrated mixture of elements from all the classical disciplines of science, namely, chemistry, physics, astronomy, and geology; even aspects of biology are included, as in the branch known as biogeochemistry. And yet the field is rightly called cosmochemistry and not, say, cosmophysics, geological astronomy, or cosmic geology. The interdisciplinarity of the field is challenging from a historiographical point of view. If one approaches it from the perspective of history of geology, or from the perspective of history of chemistry, one is likely to miss many important points that characterize the field. The historian with a narrow disciplinary perspective will easily miss the connections to nuclear science, astronomy, planetary science, cosmology, oceanography, and the environmental sciences. The point is simply that the high degree of interdisciplinarity necessitates a perspective that goes beyond traditional disciplinary boundaries.

The paper follows some of the lines of development that led from nineteenth-century geochemistry to the first phase of cosmochemistry in the 1930s, when the field was pioneered by Victor Moritz Goldschmidt in particular. "The field of geochemistry," Goldschmidt wrote, "ranges widely over the broad ground of modern science, from astrophysics and nuclear and atomic physics to geology, oceanography, and biology – mineralogy, crystallography, and chemistry being perhaps the most important contributors to geochemistry, while much help is also afforded by applied science in such fields as mining geology and metallurgy." [2] The foundation laid by Goldschmidt and others linked cosmochemistry to both astrophysics, cosmology and nuclear physics, and in the first two decades after 1945 planetary chemistry and interstellar chemistry were added to the list of cosmochemical subdisciplines. In this paper I focus on the formative period of cosmochemistry, largely the four decades between 1915 and 1955. The subject has a rich prehistory, including meteoritics and aspects of geo- and astrochemistry, the most important developments of which I summarize in section two. [3]

9.2
Nineteenth-Century Backgrounds

Astronomy and chemistry are usually considered very different sciences, with very different historical paths. Whereas astronomy is observational and based on the exact laws of celestial dynamics, chemistry is the archetypical experimental science. Yet speculations about interconnections between the two fields can be found far back in time. To mention but one example, in his 1807 lectures on the history of chemistry, Hans Christian Ørsted prophesied that "some day chemistry will have just as much influence on astronomy as mechanics so far." He added that "Then it will be necessary to regard external motion as a product of internal forces, and all natural science will finally become a cosmogony." [4] Ørsted's prophecy eventually became reality, but the path followed from chemistry to astronomy took a different direction than imagined by the Danish scientist. To become useful in the science of the heavens, chemistry first had to prove its worth in the study of the earth.

With roots in crystallography and chemical mineralogy, geochemistry emerged during the early decades of the nineteenth century. [5] The term "geochemistry" was coined by the eminent German chemist Christian Friedrich Schönbein in a work of 1838, in which he predicted the birth of a new science and mapped out its program. "In a word," Schönbein wrote, "a comparative geochemistry ought to be launched, before geognosy can become geology, and before the mystery of the genesis of our planets and their inorganic matter may be revealed." [6] Schönbein's program was eventually to be realized, but initially his term was not much used by either chemists or geologists. The older name "chemical geology" remained in use throughout the century and it was a matter of debate which of the two sciences, geology or chemistry, should be the dominant partner in the new interdisciplinary field. Should it be a "chemical geology," where chemistry was applied to known geological structures and processes, or a "geological chemistry," where chemical concepts and tools were primary? Should the geochemist be primarily a geologist or a chemist? [7] In practice, there was relatively little contact between geologists and chemists. Geochemistry was a small field without the status of a recognized sub- or transdiscipline and it lived somewhat insecurely between the two large sciences of geology and chemistry.

One of the main tasks of geochemistry was to determine the abundance of the elements in the earth's crust. In 1889, Frank W. Clarke, Chief Chemist to the U. S. Geological Survey, noted that the abundance of the elements diminishes with their atomic weights. [8] This and other abundance regularities were further substantiated with the publication in 1908 of Clarke's seminal *The Data of Geochemistry*. [9] The usefulness of physical chemistry to problems of geology and geochemistry became slowly recognized about the turn of the century, when small groups of geologists endeavored to transform the old-fashioned empirical geochemistry to a more physically oriented geochemical tradition. One of them, the American Charles Van Hise, urged in 1904 his fellow geologists "to take possession of the vacant ground between geology and physics and geology and chemistry." [10] To Van Hise, physical chemistry was not only a technical and conceptual resource but also served as a model of an interdisciplinary science. As he wrote in 1902:

> The order of results to be expected [from a new physical geochemistry] is illustrated by the great advances which have recently come from occupying the middle ground between astronomy and physics, and between physics and chemistry. For a long time astronomy and physics were pursued as independent sciences. The great discoveries of astro-physics have shown the advantages of their combination. Chemistry and physics for a long time were pursued as independent sciences. The rapid rise of physical chemistry has shown how wonderfully fruitful is the ground between the two. [11]

Physical chemistry was an important prerequisite of cosmochemistry, but it was not the only one. Another prerequisite was meteoritics, or the science of meteorites. [12] By the 1850s, chemical analyses of meteorites had shown that meteorites differ substantially in their mineralogical composition from terrestrial rocks, which was a strong argument in favor of the still controversial idea of the extraterrestrial origin of meteorites. On the other hand, the analyses also confirmed the material unity of the universe, that is, that the messengers from the heavens are composed of the same chemical elements as found on the earth. Later meteorite researchers emphasized the field's cross-disciplinary nature, involving not only physics, chemistry, and mineralogy, but also astronomy. The fascinating possibility was that meteorites, being messengers from space, could be analysed as probes of the chemical composition of the cosmos. In 1901, the American geologist Oliver Farrington hypothesized that although the mineral constituents of meteorites differ from those of terrestrial rocks, the relative abundance of the chemical elements in meteorites and on the earth as a whole were the same. [13] Farrington's hypothesis would be central to the later development of cosmochemistry.

I am not sure when the term "cosmochemistry" or related terms first appeared in the scientific literature, but in 1902–1903 the German physical chemist Emil Baur gave a series of lectures that were published in book form as *Chemische Kosmographie* [14]. With this he meant the totality of chemical processes in nature, an attempt to see the whole of nature under a chemical viewpoint. He included in his treatment subjects that later geo- and cosmochemists would recognize, such as stellar and solar chemistry, meteorochemistry, petrography, mineralogy, and chemical oceanography; but he also covered aspects of paleontology, fermentation processes, and physiological chemistry. Baur's kind of cosmochemistry was related to the period's "cosmic physics", only was his perspective chemical rather than physical. [15] Whereas the term cosmochemistry belongs to the twentieth century, cosmic physics – including the name (kosmische Physik) – can be traced back to the mid-nineteenth century.

Meteoritics served as a link between geochemistry and the chemistry of the wider universe. But there were other, less direct ways to explore this wider chemistry, namely, by analysing spectroscopically the light emitted by the stars. The astro-spectrochemistry that emerged in the 1860s opened up for new and exciting questions to be answered by means of the spectroscope. What is the chemical composition of the sun and the stars? Do there exist chemical elements in the stars that are not found on the earth – including elements lighter than hydrogen? Do stellar spectra indicate the complexity of the chemical atom? These were some of the questions asked by William Crookes, Norman Lockyer and other researchers within astrophysics and -chemistry. [16] A major aim of the astrospectroscopists was, in the words of William Huggins, "to discover whether the same chemical elements as those of our Earth are present throughout the universe." He concluded that this was indeed the case and that "a common chemistry ... exists throughout the universe." [17] All the same, might there not be elements in the heavens that do not exist on earth? This was what Lockyer, Crookes, and a few other scientists believed, basing their arguments on spectral lines from the stars that could not be identified with those known from laboratory measurements. Stellar spectroscopy led famously to the prediction of helium (1868), but also to a number of discovery claims of nonexisting elements such as "coronium" (1871), "nebulium" (1898), "asterium" (1900), and "archonium" (1914).

Although the new astrochemistry (or "celestial chemistry") attracted much attention, it was far from universally accepted among the chemists. After all, chemistry was a laboratory science that studied experimentally elements and their compounds; astrochemistry offered a few spectral lines that could not be experimented with, only speculated about. To most chemists, an element that could not be isolated and whose atomic weight could not be determined was not an element. All the same, spectroscopy helped transforming both chemistry and astronomy and in general brought the two sciences into closer contact. In a retrospective article of 1897, Huggins noted how "an astronomical observatory began, for the first time, to take on the appearance of a laboratory... The observatory became a meeting place where terrestrial chemistry was brought into direct touch with celestial chemistry." [18] Together with geochemistry and meteoritics, astrochemistry formed the background of twentieth-century cosmochemistry.

Photography, a science (or art) which astronomers saw as closely related to spectroscopy, was another field that forged links between chemistry and astronomy. Not only were photographic processes a major area of chemistry, astronomers also sometimes used chemical metaphors when describing the marvels of scientific photography. Thus, to the Swedish astronomer Carl Charlier the photographic plate was a less utopian version of what "The medieval alchemist dreamt of... [namely,] being able to produce an 'homunculus', a chemical product with all the properties of a man." The parallel may not be obvious but, as Charlier explained, in astronomical practice "after an hour's exposure the stars have been transferred from the sky to the photographic plate."[19]

9.3
Chemists, Element Formation, and Stellar Energy

Einstein's energy-mass relation, the Bohr-Rutherford nuclear atom, and progress in astrophysics transformed the Victorian speculations of Crookes and Lockyer into a more fruitful and respectable research program. In this process, the English theoretical physicist John Nicholson can be seen as a transitional figure. His work in atomic theory had clear links to the Crookes-Lockyer tradition and was, as he admitted, quite speculative. Nicholson believed that proto-elements such as "coronium" and "nebulium" existed in stellar atmospheres and that microphysical theories could be tested by means of astrophysics, which he – prophetically – characterized as "an arbiter of the destinies of ultimate physical theories." His project was to integrate microphysics with astrophysics and thus, indirectly, examine the chemistry of the heavens. "The astronomer... may have a field of chemical study which is closed to the chemist", he wrote in 1913, referring to those primitive elements that were assumed to exist in stars but no longer on earth.[20]

From about 1915, a new generation of scientists attempted to use the new physics to understand how elements were formed and why the stars shine. In this work, and especially with regard to the question of element formation, chemists played a role that for a period was as important as that played by physicists and astronomers. Several leading physical chemists, including Svante Arrhenius, Walther Nernst, Jean Perrin, and William Harkins, were interested in astronomy and cosmology and contributed to the new phase of astrochemistry. However, they worked individually and independently, and their research formed neither a recognized subdiscipline nor the nucleus of a scientific subcommunity.[21]

Nernst's theory of a stationary universe, developed from 1912 to his death in 1938, operated with the creation from the ether of radioactive atoms.[22] The hypothetical ether-born superradioactive and superheavy elements formed the *Urmaterie* of his universe, the matter from which all the elements were formed by disintegration and which was the source of the stars' energy. In 1928 Nernst urged the chemists "to seek by all suitable means this most important element in the earth also;"[23] and when Enrico Fermi and his group in 1934 (erroneously) reported the production of transuranic elements he took it as support of his conjecture. Fantastic as the hypothesis of superheavy cosmic elements may appear, Nernst was not alone in defending the hypothesis. It was argued by Werner Kohlhörster in the 1920s as a possible explanation of the cosmic radiation and played an important role in James Jeans's cosmic scenario. According to Jeans, the world did not start with hydrogen atoms or some more primary matter, but on the contrary with complex, transuranic atoms that gradually decomposed into radiation and lighter atoms. The view of Nernst and Jeans was severely criticized by Arthur Eddington, but at the time it had a certain credibility and, although unorthodox, could not be easily dismissed. For example, in 1923 Ernest Rutherford speculated that the radioactive elements represented the sole survivals of even heavier elements that had been common in the distant past, when the atoms were in course of formation. [24] And six years later Lord Rayleigh (Robert J. Strutt) discussed "the general question of whether the evolution of elements has proceeded from the simple to the complex, or from the complex to the simple." [25]

Other chemists and physicists found synthesis from hydrogen atoms to be a better explanation of how the elements were formed in the stars. The Chicago chemist Harkins suggested in 1917 that "the variation in the abundance of elements as found would seem to be the result of an atomic evolution... [and] related to the atomic number." [26] He hypothesized a transformation of four protons into a helium nucleus, a process that was also considered two years later by the French physical chemist (and later Nobel laureate) Jean Perrin. Harkins had a weakness for numerological speculations and was constantly involved in priority disputes, and perhaps for these reasons he is recognized only as a relatively minor figure in the history of science. But he had very important insights and should be counted as one of the pioneers of modern cosmochemistry. [27] Harkins, who had a thorough knowledge of geochemical data, was the first to combine the then infant nuclear physics with geochemical analyses of minerals and meteorites. He realized that geochemists' analyses of the composition of rocks were of little value because they referred to material of the "mere skin," which had been altered by geological forces. Harkins therefore concluded that meteorites were the best samples of the material composition of the cosmos and that the meteoritic abundance distribution of the elements would be an important clue to the structure and formation of atomic nuclei. Accordingly he devised nuclear models in order to account for the geochemical data of meteorites in particular.

In 1917 Harkins found that on the average elements of even atomic number (Z) are about 70 times as abundant in meteorites as those of odd Z; he further noted that the first seven elements in the order of their abundance are all even-numbered and make up almost 99% of the material in meteorites. [28] Four years later he elaborated his assumption that "the relative abundances of the atomic species of low atomic weight may be used as an index... of their relative stability." He now suggested several more rules, including that atoms with even A (mass number) and an odd number of electrons are extremely rare. [29] These rules were claimed to be valid for isotopes, and not merely elements. In 1931, after more data on the distribution of isotopes had been collected, he reported that even-A nuclei were much more frequent than odd-A nuclei. [30]

Harkins's program of linking nuclear physics and geochemistry was not initially followed by other researchers, but it was not without impact and was often referred to in the 1920s. The program received support from Francis Aston's invention of the mass spectrometer, which was of great importance to geochemistry because it

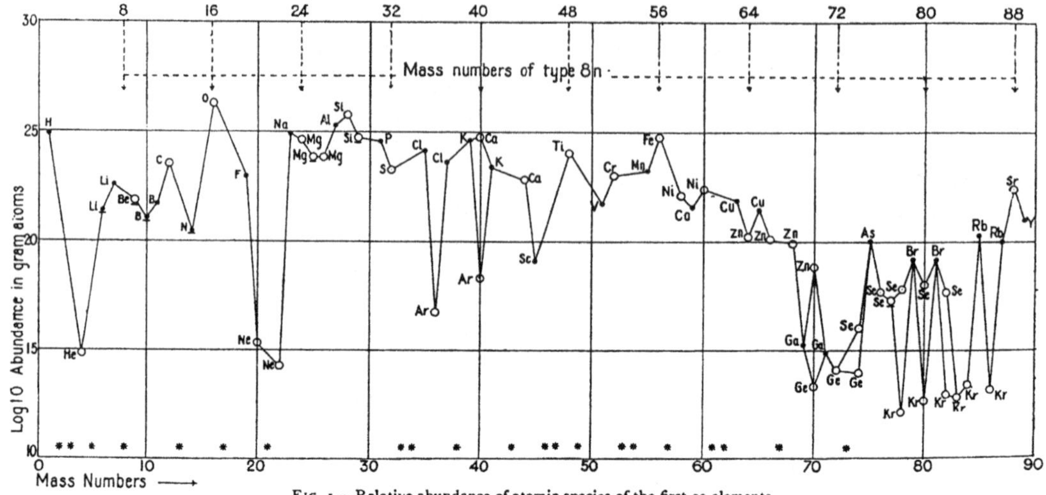

FIG. 1.—Relative abundance of atomic species of the first 39 elements.
○ Even atomic number 41 } Total, 69 species.
● Odd atomic number 28 }
✱ Missing or doubtful mass numbers.

Figure 9.1 Aston's abundance curve of 1924 [31].

opened up for the study of the abundance of individual isotopes. Aston, a physicist and Nobel laureate in chemistry of 1922, was interested in the geochemical aspects of isotope separation and produced one of the first diagrams of the relative abundance of isotopes (Figure 9.1). In accordance with Harkins he suggested that the diagram "might afford some evidence as to the relative stability of nuclei during the evolution of the atoms." [31] In another note of 1924, he speculated that the conspicuous rarity of the inert gases might be evidence for the planetesimal theory of the origin of the earth suggested by the American astronomer Thomas Chamberlin. [32]

The group of chemists with an interest in astro- and geochemistry included also Gilbert Lewis, the famous American physical chemist. In an address of 1922 he discussed the cosmic origin of radioactive elements in a manner that, he admitted, "may seem somewhat speculative in character." Like Nernst and Harkins, Lewis believed that not only had astronomy much to learn from chemistry, the latter science had also much to learn from astronomy. He argued as follows:

> While the laboratory affords means of investigating only a minute range of conditions under which chemical reactions occur, experiments of enormous significance are being carried out in the great laboratories of the stars. It is true, the chemist can synthesize the particular substances which he wishes to investigate and can expose them at will to the various agencies which are at his command; we cannot plan the processes occurring in the stars, but their variety is so great and our methods of investigation have become so refined that we are furnished an almost unbounded field of investigation. [33]

Lewis's theme, the conception of stars as huge chemical laboratories, was popular among many chemists and astronomers. For example, in 1904 the American astronomer George Hale described the sun as an "enormous crucible" in which "experiments [occur] on a scale far transcending any that can ever be performed in the laboratory." Likewise, in a 1921 grant proposal for a new solar laboratory he argued that "the cosmic crucibles in this vast laboratory of nature exhibit conditions of temperature and pressure often transcending those attainable in the laboratory, and thus present for observation experiments on an immense scale, the interpretation of which has already added much to our knowledge of physics and chemistry." [34]

The earliest phase of nuclear physics (about 1913–1925) attracted several chemists who speculated about the structure of the atomic nucleus. [35] A few of these followed Harkins in linking their speculations with geochemical data of the abundance of elements and isotopes. For example, Richard Sonder, a Swiss chemist, found several regularities in the abundance data of the elements and used these to suggest detailed electron-proton structures for all the nuclei from hydrogen to uranium. Sonder further argued that the present distribution of the elements was the result of a cosmo-genetic process from a primeval mixture of electrons and protons – a slightly modernized version of Crookes's old speculation. [36]

As another example, consider the Zurich mineralogist Paul Niggli, a leading earth scientist who made important contributions to petrology, geology, and crystal chemistry. In 1921 Niggli concluded that the abundance of magmatic elements was particularly high for atomic numbers 28, 48, 50, 80, and 82. He assumed regularities in atomic structure to lie behind and in general had a broad view of the aims and possibilities of geochemistry. "In the mineralogic-petrographic study we meet all over such questions of a general nature," he wrote. "The science of magmatic abundance is a model example of [the insight that] what is studied as separate fields by the individual sciences is, in the end, brought back into their relations." [37] Niggli found numerous relations to hold between the abundance of isotopes, their place in the periodic system, and their atomic numbers and number of nuclear electrons. He believed these numerologically based rules to be significant with respect to such problems as radioactivity and nuclear disintegration. As he wrote in 1928, his study showed "in which ways the mineralogist can contribute to the solution of these difficult problems of atomic physics." [38] For example, Niggli suggested that his rules might be understood in terms of Rutherford's new satellite model of the atomic nucleus. [39]

The American Richard Tolman, professor of physical chemistry and an authority on relativity and statistical mechanics, studied in 1922 the relative abundance of hydrogen and helium by means of chemical equilibrium theory. His work seems to have been inspired by Harkins and Wilson's suggestion of energy production from the fusion of hydrogen nuclei to helium nuclei. Although very little was known about the cosmic abundance of the two elements, Tolman's calculation clearly conflicted with what was known, namely, the marked preponderance of hydrogen over helium. He found that even at very high temperatures and very low pressures hydrogen would combine almost completely to helium. He discussed various ways to reconcile "the mutual presence of hydrogen and helium under terrestrial, solar, and stellar conditions with the tentative thermodynamic prediction of practically complete conversion of hydrogen into helium when equilibrium is attained", but found no satisfactory solution. [40] Tolman's thermodynamic treatment was based on the assumption that the helium nucleus consisted of four protons and two electrons. He mentioned as a possible way of solving the problem "to give up the hypothesis that the helium nucleus is formed from hydrogen nuclei and electrons", but dismissed the idea. Six years later a Japanese physicist, Seitaro Suzuki, reconsidered the problem addressed by Tolman, reaching the same unacceptable conclusion. Whereas Tolman only considered stellar conditions, Suzuki speculated that the observed helium-hydrogen ratio might be explained thermodynamically if the site of the equilibrium process was the early universe rather than the stars. [41] Together with the earlier works of Harkins, Suzuki's paper foreshadowed the program of "nuclear archaeology", that is, the attempt to reconstruct the history of the universe by means of hypothetical cosmic or stellar nuclear processes, and to test these by the resulting pattern of element abundance. I shall return to this theme.

Around 1930, several physical chemists applied their skills to problems of astrophysics in the manner of Tolman and Suzuki. For example, S. B. Stone, an American chemist, suggested one of the earliest theories of the origin of elements by nuclear processes – not in an astronomical or physical journal, but in the *Journal of Physical Chemistry*. [42] The early work in nuclear astrophysics assumed the reacting nuclei to be in thermodynamic equilibrium and the processes thus to be governed by tested principles of physical chemistry, namely, the equilibrium theory based on the law of mass action and the van't Hoff-Arrhenius expression for the equilibrium constant in terms of temperature and activation energy. In 1931, two physical chemists in Germany, Ladislaus Farkas and Paul Harteck, applied the method to a primeval mixture of nuclei and calculated the relative abundance for elements up to sodium. [43] Their work served as an inspiration for Carl Friedrich von Weizsäcker's important cosmological theory of 1938 in which he argued that the heavier elements were formed cosmologically, in an early hot state of the universe and not in the interior of the stars. [44]

Weizsäcker's theory shared with other theories of element formation the assumption of an equilibrium mechanism. It was the abandonment of this assumption in the 1940s that paved the way for the first successful big-bang model of the universe, proposed by George Gamow and his collaborators in 1948. That the equilibrium hypothesis might not be tenable had been suggested as early as 1931, when the two American chemists Harold Urey and Charles Bradley argued that the relative abundance of terrestrial elements could not be reconciled with the hypothesis, whatever the temperature of the equilibrium mixture. [45]

9.4 Victor Moritz Goldschmidt and the Transition from Geo- to Cosmochemistry

In the 1920s geochemistry experienced a major transformation. As the field was given a more secure theoretical foundation and new and more accurate empirical data were collected, it turned into a mature science with its own methods, problems, and visions. The leaders of the new geochemistry were Victor Goldschmidt and, in Russia, Vladimir Vernadsky and his younger contemporary Aleksandr Fersman. Fersman argued for a "chemicalization of geological thinking" and attributed to geochemistry a fundamental role because "geochemistry is speaking the universal language of atoms."[46] Although the Russian school of geochemistry was very productive and innovative, internationally its impact was rather limited.[47] Almost all the scientific works were published in Russian and much of it was unknown to Western scientists. Moreover, Vernadsky and his pupils mostly concentrated on terrestrial geochemistry, including biogeochemistry, and showed little interest in the connections to cosmochemistry, the main subject of the present essay. Although Fersman advocated that traditional geochemistry should be extended to a chemistry of the cosmos, his ideas bore no fruit. In 1923 he published a book with the title *Chemical Elements of the Earth and the Cosmos*, but it appeared in Russian only. All the same, there is no doubt that the Soviet Union was a leading nation of geochemistry in the period from about 1920 to 1950. In 1931 the Institute of Geochemistry was founded as part of the large Lomonosov Institute in Moscow. Institutes of its kind and size did not exist in the West.[48] Reflecting the nationalistic ideology of the Stalinist period, Soviet scientists sometimes claimed geochemistry to be a Soviet science. In 1938, an author wrote that "Geochemistry is, to a large measure, ours – a Soviet science."[49]

Geochemistry in the 1920s was not restricted to the study of the earth's crust and atmosphere but included in a few cases investigations of the interior of the earth. The Russian-German scientist Gustav Tammann, a leading physical chemist at the University of Göttingen, made important contributions to geophysics and proposed in 1924 an intermediary sulfide layer between the earth's outer silicate layer and its iron-nickel core. Likewise, in Zurich Niggli pioneered the physico-chemical study of magmas and thereby provided volcanology with a chemical foundation. Equally important work on the physico-chemical processes of the formation and differentiation of igneous rocks was done by Norman Bowen at the Geophysical Laboratory in Washington D.C. The works of Tammann, Niggli, Bowen, and others brought geophysics and geochemistry into closer contact.[50] However, in general geochemistry did not have a strong position in German science. In an article of 1925, Friedrich ("Fritz") Paneth complained that Goldschmidt's important works were unknown in Germany and that German textbooks in chemistry, contrary to those in England and America, ignored geochemistry.[51] This situation changed when Goldschmidt moved to Göttingen four years later.

More than anyone else, the Swiss-Norwegian chemist Victor Moritz Goldschmidt revolutionized geochemistry and is today recognized not only as the founding father of modern geochemistry but also as a pioneer of cosmochemistry.[52] Goldschmidt

Figure 9.2 Norwegian postage stamp issued in 1974 in honor of Victor Goldschmidt. Based on a portrait taken in 1923.

was the founder of important schools of geochemistry in Norway and Germany and a leading figure in the early attempts to establish geochemistry as a crossdisciplinary scientific field. He had good connections to his Russian colleagues and corresponded with Vernadsky about launching an international journal of geochemistry.[53] However, nothing came of the plan. Only after Goldschmidt's death did the first journals devoted to geochemistry appear; and it was also at that time, around 1950, that the first comprehensive textbooks in the field were published.

The Swiss-born Goldschmidt studied chemistry, mineralogy, and geology at the University of Oslo (then Christiania), where his father, Heinrich Goldschmidt, held the chair in chemistry. Although Oslo was at the time at the periphery of scientific Europe, it held a strong position in mineralogy and geology. Young Goldschmidt benefited from the fertile research environment that existed around Waldemar Brøgger and Johan Vogt, two of Norway's important earth scientists. In 1914, only 26 years old, he was appointed professor of mineralogy, a position he held for the next 15 years, until he accepted a position in Göttingen (Figure 9.2). Goldschmidt served as head of the University's mineralogical institute, but as a Jew he decided in 1935 to leave Germany and return to Oslo, where he became director of the Geological Museum. When the Germans occupied Norway he was temporarily imprisoned and in 1942 narrowly escaped the concentration camp by crossing the border into Sweden with the help of the Norwegian underground movement.

In the 1920s Goldschmidt and his collaborators focused on the relation between crystal structure and the size of atoms or ions in order to establish general laws of the distribution of elements in crystals, or what he called *Verteilungsgesetze.*[54] Among other results, they published the first table of atomic radii and discovered the phenomenon known as the lanthanide contraction. In 1924, Goldschmidt and his assistant Lars Thomassen studied the relative abundance of the rare earth elements and confirmed Harkins's rule that elements of odd atomic number are less abundant than their neighbors of even atomic number (Figure 9.3). All this was possible only because the Oslo scientists made extensive use of X-rays, crystallographically to obtain data on the atomic and ionic radii and spectrographically to determine the concentration of rare elements in minerals. Goldschmidt realized that in order to obtain a full picture of the relative abundance of the elements, the rare elements were no less important than the more common ones.

The same insight was reached in Copenhagen, where the X-ray spectrographic discovery of hafnium in 1922 caused one of the discoverers, George de Hevesy, to examine the distribution of the elements and their isotopes in terrestrial and meteoritic sources. Until the mid-1930s, when Hevesy changed to biochemistry and medical biology, geo- and cosmochemistry was his main interest. In his case, the interest derived from the light such studies could cast on problems of nuclear structure, namely, by using the relative abundance of the nuclides as a measure of their stability of the atomic nuclei. "It is this consideration", Hevesy stated in his 1932 Baker Lecture at Cornell University, "which gives us a new stimulus to study the abundance of the elements in terrestrial and meteoric matter as well as in solar and stellar atmospheres." [55] In collaboration with Hilde Levi, Hevesy pioneered in 1935 the technique of neutron activation to analyse rare earth elements and thereby detected traces of elements that had escaped Goldschmidt's analysis. [56]

Hevesy was not the only chemist whose work in X-ray spectroscopy led him to extended excursions into geochemistry. The same was the case with Ida and Walter Noddack, the two German chemists who in 1925 had discovered the element rhenium (Z = 75) by means of X-ray spectroscopic methods and, controversially, claimed the discovery of masurium (Z = 43, later technetium). This led the Noddacks to apply their analytical skills to the composition of meteorites, which they realized were much more representative of the cosmic abundance of the elements than rocks from the earth's crust. Using as data material based on the evaluation of 1600 terrestrial and meteoritic assays they contributed important if sometimes controversial data on the geo- and cosmochemical distribution of the elements. In their 1930 article, they presented the elemental abundance of 71 elements in meteorites. [57] Much like Hevesy had examined the geochemical abundance of hafnium, so the Noddacks investigated the geological and meteoritic abundance of rhenium. [58]

Hevesy's close friend and collaborator, the distinguished Austrian (and later British) chemist Friedrich Paneth, was yet another radiochemist who turned to geo- and cosmochemistry in the 1920s. He started a research program on meteoritic chemistry that made him an authority in the field of meteorite science and one of the founders of modern cosmochemistry. [59] It is also worth to mention that Hevesy's old rival from the controversy over element 72, the Frenchman Alexandre Dauvillier, turned from X-ray spectroscopy to astro- and cosmochemistry. In 1955 he wrote one of the first books ever on cosmochemistry. [60]

Goldschmidt was first nominated for a Nobel Prize in chemistry in 1928. The nominator was Brøgger, his old teacher, who praised Goldschmidt's fundamental contributions to geochemistry and crystal chemistry, and in particular emphasized his work on the rare earth elements. [61] The following year he was nominated by Walter Hieber, a Göttingen chemist, and in 1931 by Edwin Blanck, Alfred Cohen,

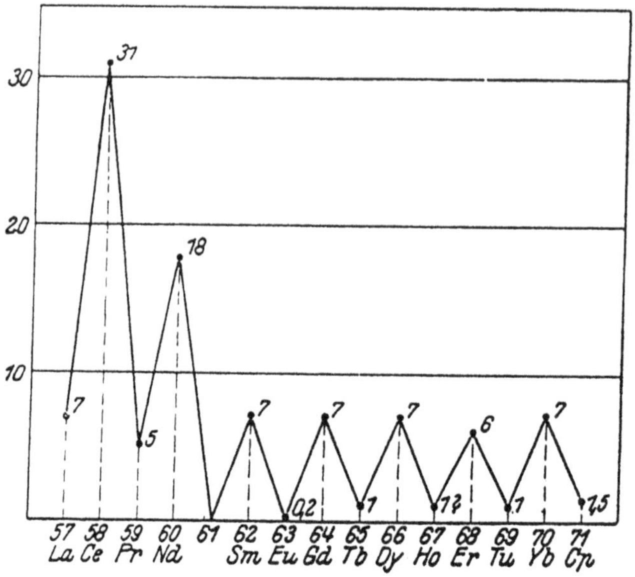

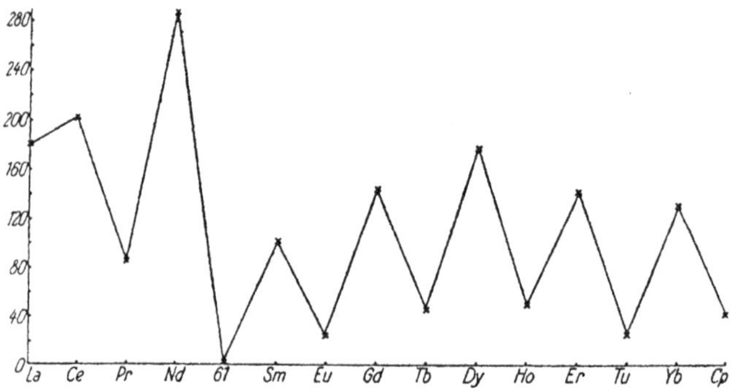

Figure 9.3 The relative abundance of the rare earth elements according to Goldschmidt and Thomassen (above) and Ida Noddack (below[57]). The Noddacks found a much higher abundance of neodymium than the Norwegian researchers, but their data were not generally accepted. Both diagrams illustrate the Oddo-Harkins rule very clearly.

Max Planck, Adolf Deissmann, and Arthur Kötz. Fritz Haber nominated him for the 1933 prize and again (together with G. N. Lewis) for the 1934 prize. Finally, Otto Ruff, a chemist from Breslau, nominated Goldschmidt for the 1935 prize. The German nominators seem to have realized that Goldschmidt, a professor of mineralogy, might not qualify for a chemistry prize, but argued that his work was highly important also from the perspective of inorganic chemistry. According to

Hieber, Goldschmidt's contributions to geochemistry were comparable with "the modern direction in biochemistry, where prominent spokesmen during the last years have been awarded the Nobel prize." [62] If biochemists could be awarded the Nobel prize, why not geochemists? In 1934, the Nobel Committee for chemistry prepared a special report on Goldschmidt. The Swedish expert, Arne Westgren, concluded that Goldschmidt's geochemical work was "primarily of geological interest" and "only secondarily of chemical interest." [63] Although Westgren found Goldschmidt's contributions to crystal chemistry to be important, these had not resulted in any major discovery and consequently he recommended not to honor Goldschmidt with the prize. There seems little doubt that, as seen from Stockholm, geochemistry – contrary to biochemistry – was not considered a proper branch of chemistry.

From the late 1920s, Goldschmidt and his collaborators started on an ambitious and time-consuming program of extending and improving existing tables of terrestrial abundance data of the chemical elements. At the same time, they combined them with solar, stellar, and meteoritic data in order to obtain a reliable picture of the cosmic distribution of elements. [64] Together with his associate Claus Peters, he investigated the geochemical properties of several elements, including boron, beryllium, germanium, and the noble metals. Goldschmidt's main interest was in the chemical composition of the earth, but he realized, as he later phrased it, that "[o]nly through comparison with the universe as a whole can we gain an understanding of the geochemical evolution of the earth." [65] In most of his work of the 1930s Goldschmidt relied on a new method of optical arc-spectrography that he and his collaborators developed in Göttingen. The optical technique was much more sensitive than the X-ray method and thereby eliminated in many cases the laborious chemical concentration of the elements from rock samples. The fruits of Goldschmidt's research program were published in 1938 and formed the basis of a lecture he gave before the Chemical Society in London in March 1937. [66] Contrary to most earlier geochemical work, Goldschmidt's studies were concerned not only with the abundance of elements but also with isotopes. He followed the progress in nuclear physics and considered, in agreement with his friend Hevesy, his compilation of data to be a tool to understand nuclear structure. He summarized the data in graphs that showed not only the variation of element abundance with the atomic number but also with the mass and neutron numbers (Figure 9.4).

Goldschmidt's cosmochemical data directly influenced the development of nuclear theory and cosmology, two sciences which at the time were widely seen as unconnected. The data came to function as the most important element of the nuclear-archaeological program the aim of which was to reproduce the cosmic distribution of nuclear species either by stellar processes or by processes occurring in a hypothetical prestellar past, that is, cosmologically. As early as 1926, Goldschmidt had noted the abnormal rarity of lithium, beryllium, and boron and suggested that it was a phenomenon to be explained by nuclear physics, possibly because the elements were difficult to build up through nuclear processes in the stars. [67] In 1938 he returned to the problem of a nuclear-physical basis for the distribution of the elements. He believed the solution was to be found in the new

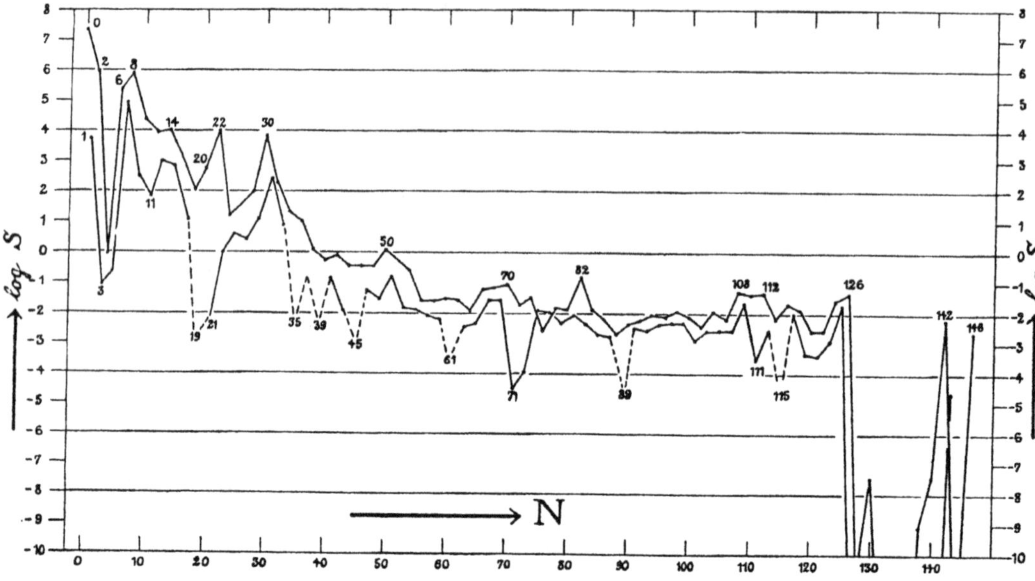

Figure 9.4 Goldschmidt's 1938 summary data of the cosmic logarithmic abundance of the elements as a function of their neutron numbers [66].

field of nuclear astrophysics which "makes possible conceptions of the origin of all the atomic species of the elements from hydrogen." [68] Goldschmidt considered Weizsäcker's new theory a promising answer but argued that it was unable to explain the sudden decrease in abundance following iron. [69] He suggested a revised version of Weizsäcker's theory in which the formation of elements between boron and iron was the result of collisions with fast (non-thermal) protons and deuterons, the reactivity being governed by the potential barriers of the target nuclei.

The cosmological version of the nuclear-archaeological program first appeared in Weizsäcker's 1938 theory, which included a reference to Goldschmidt's new abundance table. The table, or modifications of it, played a significant role in all later theories of physical cosmology. Thus, in an important pre-big-bang work of 1942, Subrahmanyan Chandrasekhar and Louis Henrich refined Weizsäcker's method and computed theoretical abundance curves based on the equilibrium assumption. "Our object," they stated, "is to compare the computed abundance-curves with the observed relative abundances of the stable nuclei, as given by V. M. Goldschmidt." [70] They found a promising agreement with Goldschmidt's data for elements up to argon, but for the heavier elements their method failed miserably. Goldschmidt's abundance curve showed that, apart from local variations, for atomic weights larger than about 100, all elements have nearly the same abundance, and this feature could not be reproduced by the Chandrasekhar-Henrich method. The failure was soon interpreted as a failure of the equilibrium approach and within a few years it would lead the way to the big-bang, non-equilibrium theory of Gamow and his associates Ralph Alpher and Robert Herman.

I shall not go into the details of the history of nucleosynthesis and its role in the development of modern cosmology,[71] and merely want to emphasize the predominant role that the cosmochemical tables played in this area of research. By 1950, all cosmologists and astrophysicists agreed that a good theory of the universe should be able to account for the formation and cosmic distribution of the elements, that is, to explain how the various nuclear species heavier than hydrogen were or had been synthesized. The better the agreement between the predicted species and those found empirically, the better the theory. This test, as it were, became an important part in the controversy between the big-bang theory and the rival steady-state theory that allowed only nuclear synthesis in existing objects such as stars and novae. For a decade Goldschmidt's compilation of data was the mark book, either in its original version or the revised version published by Harrison Brown in 1949.[72] A more detailed picture of the cosmic distribution of elements became available in 1956 when Urey and the German-American physical chemist Hans Suess published new abundance data based on corrections of certain assumptions made by Goldschmidt and Brown.[73]

9.5 Geochemistry and the Shell Model of Nuclear Structure

Element abundance data were useful not only in astrophysics and cosmology but also in the attempts to understand the structure of the atomic nucleus.[74] As mentioned, this line of reasoning was adopted by Harkins as early as 1917, of course based on a highly inadequate picture of the nucleus. It was only after 1932, with the discovery of the neutron as a nuclear component, that it was realized that not only is the atomic mass number related to isotopic abundance, but so are the proton and neutron numbers individually. Cosmochemical data played an important part in the development of the shell model, first proposed by Walter Elsasser and Kurt Guggenheimer in 1933–34 but only turned into a precise quantitative theory in the late 1940s.[75] Guggenheimer, a physical chemist, used isotopic abundance data as evidence of closed nuclear shells with nucleon numbers 50 and 82.

Maria Goeppert Mayer came to nuclear theory through an attempt to understand primordial element formation by means of hypothetical so-called polyneutrons that were assumed to have populated the early universe. Together with Edward Teller she studied how heavy elements could be formed from the polyneutrons and found that the mechanism led to a distribution of heavy isotopes in rough agreement with observations. As she recalled in her 1963 Nobel lecture, "When Teller and I worked on a paper on the origin of the elements, I stumbled over the magic numbers. We found that there were a few nuclei which had a greater isotopic as well as cosmic abundance than our theory ... could possibly explain."[76] It soon turned out that the polyneutron theory was wrong – it would lead to a world consisting mostly of heavy elements – but it had the effect of stimulating Mayer to propose the shell theory of nuclei in 1948. In her first paper on the subject she quoted isotopic abundance data in favor of the theory and referred to certain regularities in Goldschmidt's table that could be explained by her theory.

The shell model was developed independently by Hans Jensen, Otto Haxel, and Hans Suess in Germany, and also in this case Goldschmidt's data were important. Suess, a young physical chemist, had become interested in cosmochemistry during the war and happened to spend some time in occupied Norway together with Harteck and Jensen. [77] There they discussed the abundance values with Goldschmidt; and Suess realized that nucleon numbers 50 and 82 were particularly prominent in the cosmic distribution of nuclei. [78] The recalculation of the abundance of individual isotopes that Suess performed in 1947 served as an important source for the shell model proposed by the three scientists the following year. [79] Goldschmidt's role in the development of the shell model was not limited to his cosmic abundance data. In the last part of his 1938 treatise, he investigated in considerable detail how the structure of atomic nuclei could be understood in order to explain the cosmochemical data. [80] Based on arguments of a somewhat numerological nature, he suggested a kind of shell model and reached the conclusion that the neutron numbers 20, 28, 50, and 82 were what later would be called "magic". In the later shell model, the magic numbers were shown to be 20, 50, 82, and 126 neutrons or protons. Although he did not base his arguments on quantum theory, Goldschmidt should be recognized as one of the early contributors to the nuclear shell model.

9.6 Chemistry in Space

A rational deduction of elemental abundance from solar and stellar spectra had to be based on quantum theory, and the necessary foundation was laid with the Indian physicist Meghnad Saha's theory of 1920. Saha, who as part of his postdoctoral work had stayed with Nernst in Berlin, combined Bohr's quantum theory of atoms with statistical thermodynamics and chemical equilibrium theory. Making an analogy between the thermal dissociation of molecules and the ionization of atoms, he carried the van't Hoff-Nernst theory of reaction-isochores over from the laboratory to the stars. Although his work clearly belonged to astrophysics, and not chemistry, it relied heavily on theoretical methods introduced by and associated with physical chemistry. This influence from physical chemistry, and probably from his stay with Nernst, is clear from his 1920 paper where he described ionization as "a sort of chemical reaction, in which we have to substitute ionization for chemical decomposition." [81] The influence was even more evident in a second paper of 1922 where he extended his analysis. [82]

It was, in any case, Saha's theory and its development by Ralph Fowler and Edward Milne that eventually led to a solution of the stellar abundance problem. Until the mid-1920s, it was generally assumed that the stars have roughly the same chemical composition as the earth, with iron being the most abundant element.

During the last years of the decade Albrecht Unsöld in Germany and Henry Norris Russell in the United States reanalysed spectral data and demonstrated that, contrary to previous expectations, hydrogen is by far the most abundant element in the sun's atmosphere. [83] The result indicated that hydrogen might also be the predominant element in the interior of the sun and most other stars, which was confirmed by calculations made by the Danish astronomer Bengt Strömgren in 1932. During the 1930s, knowledge of the chemical composition of the stars, and hence of the world, thus changed radically. The work of Unsöld, Russell and others not only proved the preponderance of hydrogen, but also led to relative abundance data of other elements, including C, N, Si, Na, Ca, Sr, and Fe. [84] In his 1938 summary of the composition of the sun's atmosphere, Unsöld concluded that for atomic numbers larger than ten, there was good agreement between the astronomers' calculations based on spectra and the composition of meteorites as found by Goldschmidt and the Noddacks. [85] Likewise, Strömgren's improved 1940 calculation of the composition of the sun's atmosphere made use of Goldschmidt's meteorite data and concluded that the agreement between these and the solar values was "rather striking." [86]

In addition to the sun and other stars, also the chemical composition of comets was examined spectroscopically, first in the visible and ultraviolet range and later in the infrared and radio regions. According to a survey of 1943, molecules identified in cometary gases included OH, CO, CO^+, C_2, CH_2, CN, CH, and NH. [87] The form of stellar hydrogen was assumed to be either atomic or dissociated into electrons and protons, but in 1939 the German-American astronomer Rupert Wildt showed that negative H-ions exist in the cooler atmosphere of most stars. [88] Such negative ions are not part of ordinary chemistry, because their electron affinities are small compared with the atomic ionization potential; but, using standard methods of physical chemistry, Wildt showed that they can well exist in the solar atmosphere and will contribute to the sun's opacity. His result was considered very important, and according to an historian of astronomy it had the effect that "almost overnight the whole course of solar atmospheric theory was changed." [89] Wildt was also one of the early contributors to planetary chemistry, that is, the study of chemical compounds in the atmospheres of the planets by means of spectral analysis. Large amounts of CO_2 were found in the atmosphere of Venus in the early 1930s, and later in the decade NH_3 and CH_4 were detected on the large planets Jupiter and Saturn. In 1937 Wildt studied the photochemical decomposition of and reactions between molecules in planetary atmospheres. These included the formation of higher hydrocarbons during the photochemical decomposition of methane and the possible formation of formaldehyde (HCHO) from the reaction between carbonmonoxide and photochemically produced hydrogen atoms. [90]

Wildt was a pioneer of astronomical chemistry and wrote in 1940 a broad-ranging article on what he saw as a new field and for which he coined the word cosmochemistry. "The question can be raised," he wrote, "whether it would not be a timely expansion of the notion of geochemistry to adopt the term cosmochemistry to designate the science which shall deal with matter under all cosmic conditions." [91] Wildt realized that most chemists would have troubles in admitting cosmochemistry as chemistry, and pointed out that "the component 'chemistry' should be understood in its broadcast implications, meaning the science of matter in all its manifestations." He foresaw "a unified theory of matter, in which traditional chemistry will be merged with the physics of matter, and which may be called chemistry in the metaphorical sense alluded to before." His cosmochemistry was closely related to cosmology, in the sense that it complemented the geometrical approach by treating the universe *sub specie materiae*, as he phrased it. In a later review of planetary science, Wildt emphasized the importance of astro-geochemistry and gave much space to geochemistry, "the task of [which]... is to determine quantitatively the chemical composition of the Earth and to ascertain the laws governing the distribution of each element throughout the globe."[92]

9.7 Chemical Cosmogony and Interstellar Molecules

Whereas Wildt's approach to planetary chemistry was primarily astronomical, other scientists adopted a more geochemical approach in order to understand the origin and nature of the planets. Among the leading figures in American postwar cosmochemistry was Harold Urey, who was well prepared from his isotope research in the 1930s. In 1934 he had first entered meteorite research with an investigation of the ratio of oxygen isotopes in stone meteorites.[93] The center of American cosmochemistry was the University of Chicago, where Urey forged the new field in collaboration with Suess, Harrison Brown, Clair C. Patterson, Mark Inghram, and others. Brown, a nuclear chemist who turned to geo- and cosmochemistry, published with his student Patterson an important theory of meteorites. The Brown-Patterson theory was criticized by Irving Klotz, a physical chemist, who used arguments from chemical thermodynamics to raise doubts about the theory.[94] It was one among many cases in which physical chemistry had a decisive influence on planetary science.

Around 1950 Urey focused on "chemical cosmogony," that is, the origin of the planets and the moon from the perspective of physical chemistry.[95] In Urey's influential theory of the origin of the earth chemical reasoning played a decisive role, although of course in balance with considerations of geological and geophysical evidence. In one of the first issues of *Geochimica et Cosmochimica Acta* he described the role of chemistry as follows:

> The chemistry of the early history of the earth and other planets depends upon the cosmic abundance of the elements, the temperature of the accumulating planets and the gravitational fields which permit or prevent the escape of gaseous molecules and the properties of chemical substances under the conditions existing. Because of the detailed character of chemical evidence considerable information in regard to the early history can be deduced.[96]

Urey's theory was closely related to an earlier theory of the formation of the planets proposed by the German physical chemist Arnold Eucken in 1944.[97] However, published at an unfortunate time and place, Eucken's theory of the chemical development of contracting gaseous spheres was unknown to most scientists, Urey included. According to Urey, "The writer read his [Eucken's] paper only after the ideas of his paper were well developed and had he read it earlier would probably never have written this paper at all." [98] However, Urey concluded that the mechanism of planetary formation was chemical fractionation and not condensation, as proposed by Eucken. Urey's interest in the origin of the moon made him a pioneer of "selenochemistry," a field which later changed into a laboratory science when it became possible to study lunar samples obtained from the Soviet and American space programs. In the mid-1950s Urey became involved in a bitter controversy with the Dutch-American astronomer Gerard Kuiper concerning the chemical evolution of the solar system. In this controversy, methodological differences between the disciplines of astronomy and chemistry played an important role. "[Kuiper] would not regard anything that a mere chemist would say as important," Urey angrily wrote. [99].

In the 1930s, another arena of cosmochemical interest was discovered, namely the vast spaces between the stars. It had long been discussed whether there was diluted interstellar matter, but the general opinion was that the spaces between the stars were empty, hence without obscuration effects on starlight. Only in 1928 was the first evidence for interstellar calcium ions presented, and two years later the Swiss-American astronomer Robert Trümpler argued convincingly for the existence of obscuring matter in interstellar space. [100] This turned out to consist of atoms, ions, molecules, and radicals, some of them in a different chemical state than the species known from the laboratory. It was long uncertain whether molecules existed, but between 1937 and 1941 a few diatomic molecules (CH, NH, CN, CH^+) were discovered – in the sense that their spectral lines were identified. General physical arguments indicated that these and other simple molecules were fairly frequent in interstellar space. The identification of a hitherto unidentified spectral line with a rotational transition in CH, made in 1937 by the Belgian scientists Pol Swings and Léon Rosenfeld, is sometimes taken as the "discovery" of interstellar molecules. [101] In June 1941, Otto Struve organized a symposium on the interstellar medium at Yerkes Observatory with the attendance of a small number of astronomers, physicists, and chemists. [102] The attendees included Robert Mulliken and Gerhard Herzberg, both of whom would later be awarded the Nobel Prize in chemistry (in 1966 and 1971, respectively). The prediction in 1940 of cyanogen radicals (CN) made by a Canadian astrophysicist, Andrew McKellar, has an important place in the history of cosmology because the prediction and its subsequent confirmation turned out to be closely related to the famous discovery of the cosmic microwave background radiation in 1965. However, the significance of McKellar's observation was only fully appreciated after Arno Penzias's and Robert Wilson's discovery of the 2.7 K background radiation. [103]

The real breakthrough of interstellar chemistry belongs to the post-war period and depended on the development of radioastronomy and microwave technologies that opened up quite new perspectives in celestial chemistry. However, since this modern phase only started in 1963 (with the detection of the radio signature of OH), it does not belong to the chronology of the present work. It suffices to note that today more than 120 molecules and radicals have been detected in interstellar space and their unusual chemical reactions studied in considerable detail. [104]

9.8 The Emergence of Cosmochemistry

In spite of the efforts and visions of geochemical pioneers such as Vernadsky, Goldschmidt, Paneth, and the Noddacks, still at Goldschmidt's death in 1947 geochemistry had scarcely become a mature and recognized scientific discipline. Not only was it sandwiched between geology and chemistry, it was also uncertain what the field was or should be, more exactly. The old problem of geochemistry versus chemical geology had not been settled to the satisfaction of all chemists and geologists.

According to the Noddacks, geochemistry was a subdiscipline of chemistry, simply "the science of our earth's chemical composition." In a lecture of 1936, they stated that the kind of questions asked by the geochemist was essentially the same as those asked by chemists investigating other and much smaller objects: "This 'chemical' way of asking questions separates geochemistry from its neighbor fields, those of geophysics, geology, and mineralogy, with which geochemistry has otherwise much in common with respect to methods and aims." [105] However, not everyone agreed, and for some time the field's disciplinary status remained a subject of debate. In 1944, while a refugee in England, Goldschmidt was awarded the Wollaston medal, the highest honor of the Geological Society of London. Using the occasion to reflect upon geochemistry's disciplinary status, he saw the award as a recognition of "that branch of science which I have tried to develop during the last quarter of a century, namely, modern geochemistry, based upon atomic physics and atomic chemistry." Goldschmidt proceeded:

> One may wonder whether the new geochemistry is still a branch of geology or if it belongs rather to chemistry or even physics. I think it serves no useful purpose to construct strict lines of division, cutting through the unity of modern science. This prominent audience of geologists will agree with me that geology always has been, and still is, in a central position in the general development of pure and applied science... Another fundamental problem [apart from the evolution of organisms], the composition, structure, and distribution of matter, an object for modern geochemistry, is as closely connected with geology, even if it demands a kind of "combined operations" of geology with chemistry, physics, and other branches of science. Already, too, we can visualize the work of geochemistry as being closely connected with astrophysics and nuclear physics, leading up to the final problem of the origin and evolution of matter itself. [106]

In 1947, the same year that Goldschmidt died, the Finnish geochemist Kalervo Rankama pointed out to his American colleagues that there was no consensus about the concept of geochemistry and that there seemed to be "some fundamental divergence between the American and European concepts of geochemistry." [107] As Rankama saw it, the Americans had not truly understood Goldschmidt's definition of the field as consisting of three parts: "(1) The determination of the abundance of the elements and of the atomic species in the earth. (2) The description of the distribution of the individual elements in the different spheres of the earth, and in minerals and rocks. (3) The detection of the laws dominating the abundance and distribution of the elements." [108] Following his compatriot, Thure Sahama, professor of geochemistry at the University of Helsinki and a former collaborator of Goldschmidt, he stressed that geochemistry, contrary to chemical geology, "must maintain the chemist's attitude towards the problems investigated." Moreover, in accordance with Goldschmidt's program Rankama argued that geochemistry was intimately linked with "cosmochemistry... [which is] the natural end product of geochemistry, expanded and applied to the problems of the Universe." [109]

Modern cosmochemists consider their science in much the same perspective as Goldschmidt and Rankama outlined more than half a century ago. Thus, according to a recent textbook, the cosmochemist has two basic tasks: "The first is to determine the chemical composition in the material universe... The second task is to understand the reasons for the compositions that are found." [110] Goldschmidt would have completely agreed.

The chemistry-geology boundary question reappeared in Rankama and Sahama's influential 1950 textbook on geochemistry. Prefaced March 1949, the book was based on Sahama's *Geokemia*, published in Finnish in 1947. The two Finnish geochemists wrote: "The chemical geologist examines his problems from the viewpoint of a geologist. His is a geological material, and, when interpreting his observations, the analytical results, etc., he always has their geological application in mind. For the geochemist, on the other hand, the geological observations, though based on the use of geological material, represent only a certain part of his results, which are intended to establish laws governing the abundance and distribution of the elements." [111] As to the new concept of cosmochemistry, they ascertained that "geochemistry is but a branch of the general chemistry of the Universe, called cosmochemistry." [112] The Goldschmidt-Rankama-Sahama conception of geochemistry (Figure 9.5) was soon accepted and became the basis of the new and vigorous geo- and cosmochemistry both in America and Europe.

Among the early signs of professionalism was the founding in 1950 of *Geochimica et Cosmochimica Acta*, the first international journal devoted to geochemistry in its modern sense. According to the editors of the new journal, its aim was "to publish original research papers on geochemistry and cosmochemistry, such as have hitherto been scattered over a wide range of geological, mineralogical, chemical, and astronomical periodicals." They believed time was ripe for such unification, for "the chemistry of the Earth and of the Cosmos has become a branch of science independent enough to have a journal of its own." [113] *Geochimica et Cosmochimica Acta* absorbed much work that previously had appeared scattered around in journals devoted to, for example, inorganic chemistry, geophysics, astrophysics, oceanography, and meteoritics [114]. It started modestly with 338 pages in its first year but quickly rose in importance, readership, and length. (The 1997 volume consisted of

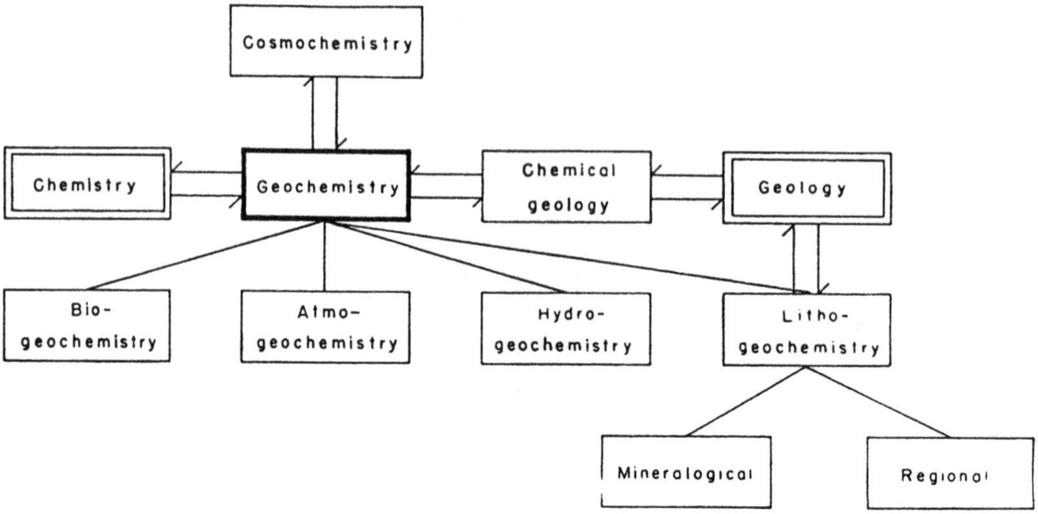

Figure 9.5 Rankama and Sahama's conception of the relations of geochemistry and cosmochemistry to their allied sciences [111, 114].

24 issues, with a total of 5,430 pages.) In addition to the appearance of textbooks and journals, the 1950s also saw the emergence of the first geochemical societies. The earliest and most important of these was the international but American-based *The Geochemical Society*, founded in 1955 with Earl Ingerson as its first president. [115] From 1957, *Geochimica et Cosmochimica Acta* became the official journal of the Geochemical Society.

To any emerging scientific discipline belongs a historical (or quasi-historical) tradition and, if possible, a founding father. In the case of geo- and cosmochemistry, Goldschmidt was chosen as the father figure. This was undoubtedly justified on scientific grounds, for Goldschmidt was a pioneer not only of modern geochemistry but also of the extension of geochemical perspectives to cover cosmic aspects. In a paper of 1950, Harrison Brown, one of the new generation of cosmochemists, paid tribute to "the late V. M. Goldschmidt, who perhaps more than any other man can be considered the father of modern geochemistry." [116] In a period marked by the Cold War, it cannot have been a disadvantage that he was a Swiss-Norwegian cosmopolitan Jew with no political commitments to either the capitalistic West or the communist East. In 1988, the geo- and cosmochemical community established a series of prestigious Goldschmidt conferences, and the highest award of the Geochemical Society is named the Goldschmidt award.

9.9 Conclusion

By 1970 at the latest, cosmochemistry had become an established field of science, organizationally based since 1967 in the International Association of Geochemistry and Cosmochemistry. Working backwards from this fact, I have sketched a possible history of cosmochemistry from its backgrounds in geochemistry, meteorite science, astrophysics, and nuclear physics. I have stressed the extreme interdisciplinary nature of the field, but also that the suffix "-chemistry" is justified in so far that progress in cosmochemistry have relied crucially on methods and theories of classical physical chemistry. Whether or not the cosmochemists were trained in chemistry – and many were not – they used chemical reasoning, methods, and instruments. Moreover, the abundance of chemical elements has historically been the single most important problem of the field; and more recently more traditional chemical objects, such as molecules, ions, and radicals, have become part of cosmochemistry.

All the same, geo- and cosmochemistry are too tightly interwoven with geology, physics, and astronomy to count as simply chemical subdisciplines. Indeed, today cosmochemistry is widely seen as belonging to the astronomical rather than the chemical sciences. According to Charles Cowley, a modern cosmochemist, "cosmochemistry is a branch of astronomy and geology for which physics and chemistry are ancillary disciplines." [117] Moreover, the term cosmochemistry is often used in the sense of "the study of the material composition of the universe," which may have nothing to do with chemistry as usually understood. Works dealing with the composition of the universe shortly after the big bang – say, a quark soup or a mixture of nucleons, electrons, photons, and neutrinos – are sometimes described as cosmochemical. [118] Presumably, such a label will surprise most chemists.

The lack of a clear disciplinary identity may be part of the reason why geochemistry and cosmochemistry are not normally recognized as belonging to the chemical landscape; and this may again explain why they have been, with a few exceptions, ignored by historians of chemistry. Cosmochemistry is not really astronomy, not really physics, not really geology, and not even really chemistry. But of course traditional disciplinary boundaries and problems should not govern the historian's interests and priorities. I hope that I have indicated in this essay that we have in cosmochemistry a fascinating field that is in need of historical and conceptual analysis. Whether such analysis is made by historians of chemistry, or by historians with a different background, is irrelevant.

References and Notes

1 For aspects of interdisciplinarity and the emergence of new disciplines, including agricultural chemistry and physical chemistry as case studies, see G. Lemaine et al. (eds.), *Perspectives on the Emergence of Scientific Disciplines* (The Hague, 1976). See also the interesting analysis in T. Lenoir, *Instituting Science: The Cultural Production of Scientific Disciplines* (Stanford, 1997).

2 V. M. Goldschmidt, *Geochemistry* (Oxford, 1954), 10.

3 A fuller treatment is given in H. Kragh, "The chemistry of the universe: Historical roots of modern cosmochemistry," *Annals of Science* 57 (2000): 353–368.

4 H. C. Ørsted, "Reflections on the history of chemistry," in K. Jelved, A. Jackson, and O. Knudsen (eds.), *Selected Scientific Works of Hans Christian Ørsted* (Princeton, 1998), 243–260, on 253. Ørsted's essay was originally published in Danish and also appeared in *Journal für die Chemie und Physik* 3 (1807): 194–231.

5 There is no good history of geochemistry and the field is rarely included in histories of chemistry. For introductions, see E. Dudich, "From alchemy through geochemistry to cosmochemistry," *Organon* 24 (1988): 261–271; A. A. Manten, "Historical foundations of chemical geology and geochemistry," *Chemical Geology* 1 (1966): 5–31; and B. Fritscher and F. Henderson (eds.), *Toward a History of Mineralogy, Petrology, and Geochemistry* (Munich, 1998).

6 C. F. Schönbein, "Über die Ursache der Farbenveränderungen welche manche Körper unter dem Einfluss der Wärme erleiden," *Annalen der Physik und Chemie* 45 (1838): 263–281, on 281.

7 This is the subject of W. H. Brock, "Chemical geology or geological chemistry?" in L. J. Jordanova and R. S. Porter (eds.), *Images of Earth: Essays in the History of the Environmental Sciences* (Chalfont St. Gilles, England, 1979), 147–170, which focuses on the 1866–1868 debate between the geologist David Forbes and the geochemist Thomas Sterry Hunt. See also Kragh, "The chemistry of the universe."

8 F. W. Clarke, "The relative abundance of the chemical elements," *Bulletin of the Philosophical Society of Washington* 11 (1889): 131–142.

9 F. W. Clarke, "The data of geochemistry," *Bulletin of the U. S. Geological Survey*, no. 330 (1908). The data were extended in F. W. Clarke and H. S. Washington, "The composition of the earth's crust," U. S. Geological Survey, paper no. 127 (1924), which for a decade was the standard compilation of the geochemical composition of the lithosphere.

10 Quoted in B. Fritscher, "The fabrication of rocks: The geophysical laboratory and the production of modernity in mineralogy and geochemistry," 375–97 in Fritscher and Henderson, *Toward a History of Mineralogy, Petrology, and Geochemistry*, 381. On physical chemistry and American geochemistry around 1900, see also J. W. Servos, "The intellectual basis of specialization: geochemistry in America, 1890–1915," in J. Parascandola and J. Whorton (eds.), *Chemistry and Modern Society: Essays in Honor of Aaron J. Ihde* (Washington, D. C., 1983), 1–19; and J. W. Servos, "To explore the borderland: The foundation of the Geophysical Laboratory of the Carnegie Institution of Washington," *Historical Studies in the Physical Sciences* 14 (1983): 147–185.

11 Quoted in J. W. Servos, *Physical Chemistry from Ostwald to Pauling: The Making of a Science in America* (Princeton, 1990), 234.

12 For a comprehensive history of meteorite research, see J. G. Burke, *Cosmic Debris: Meteorites in History* (Berkeley, 1986). On early meteorite chemistry, see also M. F. Romig, "The scientific study of meteors in the 19th century," *Meteoritics* 3 (1966): 11–25; D. W. Sears, "Sketches in the history of meteoritics, 1: The birth of the science," *Meteoritics* 10 (1975): 215–225; D. W. Sears and H. Sears, "Sketches in the history of meteoritics, 2: The early chemical and mineralogical work," *Meteoritics* 12 (1977): 27–46, and W. Czegka, "Der Einfluss geochemischer Untersuchungsmethoden auf die Entwicklung der frühen Meteoritenkunde (1799–1803)," in Fritscher and Henderson, *Toward a History of Mineralogy, Petrology, and Geochemistry*, 177–206. Valuable as a general resource of meteoritics and cosmochemistry is also S. F. Mason, *Chemical Evolution: Origin of the Elements, Molecules, and Living Systems* (Oxford, 1992).

13 O. C. Farrington, "The constituents of meteorites," *Journal of Geology* 9 (1901): 522–532.

14 E. Baur, *Chemische Kosmographie* (Munich, 1903). From 1907–11, Baur was professor of physical chemistry in Brunswick, and since 1911 at ETH in Zurich. He published on a variety of topics ranging from geochemistry over electrochemistry to biochemistry.

15 Around the turn of the century "cosmic physics" was a popular attempt to bring astrospectroscopy, geophysics, meteorology, geomagnetism, and oceanography under

the unified domain of physics. The main work of the genre was S. Arrhenius, *Lehrbuch der kosmischen Physik* (Leipzig, 1903). In spite of Arrhenius's background, his massive two-volume work did not include chemical aspects to any extent. On Arrhenius and cosmic physics, see G. Holmberg, *Reaching for the Stars: Studies in the History of Swedish Stellar and Nebular Astronomy 1860–1940* (Lund, Sweden, 1999), 171–181.

16 For historical analysis of the tradition, see W. H. Brock, "Lockyer and the chemists: The first dissociation hypothesis," *Ambix* 16 (1969): 81–99; W. V. Farrar, "Nineteenth-century speculations on the complexity of the chemical elements," *British Journal for the History of Science* 2 (1965): 297–323; R. F. Hirsh, "The riddle of the gaseous nebulae," *Isis* 70 (1979): 197–212, and A. J. Meadows, *Science and Controversy: A Biography of Sir Norman Lockyer* (London, 1972). A broader perspective is provided in C. K. Jørgensen and G. B. Kauffman, "Crookes and Marignac - A centennial of an intuitive and pragmatic appraisal of 'chemical elements' and the present astrophysical status of nucleosynthesis and 'dark matter'," *Structure and Bonding* 73 (1990): 227–253.

17 W. Huggins, *The Scientific Papers of Sir William Huggins* (London, 1909), 49, in a footnote to a paper originally published in 1864.

18 Quoted in M. J. Crowe, *Modern Theories of the Universe. From Herschel to Hubble* (New York, 1994), 185.

19 Quotation from 1887, here cited from Holmberg, *Reaching for the Stars*, 50.

20 J. W. Nicholson, "The physical interpretation of the spectrum of the corona," *The Observatory* 36 (1913): 103–112, on 103 and 105. See also R. McCormmach, "The atomic theory of John William Nicholson," *Archive for History of Exact Sciences* 3 (1966): 160–185.

21 For Arrhenius as a cosmochemist, see his "Physikalisch-chemische Gesetzmässigkeiten bei den kosmisch-chemischen Vorgängen," *Zeitschrift für Elektrochemie* 28 (1922): 405–411.

22 W. Nernst, *Das Weltgebäude im Lichte der neueren Forschung* (Berlin, 1921). On Nernst's theory, see H. Kragh, "Cosmology between the wars: The Nernst-MacMillan alternative," *Journal for the History of Astronomy* 26 (1995): 93–115, where further references to the sources are given.

23 W. Nernst, "Physico-chemical considerations in astrophysics," *Journal of the Franklin Institute* 206 (1928): 135–142, on 138.

24 E. Rutherford, "The electrical structure of matter," *British Association for the Advancement of Science, Report* (1923): 1–24. This work is not included in J. Chadwick (ed.), *The Collected Papers of Lord Rutherford of Nelson*, 3 vols. (London, 1963).

25 Lord Rayleigh, "Some problems of cosmical physics, solved and unsolved," *British Association for the Advancement of Science, Report* (1929): 38–50, on 49.

26 W. D. Harkins, "The evolution of the elements and the stability of complex atoms," *Journal of the American Chemical Society* 39 (1917): 856–879, on 876 and 878. Together with his student Ernest Wilson, Harkins suggested in 1915 that the hydrogen-to-helium fusion was the source of stellar energy. W. D. Harkins and E. D. Wilson, "The structure of complex atoms. The hydrogen-helium system," *Journal of the American Chemical Society* 37 (1915): 1367–1383. For the historical context, see D. M. Siegel, "Classical-electromagnetic and relativistic approaches to the problem of nonintegral atomic masses," *Historical Studies in the Physical Sciences* 9 (1978): 323–360.

27 On Harkins, see R. S. Mulliken, "William Draper Harkins, 1873–1951," *Biographical Memoirs of Members of the National Acadamy of Science* 47 (1975): 49–81 and G. B. Kauffman, "William Draper Harkins (1873–1951): A controversial and neglected physical chemist," *Journal of Chemical Education* 62 (1985): 758–761. In a letter to Bertram Boltwood of February 28, 1921, Rutherford described Harkins as "moderately sound" and "a man of intelligence," but added that "I wish he did more experimenting and spent less time in theorising and in endeavouring to cover every possible idea." Quoted in L. Badash (ed.), *Rutherford and Boltwood: Letters on Radioactivity* (New Haven, 1969), 343.

28 Harkins's rule of the dissymmetry between even-odd nuclei is sometimes referred to as the Oddo-Harkins rule. In 1914 the Italian chemist Guiseppe Oddo suggested that elements with atomic weights a multiple of

that of helium (A = 4) were particularly stable, which would explain the marked abundance of elements such as carbon (A = 12) and oxygen (A = 16). G. Oddo, "Die Molekularstruktur der radioaktiven Atome," *Zeitschrift für anorganische und allgemeine Chemie* 87 (1914): 253–268.

29 W. Harkins, "The constitution and stability of atom nuclei," *Philosophical Magazine* 42 (1921): 305–339, on 310. See also W. Harkins, "Isotopes: Their number and classification," *Nature* 107 (1921): 202–203, which includes what is probably the first diagram of the abundance of isotopes as a function of the atomic number. Like all other physicists at the time, Harkins believed that atomic nuclei consisted of protons and electrons. The number of electrons corresponds to the quantity A-Z, later identified with the neutron number.

30 W. Harkins, "The periodic system of atomic nuclei and the principle of regularity and continuity of series," *Physical Review* 38 (1931): 1270–1288.

31 F. W. Aston, "Atomic species and their abundance on the earth," *Nature* 113 (1924): 393–395, on 394.

32 F. W. Aston, "The rarity of the inert gases on the earth," *Nature* 113 (1924): 786. For Aston's interest in isotopic geochemistry, see also Aston, *Isotopes* (London, 1923), where he praised Harkins's "very valuable discussion" (on 111). The history of the planetisemal theory is covered in S. G. Brush, *A History of Modern Planetary Physics*. Vol. 3. *Fruitful Encounters: The Origin of the Solar System and of the Moon from Chamberlin to Apollo* (Cambridge, 1996), 22–67.

33 G. N. Lewis, "The chemistry of the stars and the evolution of radioactive substances," *Publications of the Astronomical Society of the Pacific* 34 (1922): 309–319, on 309. For Lewis's continued interest in geo- and cosmochemistry, see G. N. Lewis, "The genesis of the elements," *Physical Review* 46 (1934): 897–901.

34 Quoted in R. H. Kargon, *The Rise of Robert Millikan: Portrait of a Life in American Science* (Ithaca, New York, 1982), 95 and 114.

35 See R. H. Stuewer, "The nuclear electron hypothesis," in W. R. Shea, *Otto Hahn and the Rise of Nuclear Physics* (Dordrecht, 1983), 19–68, on 22, who notes that "physical chemists were particularly attracted to nuclear speculation during this period and . . . their imaginations were particularly unbridled."

36 R. A. Sonder, "Zum Bau der Atomkerne," *Zeitschrift für Krystallographie und Mineralogie* 57 (1923): 611–641. A more elaborate geochemical discussion was presented in R. A. Sonder, "Über die Häufigkeitszahlen der Elemente und die Existenz eines neuen periodischen Systems," *Zeitschrift für anorganischen und allgemeine Chemie* 192 (1930): 257–285. Sonder claimed that his theory enabled him to predict the abundance distribution of the elements. When Ida and Walter Noddack published their data (n. 57 below), he argued that they supported his theory of a periodical abundance variation. See R. A. Sonder, "Über die Häufigkeitszahlen der Elemente und das Vorhandensein einer Kernperiodizität," *Die Naturwissenschaften* 18 (1930): 939–940. For more details, see H. Kragh, "An unlikely connection: Geochemistry and nuclear structure," *Perspectives in Physics* 2 (2000): forthcoming.

37 P. Niggli, "Das Magma und seine Produkte," *Die Naturwissenschaften* 9 (1921): 462–471, on 470.

38 P. Niggli, "Geochemie und Konstitution der Atomkerne," *Fennia* 50 (1928): no. 6, on 24. Nuclear and atomic physicists seem to have ignored the speculations of Sonder, Niggli, and the few other scientists who worked in the same tradition.

39 On Rutherford's theory, see R. H. Stuewer, "Rutherford's satellite model of the nucleus," *Historical Studies in the Physical Sciences* 16 (1986): 321–352.

40 R. C. Tolman, "Thermodynamic treatment of the possible formation of helium from hydrogen," *Journal of the American Chemical Society* 44 (1922): 1902–1908, on 1905.

41 S. Suzuki, "On the thermal equilibrium of dissociation of atom-nuclei," *Proceedings of the Physico-Mathematical Society of Japan* 10 (1928): 166–169.

42 S. B. Stone, "The origin of the chemical elements," *Journal of Physical Chemistry* 34 (1930): 821–841.

43 L. Farkas and P. Harteck, "Thermodynamische Bemerkungen zur Entstehung der Elemente," *Die Naturwissenschaften* 19 (1931): 705–706. The Austrian-born Paul

Harteck did important work in quantum- and radiochemistry and also played a leading role in Germany's atomic energy program during World War II. See P. F. Dahl, *Heavy Water and the Wartime Race for Nuclear Energy* (Bristol, 1999). After the war, when Harteck settled in the United States, he focused on atmospheric and interstellar chemistry. See A. Farkas, "Paul Harteck: The triumphant decade 1925–1934," *Ambix* 36 (1989): 91–102.

44 C. F. von Weizsäcker, "Über Elementumwandlungen im Innern der Sterne, II," *Physikalische Zeitschrift* 39 (1938): 633–646, translated in K. R. Lang and O. Gingerich (eds.), *A Source Book in Astronomy and Astrophysics 1900–1975* (Cambridge, Mass., 1979), 309–319. See also H. Kragh, *Cosmology and Controversy: The Historical Development of Two Theories of the Universe* (Princeton, 1996), 97–101.

45 H. C. Urey and C. A. Bradley, "On the relative abundances of isotopes," *Physical Review* 38 (1931): 718–724.

46 Quoted from Dudich, "From alchemy through geochemistry to cosmochemistry," 267. On Vernadsky's contributions, see I. V. Krut, I. M. Zabelin, and I. A. Fedoseyev, "Differentiation and integration of earth sciences in the work of Vernadsky," *Acta Historiae Rerum Naturalium Nec Non Technicarum* 14 (1981): 397–420 and K. E. Bailes, *Science and Russian Culture in an Age of Revolutions: V. I. Vernadsky and his Scientific School, 1863–1945* (Bloomington, Indiana, 1990).

47 This holds more for Fersman than for Vernadsky. One of Vernadsky's major works was published in French and in 1930 it was translated into German: V. I. Vernadsky, *La Géochimie* (Paris, 1924).

48 S. I. Tomkeieff, "Geochemistry in the USSR," *Nature* 154 (1944): 814–816.

49 Quoted in P. Krüger, "Victor Moritz Goldschmidt (1888–1947) und die Sowjetische Geochemie," *Zeitschrift für angewandte Geologie* 29 (1983): 347–355, on 352.

50 G. A. Tammann, "Tammann, Gustav Heinrich Johann Apollon," *Dictionary of Scientific Biography* 13 (1976): 242–248. C. E. Tilley, "Norman Levi Bowen," *Biographical Memoirs of Fellows of the Royal Society* 3 (1957): 7–22.

51 F. A. Paneth, "Die geochemischen Arbeiten von V. M. Goldschmidt," *Die Naturwissenschaften* 13 (1925): 805–809.

52 The best account of Goldschmidt's life and work is B. Mason, *Victor Moritz Goldschmidt: Father of Modern Geochemistry* (San Antonio, Texas, 1992). See also C. E. Tilley, "Victor Moritz Goldschmidt," *Obituary Notices of Fellows of the Royal Society* 6 (1948): 51–66 and *Applied Geochemistry* 3 (1988): 359–422, a commemorative issue on the centenary of Goldschmidt's birth.

53 Krüger, "Victor Moritz Goldschmidt." The establishment of geochemistry as a scientific discipline is discussed in M. Guntau, "Zur Entstehung der Geochemie als wissenschaftlicher Disziplin," in Fritscher and Henderson, *Toward a History of Mineralogy, Petrology, and Geochemistry*, 327–360.

54 V. M. Goldschmidt, "The distribution of the chemical elements," *Transactions of the Faraday Society* 25 (1929): 73–86.

55 G. Hevesy, *Chemical Analysis by X-rays and its Application* (New York, 1932), 298. See also G. Arrhenius and H. Levi, "The era of cosmochemistry and geochemistry, 1922–1935," in G. Marx (ed.), *George de Hevesy Festschrift* (Budapest, 1988), 11–36.

56 G. Hevesy and H. Levi, "Artificial radioactivity of dysprosium and other rare earth elements," *Nature* 136 (1935): 103. G. Hevesy and H. Levi, "Action of slow neutrons on rare earth elements," *Nature* 137 (1936): 185.

57 I. Noddack and W. Noddack, "Die Häufigkeit der chemischen Elemente," *Die Naturwissenschaften* 18 (1930): 757–764. I. Noddack, "Die Häufigkeit der seltenen Erden in Meteoriten," *Zeitschrift für anorganische und allgemeine Chemie* 225 (1935): 337–364. Until 1926, Ida Noddack carried her maiden name, Tacke.

58 I. Noddack and W. Noddack "Die Geochemie des Rheniums," *Zeitschrift für physikalische Chemie* A 154 (1931): 207–244. I. Noddack and W. Noddack, *Aufgaben und Ziele der Geochemie* (Freiburg, 1937).

59 F. A. Paneth, *The Origin of Meteorites* (Oxford, 1940). See also H. Dingle and G. R. Martin (eds.), *Chemistry and Beyond: A Selection from the Writings of the Late Professor F. A. Paneth* (New York, 1964).

60 A. Dauvillier, *Cosmologie et Chimie* (Paris, 1955). Dauvillier was trained in physics and his book focused on the physical and astronomical rather than the chemical features of the universe.

61 W. Brøgger to Nobel Committee, December 20, 1928, Pro Memoria 1929. Nobel Archive, Stockholm. I thank the Royal Swedish Academy of Sciences for access to the Nobel sources.

62 W. Hieber to Nobel Committee, January 10, 1930, Pro Memoria 1930. Nobel Archive.

63 Report, May 26, 1934, 20 pp. Pro Memoria 1934. Nobel Archive.

64 For a summary of Goldschmidt's results shortly after his arrival in Göttingen, see V. M. Goldschmidt, "Geochemische Verteilungsgesetze und kosmische Häufigkeit der Elemente," *Die Naturwissenschaften* 18 (1930): 999–1013.

65 Goldschmidt, *Geochemistry*, 69. On Goldschmidt's cosmochemical program, see H. E. Suess, "V. M. Goldschmidt and the origin of the elements," *Applied Geochemistry* 3 (1988): 385–391.

66 V. M. Goldschmidt, "Geochemische Verteilungsgesetze der Elemente IX. Mengenverhältnisse der Elemente und der Atomarten," *Skrifter av det Norske Videnskaps-Akademi i Oslo, Mat.-Nat. Klasse*, 1938, no. 4, and V. M. Goldschmidt, "The principles of distribution of chemical elements in minerals and rocks," *Journal of the Chemical Society* (1937): 655–673, of which the first work was promptly translated into Russian. See also Goldschmidt, *Geochemistry*, published posthumously under the editorship of Alex Muir.

67 V. M. Goldschmidt, "Probleme und Methoden der Geochemie," *Gerlands Beiträge zur Geophysik* 15 (1926): 38–50. Goldschmidt's suggestion received theoretical support with the first quantum-mechanical calculation of stellar nuclear processes which indicated that elements with atomic numbers 3, 4, and 5 were difficult to build up by proton bombardment. R. d'E. Atkinson and F. G. Houtermans, "Zur Frage der Aufbaumöglichkeiten in Sternen," *Zeitschrift für Physik* 54 (1929): 656–665.

68 Goldschmidt, "Geochemische Verteilungsgesetze der Elemente," 131.

69 C. F. von Weizsäcker, "Über Elementumwandlungen im Innern der Sterne, I," *Physikalische Zeitschrift* 38 (1937): 176–190, where the German physicist attempted to build up the lighter elements from proton capture reactions and the heavier elements from neutron capture reactions.

70 S. Chandrasekhar and L. Henrich, "An attempt to interpret the relative abundances of the elements and their isotopes," *Astrophysical Journal* 95 (1942): 288–298, on 294.

71 For these details, see Kragh, *Cosmology and Controversy*, 109–132 and 295–305, R. J. Tayler, "The origin of the elements," *Reports on Progress in Physics* 29 (1966): 490–538, and R. V. Wagoner, "Deciphering the nuclear ashes of the early universe: A personal perspective," in Bruno Bertotti et al. (eds.), *Modern Cosmology in Retrospect* (Cambridge, 1990), 159–185.

72 H. Brown, "A table of relative abundances of nuclear species," *Reviews of Modern Physics* 21 (1949): 625–634.

73 H. E. Suess and H. C. Urey, "Abundances of the elements," *Reviews of Modern Physics* 28 (1956): 53–74. Of course, the data have been continually updated and since 1956 a series of new tables have appeared.

74 For further details, see Kragh, "An unlikely connection: Geochemistry and nuclear structure."

75 For the development of the nuclear shell model, see P. Zacharias, "Zur Entstehung des Einteilchen-Schalenmodells," *Annals of Science* 28 (1972): 401–411; K. E. Johnson, "Independent-particle models of the nucleus in the 1930s," *American Journal of Physics* 60 (1992): 164–171 (1992); and M. Mladjenovic, *The Defining Years in Nuclear Physics, 1932–1960s* (London, 1998), 287–305. As mentioned in Section 3, Niggli found in 1921 that elements with Z = 28, 48, 50, 80, and 82 were particularly frequent in the earth's crust. This may be considered an anticipation of the "magic numbers" that later turned up in nuclear theory.

76 M. Goeppert Mayer, "The shell model," in *Nobel Lectures. Physics, 1963–1970* (Amsterdam, 1972), 20–37, on 21. For the polyneutron theory, see Kragh, *Cosmology and Controversy*, 123–125.

77 Suess, Jensen, and Harteck were in Norway in order to produce heavy water for the German uranium project. See Dahl, *Heavy Water and the Wartime Race for Nuclear Energy*, 162–164.

78 Suess, "V. M. Goldschmidt and the origin of the elements."

79 H. E. Suess, "Über kosmische Kernhäufigkeiten, I: Einige Häufigkeitsregeln und ihre Anwendung bei der Abschätzung der Häufigkeitswerte für die mittelschweren und schweren Elemente," *Zeitschrift für Naturforschung* 2a (1947): 311–321.

80 Goldschmidt, "Geochemische Verteilungsgesetze der Elemente," 141–148.

81 M. Saha, "Ionization in the solar chromosphere," *Philosophical Magazine* 40 (1920): 472–488, on 476. Reprinted in Lang and Gingerich, *A Source Book in Astronomy and Astrophysics 1900–1975*, 236–242.

82 M. Saha, "On the temperature ionization of elements of the higher groups in the periodic classification," *Philosophical Magazine* 44 (1922): 1128–1139.

83 R. C. Kenat and D. H. DeVorkin, "Quantum physics and the stars III: Towards a rational theory of stellar spectra," *Journal for the History of Astronomy* 21 (1990): 157–186; K. Hufbauer, *Exploring the Sun: Solar Science since Galileo* (Baltimore, 1991), 101–106.

84 For a historical survey of stellar abundance analysis, see J. B. Hearnshaw, *The Analysis of Starlight: One Hundred and Fifty Years of Astronomical Spectroscopy* (Cambridge, 1986), 428–464.

85 A. Unsöld, *Physik der Sternatmosphären, mit besonderer Berücksichtigung der Sonne* (Berlin, 1938).

86 B. Strömgren, "On the chemical composition of the solar atmosphere," 218–257 in *Festschrift für Elis Strömgren* (Copenhagen, 1940), on 257. See also B. Strömgren, "Scientists I have known and some astronomical problems I have met," *Annual Review of Astronomy and Astrophysics* 21 (1983): 1–11.

87 P. Swings, "Cometary spectra," *Monthly Notices of the Royal Astronomical Society* 103 (1943): 86–111.

88 R. Wildt, "Electron affinity in astrophysics," *Astrophysical Journal* 89 (1939): 295–301, reprinted in Lang and Gingerich, *A Source Book in Astronomy and Astrophysics 1900–1975*, 264–267.

89 Hearnshaw, *The Analysis of Starlight*, 404.

90 R. Wildt, "Photochemistry of planetary atmospheres," *Astrophysical Journal* 86 (1937): 321–336. See also the review in R. Wildt, "The geochemistry of the atmosphere and the constitution of the terrestrial planets," *Reviews of Modern Physics* 14 (1942): 151–159.

91 R. Wildt, "Cosmochemistry," *Scientia* 67 (1940): 85–90, reprinted in H. Shapley (ed.), *Source Book in Astronomy 1900–1950* (Cambridge, Mass., 1960), 394–401.

92 R. Wildt, "Reports on the progress of astronomy: The constitution of the planets," *Monthly Notices of the Royal Astronomical Society* 107 (1947): 84–102, on 97.

93 S. H. Manian, H. C. Urey, and W. Bleakney, "An investigation of the relative abundance of oxygen isotopes $O^{16} : O^{18}$ in stone meteorites," *Journal of the American Chemical Society* 56 (1934): 2601–2609.

94 R. E. Doel, *Solar System Astronomy in America: Communities, Patronage, and Interdisciplinary Research, 1920–1960* (Cambridge, 1996), 88–90.

95 H. C. Urey, "The origin and development of the earth and other terrestrial planets," *Geochimica et Cosmochimica Acta* 1 (1951): 209–277, 263–268. H. C. Urey, *The Planets: Their Origin and Development* (New Haven, 1952). On Urey and chemical cosmogony, see Brush, *A History of Modern Planetary Physics*, 144–159 and Doel, *Solar System Astronomy in America*, 92–108. See also the contributions in H. Craig, S. L. Miller, and G. J. Wasserburg (eds.), *Isotopic and Cosmic Chemistry. Dedicated to Harold C. Urey on his Seventieth Birthday* (Amsterdam, 1964).

96 Urey, "The origin and development of the earth and other terrestrial planets," 232.

97 A. Eucken, "Physikalisch-chemische Betrachtungen über die früheste Entwicklungsgeschichte der Erde," *Nachrichten, Akademie der Wissenschaften zu Göttingen, Math.-Phys. Klasse* (1944): 1–25; A. Eucken, "Über den Zustand des Erdinnern," *Die Naturwissenschaften* 32 (1944): 112–121.

98 Urey, "The origin and development of the earth and other terrestrial planets," 274.

99 Quoted in Doel, *Solar System Astronomy in America*, 143, where the controversy is described in detail.

100 D. Seeley and R. Berendzen, "The development of research in interstellar absorption, c. 1900–1930," *Journal for the History of Astronomy* 3 (1972): 52–64.

101 P. Swings and L. Rosenfeld, "Considerations regarding interstellar molecules," *Astrophysical Journal* 86 (1937): 483–488.

102 G. Herzberg, "Historical remarks on the

discovery of interstellar molecules," *Journal of the Royal Astronomical Society of Canada* 82 (1988): 115–127.

103 For the connection between the CN-prediction and the cosmic background radiation, see Kragh, *Cosmology and Controversy*, 134–135, 345–347.

104 For the period until 1971, see D. M. Rank, C. H. Townes, and W. J. Welch, "Interstellar Molecules and Dense Clouds," *Science* 174 (1971): 1083–1101.

105 Noddack and Noddack, *Aufgaben und Ziele der Geochemie*, 3.

106 Quoted from Mason, *Victor Moritz Goldschmidt*, 101–102. One should keep in mind that Goldschmidt's audience consisted of geologists, not chemists.

107 K. Rankama, "What is geochemistry?" *American Journal of Science* 245 (1947): 458–461, on 458. Rankama held a position as Research Associate at the University of Chicago.

108 Rankama, "What is geochemistry?" 460.

109 Rankama, "What is geochemistry?" 461.

110 C. R. Cowley, *An Introduction to Cosmochemistry* (Cambridge, 1995), 1.

111 K. Rankama and T. G. Sahama, *Geochemistry* (Chicago, 1950), 6.

112 Rankama and Sahama, *Geochemistry*, 14. Cosmochemical and -physical aspects appeared prominently in the book, which included not only a substantial chapter on nuclear physics but also discussed very recent astrophysical and cosmological works in relation to the formation and distribution of chemical elements. Interestingly, the authors referred approvingly to the new big bang theory of Gamow and his collaborators.

113 *Geochimica et Cosmochimica Acta* 1 (1950): 1.

114 The first journal devoted entirely to meteorite research seems to have been *Meteoritika*, a Russian periodical that was published irregularly since 1939. In 1944 it was followed by the Dutch *De Meteoor*, and in 1953 appeared *Meteoritics*, the journal of the (American) Meteoritical Society founded twenty years earlier. For the history of this society, see Ursula B. Marvin, "The Meteoritical Society: 1933 to 1993," *Meteoritics* 28 (1993): 261–314.

115 E. Ingerson, "The Geochemical Society," *GeoTimes* 6, no. 7 (1962): 8–14, 39.

116 H. Brown, "The composition of our universe," *Physics Today* 3 (April 1950): 6–13, on 11. However, not all cosmochemists may agree in Goldschmidt's paternity. Thus, in a recent encyclopaedia it is stated that "It is generally acknowledged that the emergence of cosmochemistry as a separate discipline stems from the seminal work by Urey, Suess, and Brown in the late 1940s and early 1950s." B. Fegley, "Cosmochemistry," in J. H. Shirley and R. W. Fairbridge (eds.), *Encyclopedia of Planetary Sciences* (London, 1997), 169–177, on 169.

117 Cowley, *An Introduction to Cosmochemistry*, 3.

118 As in R. A. Alpher, "Cosmochemistry and the early universe," in E. Harper, W. C. Parke, and G. D. Anderson (eds.), *The George Gamow Symposium* (San Francisco, 1997), 50–68.

Part III
Solid State Chemistry and Biotechnology

10.
Between the Living State and the Solid State: Chemistry in a Changing World

Peter J. T. Morris

The history of materials science has been characterized by co-operation and conflict. On one hand, understanding and controlling the complex behavior of materials has necessitated collaboration between different branches of chemistry and across different scientific disciplines. In time, different sub-disciplines have fused to form the new discipline of materials science. Academia and industry have worked together to develop better materials and improved versions of existing ones. On the other hand, there have been disputes about the nature of polymers, priorities, and patents. The chapters in this section give us a good overview of these opposing characteristics of materials science. Yasu Furukawa describes the origins of polymer science from its two parent disciplines, organic chemistry and chemical physics. Herman Mark was a key player in the creation of polymer science and Mary Jo Nye examines the career of Mark's colleague, Michael Polanyi. In the period between 1914 and 1932, Polanyi carried out path-breaking work on the theory of absorption and on the X-ray diffraction of solids. In both cases, however, his contributions were overshadowed by others and disillusioned, Polanyi turned increasingly towards the social sciences and the philosophy of science. After World War II, the well established discipline of metallurgy was transformed and came much closer to the theory and methods of polymer science. The two disciplines now shared a common interest in the relationship between the structure of materials, their properties and how they were processed. Bernadette Bensaude-Vincent traces the formation of materials science in the context of the development of composite materials and the introduction of biomimetics. Materials science is certainly new, even if its roots can be traced back thousands of years. In chapter 11, Nicolas Rasmussen argues against the apparent novelty of biotechnology, by showing how "the biotech revolution" began with the development of hormone chemistry (including vitamins and auxins) in the 1930s, followed by penicillin a decade later.

10.1
Biotechnology and the Myth of a Recent "Biotech Revolution" [1]

Chapter 11 concerns episodes within that disputed territory which lies between chemistry and biology, whether described as physiology, biochemistry, or molecular biology. His chapter confronts the widely accepted interpretation of the history of biology in this century which is in terms of discontinuity and revolution. He argues instead that the biotechnology revolution is a myth fostered by exaggeration and the desire to attract attention, recognition, and reward, whether in the market place of venture capitalism or in academia. Early in his chapter, Rasmussen sets out very clearly three historical arguments, each of which attacks popular assumptions concerning the "Biotechnology Revolution." Using as examples, the careers of two plant physiologists – James Bonner at Caltech and Ezra Kraus at the University of Chicago – he shows how projects in biotechnology today had their origins in the 1930s, and how these projects involved collaborative work between academics and industrial concerns. Also he claims that "efforts to reengineer life" go back to the inter-war period.

The first two points are established admirably on the basis of the chosen examples. As for reengineering life, clearly this depends on how it is to be defined. Achieving this by conventional breeding techniques takes us back into the distant mists of history. Using artificial mutagenesis puts us squarely in the year 1944 when, as Rasmussen points out, OSRD contracts were awarded to George Beadle and Milislav Demerec to develop highly penicillin-productive strains of the mould. However, the use of the term engineering in the phrase "engineering life" is usually intended to refer to the manipulation of an organism by such means as allow one to predict the outcome (or at least to have a high degree of confidence in such prediction). The intervention is in this sense "rational", not a case of hit-and-miss, after which one has to select the hopeful results. The gene associated with the desired effect is introduced from a donor organism in which that effect is produced. In the absence of such techniques as we have in recombinant DNA technology, scientists in the forties and fifties bemoaned the fact that it was not possible to make "designer drugs". Rasmussen recognizes the significance of this point as far as concerns the public image of the "Biotech Revolution". Along with the recently acquired ability to patent life forms, this manipulative power, he explains, "struck a deep cultural nerve in a different way from [the] hormones and wonder drugs [of the 1930 and 1940s]."

There are two more aspects to the "Biotech Revolution" that perhaps deserve more comment. The first is the form of academic-industry liaison. It is significant that Bonner did not form an industrial company until 1980 and that there is no mention in Rasmussen's chapter of Kraus ever having formed one. In other words, academic collaboration with existing industrial companies has a long history – one that goes back to organic chemists in German universities in the nineteenth century. But how old is the custom of academic biologists forming their own "venture capital" companies and subsequently selling them to established industrial concerns? The second point concerns the scale of academic-industrial activity. As Rasmussen suggests, "the Venn diagram has become more densely populated since 1945," but it would be interesting to have some figures. Even if the reader does not go as far as Rasmussen in denying the revolutionary nature of current biotechnology, the reasons he gives for the popularity of the standard view by both supporters and opponents of the technology makes a lot of sense. His chapter very successfully draws our attention to a neglected series of episodes in plant physiology which are highly relevant to the broad history of biotechnology.

10.2 Polymer Science

Yasu Furukawa shows in his chapter how the study of polymers became interdisciplinary by examining its origins in Germany and America. He begins by examining the reaction of Hermann Staudinger against the prevailing physicalist colloid (or micellar) model of polymers between 1917 and 1930. Because his opponents used the term "polymer" loosely, Staudinger created a new word "macromolecule" to create a clear distinction between his concept of very large molecules held together by valence bonds and any other model for polymers. Although there are scientific arguments in favor of macromolecule (not all large molecules are polymers), polymer was always the preferred term in the English-speaking world. This was largely owing to the influence of Herman(n) Mark, a polymer chemist who had worked with chemists of the colloid school and who was always distrusted by Staudinger. The term "polymer science" stemmed from Mark's own interests in both chemistry and physics. Staudinger was entirely an organic chemist (with biological and botanical leanings). By contrast, Mark was a physical chemist by training, and in Vienna in the 1930s, and from 1940 in Brooklyn, Mark taught polymer chemistry as a combination of organic chemistry, physical chemistry, and physics. Physical studies of polymers (however understood) had existed since 1802, but polymer physics was a product of the 1930s and 1940s, a result of Mark's own research program and the entry into this new field of young chemical physicists eager to use their mathematical skills. Industry had began to encourage collaboration between organic and physical chemists; and this process was accelerated during World War II by the synthetic rubber research program in the United States. When the synthetic rubber program started in 1942, the large rubber companies already had rudimentary research labs and, in the absence of trained polymer scientists, they used a mixture of chemists and physicists. U. S. Rubber was very "physical": of their 18 PhDs, nine were in physics and physical chemistry, against only three in organic chemistry. At the other extreme, of the twenty PhDs working for Goodrich, at least ten were in organic chemistry, against only five in physical chemistry and none in physics. Goodyear was in-between: of their fourteen PhDs, four were in organic chemistry and five in physical chemistry, colloid chemistry, and physics.

By the 1960s, organic chemistry and chemical physics were in Furukawa's words, "complementary partners in perfecting the science of polymers". The relationship between polymer chemistry and biomolecular chemistry, was more limited;

Staudinger was interested in biology and Mark encouraged young scientists to study the structure of proteins. It might be remarked in passing that the coverage of biomolecules in polymer science textbooks is often rather perfunctory, bearing in mind the importance of the topic, and could even be described as paying lip-service to the relationship between polymer science and molecular biology. In the first of two short concluding sections, Furukawa briefly surveys the maturing of polymer science in the 1960s, showing how its increasingly specialized nature brought it into potential conflict with mainstream chemistry. For the evolution of materials science from polymer chemistry and metallurgy in the decades that followed, we have to turn to the chapter by Bernadette Bensaude-Vincent.

Finally, Furukawa records the ultimately futile attempts of Staudinger's widow, Magda Staudinger-Woit, to preserve the term macromolecular chemistry. When a monument was erected to Staudinger in Freiburg recently, with an accompanying booklet, the resident professor objected to any alteration of "polymer science" in the booklet to "macromolecular chemistry". With his German background, he obviously saw the former as something more modern and relevant, but as Furukawa demonstrates, "polymer science" is as old as "macromolecular chemistry".

Despite his own achievements, Furukawa's chapter indicates the need for future research. How did chemists and physicists work together in the 1940s and 1950s to create a new unified discipline of polymer science? It would also be useful to look at how the discipline was taught in the period between 1940 and 1965, especially in the United States. To do this, one would need to examine which courses were taught, the launch of new journals and the evolution of textbooks across various editions. As the polymer science in America started in a few key institutions, notably Brooklyn Polytechnic, it would also be valuable to look at the diffusion of the new discipline from these "seed" institutions, tracing the careers of the early alumni and co-workers such as Charles Overberger.

10.3
At the Boundaries

Mary Jo Nye's chapter is very different from the other three, in that it is concerned with the career in physical chemistry of one individual, Michael Polanyi, rather than the development of an interdisciplinary field. She begins with Polanyi's views on scientific practice, notably his distrust of the idealization of scientific detachment, which arose from his own experiences. After looking at his education in Karlsruhe, Budapest, and Berlin, Nye turns to his first major research topic, the potential theory of adsorption. Polanyi developed a theory of gas adsorption based on the concept of potential gradients. His highly original ideas, later vindicated by quantum theory, were in competition with Irving Langmuir's much simpler monomolecular layer theory, for which Langmuir won the Nobel Prize in 1932. Although her treatment of this case study is admirable, Nye does not address all the possible aspects of her comparison between Langmuir and Polanyi. Specifically, how far was what we now call PR (public relations) responsible for Langmuir winning the Nobel Prize? It was a rather uneven contest between the head of research at a huge American corporation who socialized with celebrities such as Charles Lindbergh and an obscure Hungarian physician working in a Berlin reeling from war, a debased currency and revolution. Many scientists do everyday science and get very little credit for it. A few scientists do remarkable science, and receive ample credit. A cynic might add that a few scientists do remarkable PR and win the glittering prizes.

Nye then turns to Polanyi's brilliant work on the X-ray diffraction of polymers and solids in general. In this field he was more successful, although some of his ideas were also developed independently by other scientists. Polanyi's main problem in this area was lack of support from his colleagues in Berlin. This was partly because his ideas were too advanced for the 1920s, but also reflects Polanyi's inability (or at least unwillingness) to follow the current trends.

In her conclusion, Nye looks at "rewards and recognition in the scientific community", showing how Polanyi was in several ways a marginal figure who also had the misfortune to be ahead of his time. Michael Polanyi was upset that he did not get the credit for what he obviously felt was remarkable science, partly in his view because his research was not defined as remarkable science by his community. This demonstrates the danger of working between two disciplines, but operating between two disciplines or moving from one to another has been postulated as one of the best ways to win prizes (compare Michael Chayut's paper [2] on Paul Flory, to give just one example).

It is striking that Michael Polanyi failed to carry out any decisive experiments, particularly for his adsorption theory, but also for the X-ray crystallography of solids. In experimental science, and probably rightly, the credit goes to those who formulate the crucial test of a theory. In chemistry one thinks of Lavoisier, in physics Michelson and Morley, and in polymers Carothers. This, it seems, Michael Polanyi failed to do. Rather than formulate how his theory could be distinguished from Langmuir's and hence shown to be correct, Polanyi switched to other topics. By the time he came back to his theory in the late 1920s, Langmuir had won the field. Although his initial analysis was considered by Mark and others to be preternaturally brilliant, Polanyi also failed to use his grasp of X-ray crystallography to devise a crucial test for the macromolecular theory.

Furthermore, he also failed to distinguish his theory clearly from that of Langmuir, with fatal results for his chances of success. Instead of saying the monomolecular layer theory was false, Polanyi said that Langmuir's theory was a special case of his own theory. This failure was crucial, and is comparable to his colleague Herman Mark's tendency in polymer science to see the best in both sides of an argument and to go for compromises. In one respect, as Furukawa has shown brilliantly in his chapter and at greater length in his book, [3] Mark was successful: polymer science – his own term – is now largely based on Mark's formulation rather than Staudinger's. His friendly and helpful manner was universally admired, and he is fondly remembered by all who knew him, but it was Staudinger, not Mark, who won the Nobel Prize. Successful scientists like Staudinger tend to have a kind of tunnel vision; their view is right and everyone else has to be bludgeoned into submission. It is a ruthless, somewhat bleak, approach to science, but one that brings prizes to those who embrace it, even if their narrow view is partly incorrect as Staudinger's most certainly was. Sadly, the prizes often go to those who formulate the issues most clearly or at least shout the loudest!

10.4
A Composite Field of Research

In her chapter, Bernadette Bensaude-Vincent traces the evolution of materials science in three different ways. In her opening section, Bensaude-Vincent shows how the traditional discipline of metallurgy was changed by its contacts with the emerging science of solid-state physics and also X-ray crystallography. One can see parallels with the development of polymer science outlined in Furukawa's chapter.

In the second case study, Bensaude-Vincent examines the development of reinforced materials from no less than four different angles. The study of composites was linked to a growing interest in the relationship between structure, properties, and function, with the implication that reinforced plastics represented the triumph of function over structure. Bensaude-Vincent then examines the differences between composites and reinforced plastics. Although she gives us a good description of the development of composites, not all readers will accept her contention that there was a technological shift from reinforced plastics to a more advanced, and by implication different, class of materials, called composites. Certainly modern composites are more advanced than the glass-fiber-reinforced plastics of the 1950s, but that is hardly surprising. The polyethylene of the 1990s is a great improvement on the polyethylene of the 1950s. To be sure, high-performance composites have transformed a field that was once dominated by small boats and home kits for hobbyists, but this is only an extreme case of developments that have taken place across the entire field of polymers and, indeed, other materials. Bensaude-Vincent then fleshes out her analysis of the development of composites by examining the creation of new reinforcing fibers, notably the aramids. As the engineering demands on composites grow and their manufacture becomes ever more complex, Bensaude-Vincent traces a shift away from chemicals towards a more generalized engineering-industrial R&D complex.

In her third case study, Bensaude-Vincent looks at one of the key projects of modern materials science: biomimetics, learning from nature. After decades of assuming that they could do better than nature, as in the case of synthetic rubber, materials scientists in the 1990s came to appreciate that nature can do many things better than we can. This appreciation partly came about because many biomaterials are structurally complex composites, in tune with current materials research and development. The investigation of biological systems by materials scientists also revealed the importance of self-assembly. The development of biomimetics involved collaboration with biologists, thus extending the degree of multidisciplinarity in an already heterogeneous field even further.

Bensaude-Vincent traces the evolution of materials science largely through technological developments. There are alternative ways to explore this process, for instance, by examining the definitions of the field given in undergraduate textbooks. Certainly there was a time in the past, in the late 1960s, when "materials sciences" (note the plural) was a loose umbrella term, but by the 1990s, they have now coalesced into one discipline, a process mirrored in the succeeding forewords of one influential British textbook: [4]

> *First edition* (1969)
> "The study of the science of materials has become in recent years an integral part of virtually all university courses in engineering. The physicist, the chemist and metallurgist may, rightly, claim that they study materials scientifically, but the reason for the emergence of the 'new' subject of materials science is that it encompasses all these disciplines. We hope that, in addition to providing for the engineer an introductory text on the structure and properties of engineering materials, the book will assist the student of physics, chemistry, or metallurgy to comprehend the essential unity of these subjects under the all-embracing, though ill-defined, title 'Materials science'."

> *Fourth edition* (1990):
> "Since 1985 ... there have been rapid and significant developments in materials science."

Similarly, the 1994 edition of a widely used American textbook states [5]:

> "The discipline of *materials science* [emphasis in original] involves investigating the relationships that exists between the structures and properties of materials."

How did the different disciplines (principally metallurgy and polymer science) became one discipline, especially in the period between 1965 and 1990? Clearly it has much to do with the development of new materials, but also with the creation of new materials for the burgeoning electronics and aerospace industries, hence as a service discipline. It also has much to do with the new physical theories of the solid-state that were developed in the 1940s and 1950s and with the development of new expensive instrumentation, most notably electron microscopes, that were needed by all the fusing disciplines. So it is partly an account of access to new instrumentation and techniques and a pragmatic attempt to access the resources to pay for them, although leading metallurgy faculties such as Imperial College (part of the University of London) had several electron microscopes by the mid-1960s.

To be sure, the development of materials science was closely associated with the Cold War (as the synthetic rubber program demonstrates) and the military-industrial complex. This is clearly illustrated by the close collaboration between the Royal Aircraft Establishment at Farnborough and Courtaulds in the development of carbon fibers for composites. The fusion of the different disciplines was encouraged by companies who had long abandoned such disciplinary boundaries themselves. But it was also brought about by the need for metallurgy faculties and societies to meet the challenge from new intellectual competitors, by rebranding themselves as more modern and thus relevant "materials science" faculties, and as Institutes of Materials.

The fusion of the various disciplines was gradual; most faculties went through an intermediate stage in the 1970s and early 1980s (as at Imperial College) of being faculties of metallurgy and materials science to avoid frightening away potential metallurgy students. The new faculties were very largely based on the former metallurgy faculties, perhaps reflecting their long standing (at Imperial College it goes back to the mid-nineteenth century) compared with polymer science. It is to be hoped that a historian will sooner or later carry out a close study of how academic departments, independent research units and learned societies rebranded themselves and merged in the crucial period between 1960 and 1990, examining the courses these institutions offered, their research programs, and the evolution of textbooks in this field.

10.5 Conclusion

Bernadette Bensaude-Vincent concludes her chapter with a question: "[is there] a future for chemists?" All the areas covered in this section began in chemistry in one way or another, but chemistry is in danger of being squeezed out of the picture as the fields of materials science and biotechnology mature. Chemists have been adept at retaining a foothold in both fields by being flexible and emphasizing the advantages chemistry can bring to their development. Worldwide, chemical societies and chemistry faculties are scrambling to retain a viable position in both areas. This process involves a redefinition of chemistry that is both broad in scope but also sees it as a service discipline. Ultimately, chemistry may fracture along the fault-line between the solid state and the living state, and thus become part of two "mega-sciences", materials science on one hand and biomolecular sciences on the other. While this process is a result of current developments, it would also echo chemistry's medieval roots in assaying and medicine.

References and Notes

1 The comment on Nicolas Rasmussen's paper has been written by Robert Olby, whose generous willingness to contribute to this introduction is gratefully acknowledged.

2 Michael Chayut, "New sites for scientific change: Paul Flory's initiation into polymer chemistry," *Historical Studies in the Physical Sciences* 23 (1993): 193–218.

3 Yasu Furukawa, *Inventing Polymer Science: Staudinger, Carothers and the Emergence of Macromolecular Chemistry* (Philadelphia: University of Pennsylvania Press, 1998).

4 Prefaces to first and fourth editions of J. C. Anderson, K. D. Leaver, R. D. Rawlings and J. M. Alexander, *Materials Science*, fourth edition (London: Chapman & Hall, 1990).

5 William Callister, *Materials Science and Engineering: An Introduction*, third edition (New York: John Wiley & Sons, 1994), 2.

11. Biotechnology Before the "Biotech Revolution": Life Scientists, Chemists, and Product Development in 1930s–1940s America

Nicolas Rasmussen

It is often said, in the many headlines the field has attracted over the past two decades, that the "rise of the biotechnology industry" is transforming medicine, agriculture, and other fields beyond recognition. Biotechnology is even sometimes placed in the same class of industrial innovation as the integrated assembly line or the steam engine. But what exactly is this new industrial revolution about? Purportedly, with recombinant DNA, rational manipulation of living things on an industrial scale became possible, for the first time permitting firms to operate in medical, agricultural, and chemical markets simultaneously. A recent *Science* magazine article by a Harvard business scholar captures this notion of the so-called "life science industry" today (Figure 11.1). When molecular biological knowledge was transferred from academia to industry in the late 1970s, other commentators say, the impact on wider society was likewise revolutionary, if somewhat sinister. Every life scientist capable of forming a relationship with a pharmaceutical or chemical company interested in DNA quickly did so. Universities, in an age of shrinking state support for basic science, eagerly encouraged mutually profitable ties between their life scientists and industry. Life scientists suddenly became entrepreneurial dealers in expertise, the story goes; the commons of biological knowledge were rapidly fenced. According to such alarmist analyses, society thus lost impartial experts to consult in policy matters, since all practitioners of molecular biology now have vested interests. Furthermore, academia itself as a haven for impartial knowledge has been fundamentally corrupted by erosion of its boundaries with industry. [1]

There is much that can be questioned about this story, let alone the many other types of fears (and hopes) that the new biotechnologies have engendered. Here I focus on the historical premises of these particular claims about biotechnology's novelty as a commercial endeavor to engineer the living world, and about the connections between academic life science and industry that are supposed to have recently emerged in order to make such ambitions reality. As I will be contending in the case of the United States (though similar arguments might be made about certain European nations as well), industrial efforts to re-engineer life are not new; such efforts were already organized over half a century ago, in the interwar period –

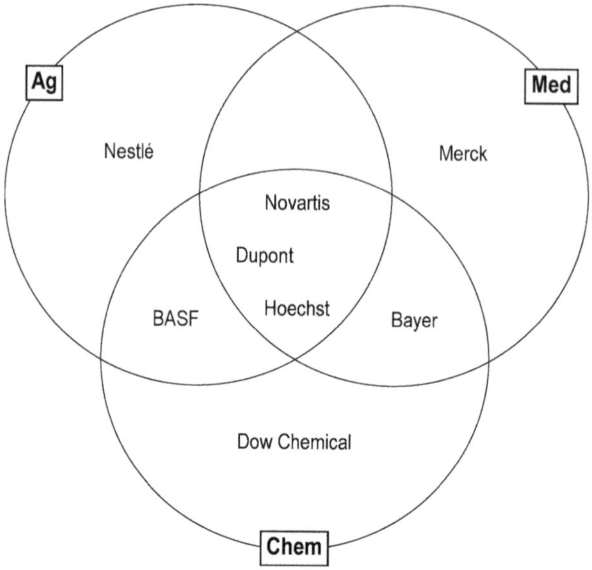

Figure 11.1 "The Life Science Industry" of the late 1990s, as conceived by a business analyst (redrawn from Juan Enriquez, "Genomics and the World's Economy," Science, 281 (1998): 925–926.

or a full century ago if we were to count agricultural genetics projects under the "engineering of life" rubric (which I will not for present purposes). I will be making three related arguments concerning early twentieth century industrial biology in America. First, not only were collaborative projects between academic life scientists and industry rather common, but in the basic nature of the exchange of goods underpinning them, and even in some respects of their form, these collaborations resembled those prevalent today more closely than is generally appreciated. Second, many of the same projects occupying today's genetic engineers were already underway in the 1930s. Some that bore fruit through the technology of an earlier day are now being carried further, while others that were abandoned have now been taken up again. Taken together, these two arguments amount to the following claim: firms with advanced life science expertise were already operating in medical, agricultural, and chemical arenas simultaneously; such firms were already working closely with academic life scientists; and they were doing much the same sorts of things as today's occupants of this economic space.

This then is a case for continuity, for evolution as opposed to revolution. In order to make it, I present several short vignettes about industrial biology projects in the 1920s-1940s from my own research and that of others, using them to populate my own "Venn" diagram. There is not space here to develop the full case for continuous development of technological projects from the interwar period right through to the 1980s. Rather, I will simply attempt to show that, in the 1920s and especially 1930s, there were academic life scientists with industrial projects, and with industrial connections, quite similar to those regarded as typical of the post-1980 era of genetic engineering. The aim in describing forerunners is not to deny change, or to deprive past science of its qualitative distinctness. I would never wish to deny that certain features of today's genetic engineering based biotechnology – particularly those related to the capacity of the living "products" of genetic engineering to reproduce themselves – raise dangers, and challenge deep cultural assumptions, differently from biotechnology of the past. Nor, as a matter of economic history, would I deny that there may be a recent quantitative increase in activity at the center of the "Venn" diagram. However, by recognizing that the pattern of commercialization of life science today has its roots or at least foreshadowings in the interwar period, we can begin to ask questions about the structural reasons underlying this pattern which would be inconceivable if we take it for granted that today's "biotechnology industry" is radically new in the ways at issue here.

My third argument, in the conclusion, proposes an explanation of what I see as the myth of a *recent* biotechnological revolution in life scientist-industrialist relations. This argument concerns historiographic issues, because underwriting the picture of recent revolution is a narrative shared by many historians of biology as well as both critics and advocates of genetic engineering, that treats molecular genetics as coextensive with molecular biology, and views current biotechnology as the triumphant arrival of molecular genetics from the lonely wilderness in which it was born in the 1930s. Certainly, it does follow that if "biotechnology" is defined strictly as the application of genetic engineering, then molecular genetics is its main ancestor, and it was not born until the 1970s. However, such a definition of biotechnology is too narrow even to accommodate universally included items such as monoclonal antibodies. But if we expand our definition of "biotechnology" to something like "the use of biological science to intervene in life processes on an industrial scale," we stand to gain a perspective that makes similarities between current biotechnology and that of half a century past clearer, and we also discover that, contrary to some historical opinion, the nascent molecular genetics of the 1930s was not particularly special among life sciences of the day in its technological orientation. Thus I shall throughout use "biotechnology" in this more inclusive, [2] if vague sense. Nevertheless, no particular definition of "biotechnology" is required for my arguments concerning the similarity of industrialist-biologist collaborations and the nature of the projects underway, in the first half of the century and in the current day. In the conclusion I make some suggestions as to the political reasons that may explain why the sort of interpretation I am advancing tends to be avoided, in favor of historically less well supported narratives of revolution and radical change.

11.1 Hormones: "Master Molecules" of Life Between the Wars

In the two decades between the first and the second world wars, pharmaceutical and chemical industries in the United States underwent fundamental changes. Both industries were liberated from technological dependence on German firms by the World War I abrogation of enemy patents, and began to do much more of their own research and development than ever before. In pharmaceuticals the change was especially pronounced, as John Swann and others have shown. Before World War I drug firms, which in America had mostly evolved from compounding pharmacies, seldom employed scientists for work other than quality control and, in a few cases, the production of biologicals such as antitoxins. But in the 1920s, ethical drug manufacturers greatly raised their research expenditures, turning increasingly to science-driven product development to compete against one another and, in line with reform movements within the medical profession and newly stringent drug regulation, emphasizing science in their marketing to differentiate themselves from traditional remedy-peddlers. Along with their investment in biology research, drug firms in the interwar period relied increasingly on patent protection for their innovative compositions, as opposed to trade secrets. There was also a growing overlap of drug and chemical industries by the end of the interwar period, with acquisition of pharmaceutical houses like Lederle by chemical firms like American Cyanamid, and with drug firms such as Pfizer moving into production of bulk biochemicals. Drug firms built up substantial in-house staffs of trained research scientists, and by the 1930s some like Merck, Lilly, and Squibb had even established quasi-independent institutes for the advancement of biomedical knowledge. Nevertheless, they relied increasingly on ties with academic biologists to keep them abreast of topics at the research frontier of life science, and as we shall see the sorts of ties that were formed closely resemble those so exercising the critics of recent biotechnology. [3]

By the late 1920s vitamins, "hormones", and other growth factors had become the subjects of biology research with the most glowing industrial promise, since they offered a key to control over vital processes. The premier "master molecules" (to borrow Evelyn Fox Keller's notion) [4] of interwar biology, hormones promised medicine not only the power to replace defective organ function but also to enhance sexual and other physiological functions of the human body. Thus, many ethical drug firms and physiologists pursued the commercialization of animal hormones, despite their lingering association with quackery from turn-of-the-century glandular transplants and extracts. By searching for a purified active ingredient from an endocrine gland by means of cutting-edge biochemistry, which could be published in scientific literature, and patenting the extraction process (or if possible, a synthetic process and/or synthetic derivative), drug firms could satisfy the new demand for "scientific" drugs and legally defend their proprietary hormone products. I will quickly sketch a few such hormone product development stories, drawing mostly on the work of other historians. Adrenalin was the first hormone to make the transition from glandular extract to pure chemical drug, and in fact was being sold by Parke-Davis under that trade name even while the chemical identification of the active molecule, epinephrine, was still uncertain. Glasgow-trained biochemist Jokichi Takamine, winner of a highly competitive race to isolate the adrenal hormone responsible for raising blood pressure and to patent his process for purifying it from adrenal tissue, licensed his patent to the firm, which began marketing it in 1901. The drug proved popular quite early among surgeons both for its ability to stop bleeding during procedures and for its quick action against asthma attacks and anaphylactic shock (all too often caused by a doctor as well, in the days of serum therapy). Parke-Davis managed to retain market dominance in this profitable product long after the expiration of the patents, through its copyright on the trade name "Adrenalin" which remained virtually synonymous with the drug well into the twentieth century. [5]

A young biochemist named Edward C. Kendall, frustrated by a brief stint at Parke-Davis because he did not feel that the firm had yet come to appreciate a role for the laboratory scientist beyond routine quality control, realized his dream of being both biologist and entrepreneur soon after his arrival in 1914 as head of the biochemistry section at the Mayo Clinic. Aided by the unusual presence of iodine in thyroid gland extract, which served as a chemical marker of the hormone activity, in 1915 Kendall had already crystallized a pure compound active in treating the thyroid deficiency condition Myxoedema. In 1916, well before his definitive 1919 paper on the chemical attributes of "Thyroxine", Kendall had patented his thyroid hormone preparation method and assigned the intellectual property to the University of Minnesota, on whose faculty he served as a result of the establishment in 1915 of the Mayo Foundation as a branch of the University's medical school. In 1919, the University granted an exclusive license for use of Kendall's patent to the Squibb pharmaceutical firm in exchange for half of the profits from the product, which Squibb successfully marketed. These royalties were deposited in a special "Thyroxine fund" from which Kendall could draw money to support his laboratory's projects. Kendall's thyroid hormone work was an important precedent in subsequent years, showing a way in which a biomedical scientist could materially benefit in a not-unseemly manner through intellectual property management, while at the same time benefiting humanity by ensuring that his discovery be reliably developed and brought to market according to the highest standards of the ethical drug industry. For example, in 1922, when Toronto physiologist John J. R. Macleod was weighing up whether to patent the insulin preparation procedure just developed by Frederick G. Banting, Charles H. Best, and James B. Collip in his laboratory, Kendall explained the satisfactory thyroxin arrangements, offering Macleod business advice and sharing his ethical perspective as well ("I can see no more reason why the man that separates the active constituent of the pancreas should not share financially as much as the man that makes a new wireless telephone"). Biochemist Harry Steenbock of the University of Wisconsin also emulated Kendall, ultimately creating the independent Wisconsin Alumni Research Foundation (WARF) to manage his Vitamin D patent and the royalties from it, when the University proved reluctant to accept and manage the biomedical patent itself. [6]

Insulin, as the hormone therapeutic with the most significant clinical impact (saving hundreds of thousands of otherwise terminal diabetics), and as one of the most astonishing drug business success stories of its day, deserves careful attention. Insulin's smooth and rapid transit from lab to experimental clinic to marketplace owes its success to a collaborative contract between the University of Toronto and the Eli Lilly firm. In 1922, following on positive results from the first clinical trials with the pancreatic substance made by Frederick Banting and biochemist collaborators in Toronto, the view prevailed that intellectual property law was necessary to protect the quality of the life-saving hormone, and the university took out a patent on their preparation method for insulin and its use in diabetes treatment. The university also began to scale up production from beef pancreas, in its own

Connaught Serum Laboratories. However, as so often occurs when chemical processes are changed from benchtop to factory scale, the uniformity and quality of the product began to suffer. Seeking a drug firm to help improve the production methods and standardize the drug, Toronto settled on Lilly. Under a contract granting the firm exclusive use of Banting's "Iletin" name for the hormone (a duplication of the Parke-Davis "Adrenalin" strategy), and a limited-term exclusive license to Toronto's patented process, in exchange for a percentage of the profit stream and a degree of control over the firm's marketing activities, Lilly and Toronto scientists began an intensive period of knowledge sharing. By the end of 1922, Lilly had improved the yield, purity, and stability of the hormone by adopting Jacques Loeb's isoelectric precipitation method for its production, and it was manufacturing enough to supply extensive clinical trials throughout North America. By 1924, when competing firms began selling equivalent preparations, "Iletin" had already become Lilly's biggest product ever and had so thoroughly become the brand of choice among physicians that the firm never relinquished its hold on this market. The experience convinced Lilly that collaborating with outside academic researchers could be very good business, and the firm pursued further arrangements of the kind. [7]

Hard on the heels of Lilly's commercial and clinical success with insulin, a number of ethical drug firms both in the United States and Europe quickly sought the scientific expertise to produce this and other hormones with a high degree of purity and reliability. In 1923, the Dutch pharmaceutical house Organon was founded as a joint venture between University of Amsterdam physiology professor Ernst Laqueur and a meat packing firm, with Laqueur supplying the scientific knowledge and trained staff, and his partners supplying the equipment, facilities, and raw materials. The first drug offered by the "start-up" firm that year was insulin. However, with the financial assistance and materials supplied by Organon, Laqueur was able to make rapid progress in what a contemporary commentator called the "elbowing and jostling and . . . jockeying for position in the neck and neck race to isolate and synthesize the much desired and long sought for hormone" secreted by the ovary, and presumed responsible for female sexuality, and the male sex hormone as well. In 1925, Organon put its first "female hormone" organ extract on the market, which soon attracted broad interest among physicians in the treatment of menopause, infertility, and a variety of psychological conditions. In the United States, Parke-Davis successfully pursued a parallel strategy, collaborating with the leading American endocrinologist Edward Doisy of St. Louis University in the production and marketing of the firm's estrogen product for a similar range of conditions, using Doisy's patented production methods and his distinctive term "Theelin" for the Parke-Davis product's brand name under a licensing agreement. It should be stressed that these eminent scientists received not only financial rewards but also, and more significantly, material and financial support for their research programs in exchange for their intellectual property, consulting, and services in standardizing hormone preparations. Doisy, who in 1929 had divided the honors in the race to purify estrogen with rival Adolf Butenandt of Germany, shared the 1943 Nobel Prize in medicine for his work on the clotting factor vitamin K (while

Butenandt shared the 1939 chemistry prize for his sex hormone work), and Laqueur is generally credited with being the first to purify testosterone in 1935. [8]

Even while pharmaceutical firms were still extracting their steroids and adrenaline from slaughterhouse waste in the early 1930s, plant hormones were making the transition from natural extracts to fully synthetic products. Auxin, the signaling molecule controlling shoot elongation, was in 1926 the first of the plant hormones to be isolated and, by the mid-1930s, easily manufactured synthetic auxins with novel biological activity were being hotly explored. We shall return to synthetic auxins below. A race to discover entirely new hormones governing other plant growth processes was also underway in the 1930s, and one of the hotbeds of this new molecule-oriented plant physiology is the California Institute of Technology (Caltech), in 1935 the institutional home of auxin's discoverer Frits Went. In that year young James Bonner joined the faculty, fresh from his own Caltech doctorate. The two main research projects Bonner initiated as a new instructor were both searches for new hormones. In one project he invented an assay for a substance from certain plant material that stimulated cell division and healing on cut surfaces; and with this assay he discovered a new wound healing hormone which he dubbed "traumatin," after the endocrinological style of the day. His second project began as an effort to develop plant tissue culture methods with chemically defined media which could be used to assay the effects of hormones and other molecules added to the culture medium. With this system, Bonner found that pea roots, which would stop growing after a few culture passages in a simple medium containing only minerals and sugar, would grow again if additionally supplied with the B vitamin thiamin (B_1). This did not in itself prove that thiamin functions naturally as a hormone in plants, since many substances which were clearly not true plant hormones (e.g., mammalian sex hormones) promoted plant tissue growth in certain bioassays. However, whether or not thiamin was a natural plant hormone, the practical implications were the same: it was possible that spraying thiamin on whole plants would stimulate growth. [9]

Bonner had obtained a sample of Merck's newly available "Betabion" crystalline synthetic thiamin (B1) by writing the research director of the firm, then a leader in vitamin synthesis. This 1936–1937 contact with Merck was perfectly timed in that the firm had a new, strong interest in plant hormones. Merck had just begun marketing synthetic auxins to horticulturalists for the propagation of cuttings, licensed under a patent of the Boyce Thompson Institute of upstate New York, under the trade name "Hormodin A," and anticipated marketing other hormone preparations under the "Hormodin" brand for a range agricultural and horticultural purposes. Thus to Merck, Bonner's discovery that thiamin might be a growth regulator in plants was not only a potential reinforcement of the firm's new plant hormone business, but also a surprising synergy with its established vitamin B business. Furthermore, the financial stakes were potentially huge, since if B1 were to become a major agrochemical, Merck's control of the patents needed to manufacture thiamin economically would have virtually guaranteed the firm's dominance in this market. After seeking expert opinion on Bonner's ideas and scientific reputation, Merck entered into a collaborative relationship with the young plant physiologist. [10]

While courting Merck, in 1937 Bonner extended his tissue culture experiments to roots from a number of different plant species, finding that most required thiamin, and also started a series of experiments with whole plants grown in defined mineral solutions with and without added thiamin. These research thrusts had strong commercial relevance, in that they aimed to establish how widespread thiamin's growth-promoting effects might be, and how applicable under greenhouse conditions. Though many economically important species, including maize, beans, and fig showed little or no effect of thiamin, some other, slower growing plants, including carob trees and the ornamental flowers camellia and cosmos, showed a strong effect. Shoot growth in these species doubled with thiamin treatment. As Bonner put it in his earliest major scientific publication on the effect, "vitamin B1 may be of some importance in practical agriculture" even if not every crop could benefit. [11] Bonner also began a trial with collaborators at the University of California, Riverside, to test the effects of prolonged thiamin treatments on oranges – an experiment with obvious commercial significance in Southern California. In some of his best greenhouse trials, Bonner obtained a doubling of the dry weight of certain plants such as cosmos treated with thiamin, but after more than a year of effort to identify the confounding variable, the effect remained highly capricious. Thiamin treatment was still far too unpredictable to become a commonplace agricultural biotechnology.

Bonner's efforts to explain and reduce the variability of thiamin's growth-promoting effect on whole plants took on new urgency in October 1938, when the national magazine *Better Homes and Gardens* magazine ran a prominent 4-page item, luridly suggesting that the hormone treatment would produce giant flowers such as plate-sized daisies. By this point Bonner was drawing funding from Merck through the Research Corporation, an independent agency which managed patents for academics and returned part of the profit stream to research funds designated by either the patent holder or licensing firm, and was reporting back to Merck research director Randolph Major on a monthly basis. [12] Perhaps due to the publicity as well as an awareness that other plant physiologists were having great difficulty reproducing his results, Bonner and Merck redoubled the effort to discipline the thiamin effect – in the more successful trials with cosmos, still a doubling or more of dry weight – and to bring it closer to market. In what bears considerable resemblance to a large-scale clinical trial of the day, [13] Bonner in early 1940 contacted plant researchers throughout North America, proposing that they participate in a cooperative standardized field trial, with cosmos seed and thiamin supplied by Merck and himself. More than two dozen scientists were recruited, and materials, together with detailed instructions for performing standard experiments and forms for reporting the data, were shipped to the collaborators. When the results came in autumn of 1940, Bonner was only able to prove a modest thiamin boost to growth, significant statistically but not practically. His own orange trials showed no difference with thiamin, as did the experiments of most plant physiologists who had attempted replication in a variety of plants. In 1943, Bonner finally saved face to some extent with a report on 85 cosmos experiments to some extent attributing the variability of response to temperature, but by this point the idea of B1 as a major agrochemical was dead. Together with the disappointing outcome and the new demands of war, the media sensation around thiamin may have contributed to Merck's retreat from the plant hormone business. Advertisements exhorting gardeners to "Amaze Your Friends" with thiamin-treated vegetables would have given plant hormones an unscientific air out of keeping with the sober pharmaceutical firm's image.[14]

Thus before the start of World War II, drug companies in America and elsewhere had a well-established pattern of tapping university-based biological science and a range of means for doing this, including arrangements with senior academic researchers in which they traded a royalty stream in exchange for a license, usually exclusive, to the academic researcher's patented biomedical discovery. Although there certainly was some discussion as to the propriety of taking out patents on life-saving medical innovations, universities too rather quickly became accustomed to benefiting in this way from their life scientists (in much the same manner as from their physical scientists and engineering faculty). Together with debate concerning the morality of these scientist-industrialist relations, resonating with those arising from recent biotechnological collaborations, there were also worries about the impact on academic science that sound virtually indistinguishable from those still being expressed a half century later. For instance in one 1937 article in the *Journal of the American Medical Association* on the increasing number of American universities gainfully managing patents of life scientist faculty members, chief editor Morris Fishbein commented on some drawbacks of the practice, even while encouraging physician-scientists to become more accepting of industry and savvy about patenting their work.

> While it is obvious that the entire trend of the times is toward the holding and control of patents . . . there are some arguments opposed to the holding of patents by universities . . .
>
> The exploitation of patents by universities places them in direct competition with one another. For example, there are now many patents concerned with the development of vitamin D. Because of financial interests involved, the results of current research are jealously guarded and probably all research is being delayed . . . [Even] workers in the same university, because of the rewards involved, may develop a competitive spirit which is likely to destroy [the] cooperation in science which is responsible for much of our current progress. Such competition, in fact, defeats the whole purpose of a university.[15]

Fishbein went on to lament efforts of some patent-holding universities even to bar scientists at other institutions from using their inventions for research purposes! The similarity of these concerns to those heard so loudly in recent years, about the impact of commercial involvement on academic life science, and the possible economic causes underlying the congruent situations of yesterday and today, will be discussed further in the conclusion.

11.2
Pharmaceuticals in Peace and War

As is well known, when the United States entered World War II the government agency responsible for war research, Vannevar Bush's Office of Scientific Research and Development (OSRD), enlisted the expertise of physical and biological scientists in the nation's universities. Some of them, especially physicists involved in atom bomb and radar development, were essentially drafted into large government-run labs, but many others were given contracts to pursue war research at their own institutions. While this is not the place to make generalizations about the impact of the Second World War on life science in America (and its frequent exaggeration), I would like to discuss at least a few examples of war research in the biological sciences which repeated prewar patterns of industrialist-biologist interaction to some extent, and thus can provide additional evidence as to the nature of these interactions and the types of projects pursued long before the current era of biotechnology. It is significant that the contract system allowed the government to treat academic researchers much as industrial sponsors had before the war. University of Pennsylvania pharmacologist A. N. Richards, chosen by Bush to head the the OSRD's life science arm (the Committee on Medical Research, CMR) when it was established in mid-1941, had decades of experience interacting with drug firms, and for most of the 1930s had worked intimately with Merck as a "general consultant" on many of that firm's research, development, and public relations problems (especially within the biomedical community). Richards was thus ideally suited to foster and propagate the type of mutually profitable relationships between academic biologists and the pharmaceutical industry in which he had become adept between the wars. [16]

The story of wartime drug development which perhaps best illustrates the way in which the American government's research contract system could simply extend and amplify the existing pattern of collaboration between academic life scientists and industrialists, is the case of the steroid hormones of the adrenal cortex. Adrenal gland extracts, like those from testes, had been marketed since the turn of the century for alleged strength-enhancing powers. Around 1930 the hunt for an adrenal hormone essential for life (unlike adrenaline) began in earnest when physiologists found that adrenal extracts could keep alive otherwise terminal adrenalectomized dogs and cats. Experiments on adrenalectomized animals with and without hormone treatments soon produced a several of lines of evidence that one role of the sought-after adrenal secretion was to counteract fatigue by releasing energy stores to the bloodstream under conditions of oxygen deficit, for instance at high altitudes. Thus, given the critical state of the air war over Britain, a May 1941 report from Allied intelligence that Nazi scientists had a new adrenal miracle drug which allowed Luftwaffe pilots to fly at 40 000 feet without oxygen, and that U-boats were shipping tons of adrenal glands from Argentina to Germany, prompted more action than skepticism. The hunt for this hormone was already quite fierce, the chief contender at this stage being Tadeus Reichstein of Zurich. Reichstein, who was collaborating with Laqueur and Organon in Holland, had in 1937 synthesized the first active steroid of the adrenal type (which, however, had little carbohydrate-mobilizing biological activity). The leading contender on the Allied side was none other than the enterprising E. C. Kendall. Since 1934 Kendall had been one of the world's chief sources of high-potency adrenal extracts through arrangements with two drug manufacturers (Parke-Davis and Wilson Laboratories), which supplied a "pilot plant factory" in his Mayo lab with over a ton of fresh frozen beef adrenals every month in exchange for a portion of the adrenaline extracted from the glands and services standardizing commercial cortical extracts by bioassay at the Mayo. The excess adrenaline Kendall sold for cash. Through his industrial connections, Kendall maintained a large and self-financing laboratory through the Depression and quickly outdistanced his American competitors, keeping abreast of Reichstein.

Upon the 1941 rumors and espionage reports about hormones, the OSRD immediately responded by organizing more than half a dozen CMR physiology contracts to test whether adrenal steroids might augment the altitude tolerance of experimental animals, and also a similar number of chemical projects. These latter involved some of Kendall's closest academic competitors in the adrenal hormone race, as well as several drug firms, in a 'collaboration' to find an industrially practical method to synthesize the adrenal steroid that Kendall had identified as most promising. (This synthetic project was necessary because the quantity of adrenal glands was inadequate to meet the needs even of patients with fatal adrenal insufficiency conditions, let alone the Air Force.) The conditions under which the chemists carried out their war research were less than fully equal and cooperative, largely because they reflected pre-war atmosphere of competition for scientific credit and intellectual property in the hormone field. When the OSRD chemistry projects were first initiated, with Richards' blessing Kendall quickly lodged a secret patent application on the process of synthesizing a key intermediary compound, with the help of the Research Corporation. Apart from Merck, which was brought in as Kendall's partner (and licensee) in the cortical steroid synthesis project by the Research Corporation, none of the others involved were given details of Kendall's new process. Indeed, to the consternation of at least some of the academic chemists working under OSRD contract on the project, but to the satisfaction of Kendall and the drug firms, in order to protect proprietary information progress reports of all participants were known in detail only by the chairman of the committee Richards put in charge, W. Mansfield Clark, who would decide when other participants needed to know of the activities of the others. As for inequality, since he accepted no money from the government for his research, which was essentially just a continuation of the project he had been pursuing since 1934, Kendall was permitted to patent his wartime work, while his academic rivals brought into the project as OSRD contractors were subject to a stringent patent clause assigning all intellectual property acquired in contract work to the government. Moreover, Kendall was fed intelligence concerning his academic rivals' chemical requisitions from Merck, from which all of certain essential starting materials had to be ordered.[17]

In mid-1943 the CMR physiology projects had shown beyond much doubt that the sugar-affecting adrenal steroids would be of little value in military aviation, and the chemists appeared not much closer to achieving synthetic production in any case.

Unsurprisingly, the physiologists and most of the chemists moved on to other projects by the start of 1944. However Kendall persisted, working ever more intimately with Merck chemists straight through the remainder of the war at an undiminished pace (independent of OSRD funds as he was). There seems to have been "whispering" and finger-pointing in Kendall's direction during the last year or two of the war, a feeling that Kendall's unwillingness to alter his competitive scientific style was one reason the adrenal collaboration had failed, by comparison with the other large biochemical projects on penicillin and antimalarials. But as Kendall was aware, Reichstein too was forging ahead despite the war, now working with the Swiss firm Ciba since Holland was under occupation. Ultimately it was the processes researched by Kendall and Merck chemists during the war which in 1947 made the large scale synthesis of the adrenal hormone Kendall called "cortisone" possible. Merck supplied Kendall's Mayo colleague rheumatologist Philip Hench the drug that in 1949 astonished the world by making bedridden arthritics walk again. [18] The Research Foundation negotiated cross-licensing agreements among Merck, Ciba, and some other holders of patents pertinent to the long and complex synthetic process, and cortisone quickly became the archetypical postwar miracle drug. Reichstein, Kendall, and Hench shared the 1950 Nobel Prize in Medicine. It clearly was not second-rate science that Kendall had been doing before and during the war, and as he was long aware, this was a type of biochemical work that could only be effectively pursued in close collaboration with industry.

At the risk of redundancy, I will briefly retell the story of how penicillin came to be mass produced in wartime America and thus commercialized, stressing the aspects of this famous event most pertinent to my argument. For despite certain idiosyncrasies related to the war which made penicillin a success story of extraordinary cooperation among normally non-communicating industrialists, government agencies, and academic researchers, no other story of product development in the 1930s and 1940s could more perfectly exemplify the already existing overlap of American pharmaceutical, chemical, and agricultural industry – both in terms of markets served and production processes – that I am trying to demonstrate. Furthermore, in considering the tactics that were necessary to make this success come to pass we can discern the outlines of the normal patterns of knowledge-sharing already described above, which in some measure needed to be overcome. Let us take up the story in mid-1941, the point at which British medical researchers Howard Florey and Norman Heatley, having just successfully carried out penicillin's first clinical trial despite pitiful quantities of home-made drug, traveled to the United States in search of a pharmaceutical firm willing to produce the material for further trials. Scheduling a Washington meeting with National Research Council head Ross Harrison, they were fortunate that Harrison introduced them to two USDA scientists familiar with industrial microbiology. Florey and Heatley were sent to the USDA's Northern Regional Research facility in Peoria, the national center for research and development of industrial-scale fungal fermentation for use of agricultural surplus and byproducts. The agricultural scientists there had recently built a new large-scale fermenter allowing submerged mold culture, that is growth through an agitated liquid medium as opposed to surface culture, and were using it to grow various fungi in media based on corn steep liquor, a plentiful byproduct of maize processing. Use of submerged fermentation in corn steep raised penicillin production per liter thirtyfold, compared with British surface culture methods. Higher-producing strains of mold collected by the USDA soon raised it even more.[19]

At the end of 1941 improvements in production methods and yields at the USDA Peoria lab were ready to be translated into drug production so that clinical trials might be possible. Reaching this stage was already the result of extraordinary, war-driven central planning, in that foreign medical researchers would never have been invited to take central roles in a pharmaceutical development project at a government agricultural research facility under other circumstances. But the OSRD and other wartime agencies had to make further direct interventions in both industry and academia to move the project toward implementation. Several major pharmaceutical firms including Merck, Squibb, and Pfizer were given free access to the Peoria findings and induced to build pilot fermentation plants to manufacture the drug. The firms would build the small plants and pursue yield improvement research at their own expense, reporting advances to the OSRD and other agencies for dissemination to the other manufacturers – leaving open, at least in principle, the possibility for the firms to patent any process technology improvements. Roughly simultaneously, the government initiated contracts for university-based chemists to synthesize penicillin. Many firms also pursued their own synthesis research programs with great enthusiasm, for chemical processes were more familiar to the drug manufacturers, and in addition could be far more easily protected by patents than fermentation production with difficult-to-patent microbes (although Bush reserved the right to require firms to license any patents on either synthetic or fermentation production, subject to reasonable royalties). By mid-1943 government-coordinated crash clinical trials using penicillin made in the pilot fermentation facilities were finished, and standard dosage regimens determined, so the time had come to begin full scale production. The War Production Board then compelled foot-dragging drug manufacturers with a heavy investment in penicillin synthesis, such as Merck, Commercial Solvents, and Squibb, to put full scale fermentation factories into operation at the end of 1943. This commitment to the agricultural scientists' fermentation process, despite reluctance of the majority of factions believing in chemical synthesis, was ultimately vindicated. Chemists did not synthesize penicillin until 1959, and fermentation remains the most efficient manufacturing process (as it is now also for steroids, vitamins, and many drugs).

Leading researchers in fundamental biology also had a major part to play in this drama, along with the USDA scientists and drug firms. As mentioned, high producing *Penicillium* strains had been introduced after a world-wide search by USDA mycologists, though ironically, the best one came from a rotten cantaloupe in the Peoria fruit market. In what might justly be counted the first true project in genetic engineering, as distinct from scientific breeding, in 1944 OSRD research contracts were initiated with several geneticists to develop a new, still better strain by deliberately inducing mutations with radiation. The contractors included Milislav Demerec of the Cold Spring Harbor labs and George Beadle of Stanford – soon to be Bonner's department head at Caltech (and in whose lab, incidentally, Bonner's

Chicago-trained plant physiologist brother David worked, again suggesting overlap in the histories of molecular biology and plant physiology). Using the new one-gene/one-enzyme hypothesis and everything known about metabolic pathways in lower plants, Beadle's and Demerec's groups each generated improved strains. Demerec's strain more than doubled penicillin yield per liter of culture, and was eagerly adopted by the drug firms starting fermentation production, serving as the progenitor for later strains developed and used by industry into the 1980s. Here central planning overcame the existing economy of scientific and pharmaceutical information flow, transferring knowledge rapidly from biomedical and agricultural researchers in both academia and government to the drug industry, and compelling drug producers to alter production infrastructure rapidly so as to accommodate the innovation. The institutional barriers for all these changes could never have been overcome so quickly without the wartime rationale for meddling with private industry, but the results illustrate well what was technically and commercially feasible for American biotechnology in the 1940s.

Not all wartime biotechnological projects had to do with chemical-biological weapons or drugs. In the course of research on biological electron microscopy, I encountered a rather different project worthy of a brief description, this one well illustrating both the state of American biotechnological research and development circa 1940, and the continuation of fairly typical interwar academic-industrialist collaboration patterns into the war years. Francis Schmitt, self-styled molecular biologist and biophysicist, came to the Massachusetts Institute of Technology (MIT) in 1941 to take over the life science division on a mission from the Rockefeller Foundation to make biology more like physics and engineering. An early purchaser of RCA's first commercial model microscope, even in his first semester at MIT Schmitt had broken new ground studying the structure and self-assembly of fibrous macromolecules such as collagen, the major protein of skin and connective tissue. Indeed his group found that collagen could be purified from tissue and brought into solution by a certain acid treatment, and then by altering ionic conditions be made to re-assemble into fibrils indistinguishable from the molecule's native form. Immediately after the attack on Pearl Harbor in December 1941 (if not before), the MIT group conceived a way in which to make this research medically and industrially valuable. Evidently aware that shipments of the Argentinean sheep from which absorbable gut suture was made were thought to be threatened by German U-boat action, Schmitt and his team proposed that the collagen solution could be employed to make artificial sutures by a process similar to the spinning of synthetic textiles. After some preliminary benchtop experiments on how to prepare and extrude liquid collagen such that it would rapidly form thin, uniform fibers, the group approached suture manufacturers Johnson & Johnson and in March 1942 won a grant from the firm (via the Johnson Research Foundation) to further the project. By April they had already demonstrated their collagen manufacturing process on a scaled-up basis to Johnson & Johnson representatives, and by November 1942 they had demonstrated the automated manufacture of collagen threads using equipment of the Pennsylvania-based American Viscose textiles firm. [20]

In December 1942 Schmitt's group accepted an OSRD contract to pursue the collagen project further, one which by special arrangement (on the grounds that the basic innovation was carried out before government involvement) carried an especially lenient intellectual property clause only granting the US government a license to Schmitt's process, rather than the usual assignment of patents to the government. Probably, future patent rights had already been assigned to Johnson & Johnson in exchange for royalties. American Viscose was granted a separate contract to develop the fabrication technology for suture manufacture from Schmitt's collagen, and Armour Laboratories was made a subcontractor under Schmitt to work with the MIT group on optimizing methods for preparing collagen solutions from the beef achilles tendon that Armour obtained plentifully from its Chicago slaughterhouses. Leather manufacturers helped Schmitt buy a second electron microscope. The synthetic sutures were tanned with chromic acid to varying extents and tested in experimental animals to gauge inflammation reaction and tissue absorption rates, and in September 1943 their performance was judged sufficient by Johnson & Johnson experts to merit approval of pilot plant production facilities by the CMR.[21] By 1944, the sutures reached successful preliminary trials with surgeons, especially at Massachusetts General Hospital. MGH surgeons also interested Schmitt in fabricating collagen sheets for use as "artificial skin" in burn patients; these were made and tried on patients, but not used clinically to any great extent. Achieving significant surgical use in wartime trials, however, were narrow tubes of pure collagen extruded by Schmitt's method. In traumatic injuries such as occur on the battlefield, the proximal end of a severed nerve would be inserted in one end of the tube and, during surgery, the distal end would be inserted in the other. Use of the collagen sheaths in suturing nerves increased the reinervation rate of severed and severely damaged limbs appreciably.[22] Although no spectacular marketplace success immediately followed these advances, in the MIT collagen project we see another example of life science bringing a technology into existence through close interaction among medical institutions and industries (MGH, Johnson & Johnson), agricultural industry (e.g., Armour, and its medical subsidiary Armour Laboratories), and chemical industry (e.g., American Viscose). And though Schmitt's war contract may have facilitated cooperation among his industrial partners, the project was initiated and pursued for nearly a year entirely by the collaborators, without involvement of any government agency. It was precisely the sort of arrangement that might have formed spontaneously, without the war.

For a final wartime vignette I will return to academic physiologists researching plant hormones, and how their war work led to agricultural herbicides. As with thiamin, before the war the practical idea influencing research into auxin and other plant hormones was that a compound that *enhanced* crop plant growth hormonally might aid agriculture. However, in the course of the war some plant physiologists seeking ways to make their hormones contribute to victory reconceived of hormones as plant-killers rather than plant growth enhancers. The academic biologist leading this group was Ezra Kraus, head of Botany at the University of Chicago. In the 1930s research at Chicago had revolved generally around the synthesis and transport of endogenous hormones, and their interaction in shaping growth with environmental conditions – maintained in the era's finest greenhouse facilities, thanks to Rockefeller Foundation funding. Like Caltech, Chicago had no agricultural school, but Kraus helped establish the new USDA Bureau of Plant Industry research facility at Beltsville, Maryland, and in the late 1930s there was a rather free flow of students and ideas between that institution and Chicago. Through 1940, all of the plant hormone studies under Kraus at Chicago and Beltsville dealt with topics like growth rate, differentiation, and flowering in crop plants, agricultural uses of auxin, and even the relationship of hormones to cancer in a plant model. There was substantial experimentation, both practical and physiological, with the new (and invariably patented) synthetic auxin analogs flowing from the aforementioned Boyce Thompson group of Percy Zimmerman and his collaborators, but there was never any suggestion of using hormones to kill. [23]

However, the idea certainly occurred to Kraus by December 1941. Barely a week after the Japanese attack on Pearl Harbor, Kraus drew up an informal proposal to enlist his hormone expertise in the war effort, and submitted it to the National Academy of Science's top secret committee on chemical and biological warfare (CBW), on which he sat. Kraus proposed investigating hormones to enhance American crop production more or less as he was already doing in Beltsville, and for offensive purposes he suggested developing hormones to destroy crops and forest cover useful to the enemy. The CBW Committee gave Kraus's offensive project approval in February 1942, together with more than a dozen other biological warfare projects ranging from production of anthrax spores and development of anthrax vaccines, to culture of the late potato blight microbe (presumably for possible use against German potatoes). Kraus and his students at both Chicago and USDA facilities set to work growing rice in vats, and testing the killing power of both hormones and traditional inorganic poisons such as arsenic. The latter, disappointingly, appeared to be more reliably effective on rice. But through two years of meager results Kraus and his team soldiered on in search of a way to make hormones useful against the Japanese staple crop until, at the start of 1944, they finally started testing some of the newer synthetic auxins discovered by Zimmerman, including 2,4-dichlorophenoxy-actetate (2,4-D). With this and the closely related compound 2,4,5-trichlorophenoxyacetate (2,4,5-T), dramatic killing of tomatoes in the greenhouse was obtained. Outdoor tests during the summer of 1944 by Kraus's USDA group showed that 2,4-D had a selective toxicity on dandelions and other broadleaf plants in grass lawns, and at about the same time a former Kraus student, just moved to the Cornell experiment station, showed 2,4,5-T to be effective against certain farm weeds. Kraus, who probably instigated the Cornell work too, was excited by these results and not at all discouraged that the hormones appeared to be poor herbicides against rice and other grasses.

Kraus, as editor of the plant physiology journal *Botanical Gazette*, quickly published the full results side by side in December 1944. Both papers cite correspondence with Kraus in 1941 for the idea that hormones could be used for weed control. [24] By this point Kraus evidently recognized the agricultural potential of 2,4-D and 2,4,5-T as herbicides, and may have been seeking to establish a priority date for conceiving this invention (a move with major intellectual property implications) and reducing it to practice despite the inconvenience that there was a war on; indeed, his work was a military secret. As Kraus was well aware, since he remained advisor on the project, in July, 1944 military herbicide research had been officially transferred to the Chemical Warfare Service facility at Camp Detrick, Maryland, and more stringent barriers to publication on the herbicidal properties of hormones were soon instituted. For the next year a group of military scientists there conducted trials on various methods for applying herbicides, and, with specially developed biological assays, tested over 1000 synthetic compounds for their potency as plant growth regulators and toxins before the war ended. The herbicides were never deployed against an enemy until Vietnam (as Agent Orange, made of 2,4-D and 2,4,5-T), although the Navy contracted with Sherwin-Williams for 2000 tons of 2,4-D in August 1945, probably for clearing bases and airfields.

Despite its continuing top-secret status, however, during 1945 the public became progressively more aware of herbicide research, thanks more to business interests rather than to plant physiologists. Many of the media items that broke the story were instigated by the American Chemical Paint Company. Franklin Jones, an industrial chemist and sales manager for the small Pennsylvania firm, evidently caught wind of the 2,4-D and 2,4,5-T work at USDA, either from Kraus's Beltsville group or from another physiologist at the plant introduction bureau in Glen Dale, Maryland who had been working independently. In late 1945 Jones received a patent on the use of these compounds as weed killers. This news caused uproar in the chemical industry, setting off a melee of intellectual property litigation that there is no space to describe here. By 1947 the nation's major agrochemical manufacturers including Dow, Sherwin-Williams, and Du Pont successfully had teamed together, with help from the USDA, Kraus, and much of the plant physiology establishment, to neutralize the American Chemical patent. The small company would not be allowed to monopolize the promising new product category, so important, along with new insecticides like DDT, to the chemical industry's reconversion to peacetime production and expansion in the agrochemical sector. Even in the beginning of 1946, as the legal battles mounted, several of the larger firms were building 2,4-D factories, and were severely depleting Camp Detrick's scientific staff in their eagerness for hormone R & D talent. The suburban lawn-owner, the first market segment targeted for the product, was bombarded with advertising. Sherwin-Williams's 1946 campaign for its "Weed-No-More" 2,4-D brand generated 175 million advertising messages alone, while the press and the firm's competitors all fueled the marketplace frenzy.

Kraus was instrumental not only in legal proceedings to block an American Chemical monopoly on 2,4-D weed killers, but also in promoting the trials by agricultural scientists which were necessary to transform hormone herbicides from a suburban convenience into a serious agricultural innovation. Not that the mainstream agricultural science community needed much prodding; but when Kraus made a show of personally ingesting one-half gram of pure 2,4-D per day for three weeks to demonstrate its safety, he must have been trying to allay somebody's concerns. Findings from agricultural stations soon spelled out how the new chemical could safely be used on monocotyledonous crops like wheat and maize, where much like an antibiotic it would selectively poison broadleaf weeds among these cultivated grasses. By 1949, US production of 2,4-D had reportedly reached 20 million pounds, enough to treat over 20 million acres. Indeed, that year 20 million acres were sprayed in the Great Plains region alone, where 2,4-D was already becoming standard practice for spring wheat production. The hormone was a terrific plant killer. Still, the original dream of using synthetic hormones to enhance and manipulate crop plant growth remained strong for Kraus. As he put it in his keynote speech for a late 1945 conference, with hormones, "I know of no single process of the living plants that cannot be brought eventually under absolute control" through further plant hormone research. [25] In late 1945 Sherwin-Williams demonstrated its faith in Kraus's vision with a $ 25000 per year, 5-year research contract with the University of Chicago, in order to share in Kraus's plant hormone research. Kraus had now, later in his career than Bonner, similarly learned how to run a research program mutually beneficial to himself and to industrial partners, in much the same manner as the endocrinologists. In 1947 he went an entrepreneurial step further, attempting to parlay a potential new method for 2,4-D -mediated sucrose production from plant starch into corporate sponsorship for a new Botany building. [26] Both his actual contract and this more grandiose scheme, which failed for technical reasons, capture well the spirit of the day. Biologist-entrepreneurs could bargain their knowledge of plant hormones for high stakes with chemical and drug firms. The war years may have introduced a larger number of American life scientists, like Kraus, to what might be called the bio-industrial habitus (or way of life), but it by no means created it. [27]

11.3
Conclusion

The products developed during the war through biologist-industrialist collaboration were of mainly the same type as those being brought to market before the war, and there is every reason to suppose that even without America's involvement in the war these same products would soon have been developed anyway in much the same manner (so long as a peacetime demand existed). This is most obvious in the case of cortical steroids, where Kendall merely continued his prewar research program to repeat the earlier medical and commercial successes of thyroxine, insulin, testosterone, and other hormone drugs. As for penicillin, government involvement evidently hastened the pharmaceutical industry's acceptance of fermentation production, and facilitated the transfer of that knowledge from USDA and academic labs to industry. Still it would be safe to assume that the ultimate outcome, the commercialization of the drug, would be the same without World War II, since the necessary circuits of information and personnel flow already existed in the interwar period. For instance, USDA microbiologist James Currie had made Pfizer a leader in fermentation production technique after World War I when he left the Peoria lab for the firm, ultimately giving Pfizer the knowledge base which made it the largest penicillin producer during the war. A similar argument can be made about the development of plant hormones as agrochemicals. Synthetic hormone products for nurseries and orchards were being marketed by Merck and also by Du Pont before the war. Kraus's wartime project no doubt accelerated the development of hormones as agricultural herbicides, but Du Pont in the late 1930s already recognized the herbicidal potential of some chlorinated phenoxyacetic acids, and in fact held a broad 1943 patent on many of these compounds (including 2,4-D) for all forms of plant growth control. Thus hormone herbicides too were merely a matter of time, with or without government contracts. Even Schmitt's wartime collagen project, which was entirely arranged among the collaborators without government help (and indeed was being funded by industry before Schmitt's OSRD grant), might have happened without the war, though in the absence of a U-boat menace there would have been less reason to seek substitutes for natural sutures. Had a demand existed, Johnson & Johnson and Armour certainly would have been interested in sponsoring Schmitt's collagen work (as United Shoe Machinery Corporation, excited by the possibility of producing leather goods by plastic manufacturing methods, became immediately after the war).

Nor was the type of interaction fostered between academic life scientists and drug firms to enable such innovations dramatically altered by the war. Rather, as I argue above and in more detail elsewhere, wartime science management agencies like the OSRD's Committee on Medical Research relied upon the tried and true framework of collaborative project-oriented research contracts which, along with research fellowships for junior scientists, had become the usual means by which American industry between the wars tapped knowledge from academic researchers at the forefront of science.[28] Graduate students and junior scientists like Bonner were given fellowships or small grants by pharmaceutical firms, sometimes indirectly through the Research Corporation or NRC, as a means of stimulating work on problems interesting to them, and to buy a "window" on action in university biology – as industry long had done with chemistry. Such contracts often contained stipulations on patent assignment or licensing rights, and many restricted the sharing of unpublished information. As we also have seen (and also as anticipated in chemical firm-chemist relations), drug firms in the interwar period learned to negotiate arrangements with senior academic life scientists as well, exchanging royalty streams for exclusive licenses to patented biomedical discoveries, exclusive use of copyrighted terms as brand names, plus technical advice and services such as hormone standardization done in university labs. Very often the proceeds of these royalties went to special research funds kept by the universities to support the work of the entrepreneurial life scientist as well as other research at the institution, to preserve a certain apparent distance between scientist and industrialist. But as we have seen, by the 1930s such increasingly widespread management of biomedical knowledge for profit by academic scientists and institutions had given rise to a perceived serious erosion of traditional patterns of sharing scientific knowledge and materials between and even within universities. Some worried that knowledge was being produced for private, not public benefit, and the boundaries between impartial academic knowledge and big business were dissolving. No doubt there has been considerable evolution in the legal details of the contracts enabling the commercialization of university-generated life science since the 1930s. No doubt too contemporary business arrangements between life scientists and small "start-up" firms in which they are often principals – an option only made commonplace by the new availability of venture capital in the 1970s – differ in some ways from arrangements with the large firms with which industrially-minded life scientists of the past were compelled to collaborate. However, large drug firms are still the ultimate client served by these life scientists, today as yesterday. The underlying pattern of resource exchange (knowledge for materials, cultural for economic capital), together with the threats to the autonomy and productivity of life science, are essentially the same as those that were exploited and decried half a century later.

Why were the patterns of interaction between pharmaceutical companies and academic life scientists in certain fields so similar across such a substantial gulf of time? Though admittedly speculative, it seems likely that the initial growth in popularity of this practice owes something to the economic exigencies of the lean Depression years, as well as to the increasing value of certain types of life science research to the pharmaceutical industry. As certain areas of science came to offer more to drug companies, and as life scientists stood to benefit from the large material resources offered by corporate sponsors, they turned to each other for mutual benefit – especially as the Depression reduced other support options for scientists. Endocrinology met the conditions promoting such alliances particularly well. The field was fiercely competitive, no doubt partly due to the rewards that could flow from a patent, and the acquisition and processing of starting materials (slaughterhouse waste, urine) on an industrial scale was a great aid to rapid progress in research. Presumably many cultural factors played their part also, for instance a resurgent Yankee pragmatism in academia since the Progressive era and an ethos of "cooperative individualism" penetrating from the business world to American science during the Republican 1920s. [29] But to me the most powerful explanation lies in on the alignment of economic forces around this type of life science. These structural, economic reasons easily account for the resurfacing (not *reappearance*, for though there may have been an eclipse for basic fields there was never a disappearance of industrial sponsorship of certain "applied" areas of life science) of this same sort of life scientist-industrialist relationship in the last decades of the twentieth century. By the mid-1970s, the pharmaceutical industry was in position to make money by adapting the production technology based on fermentation and the new methods for expressing mammalian genes in bacteria that molecular geneticists had developed. Lacking the in-house expertise, they contracted with start-up firms associated with academic centers, as Lilly did in sponsoring Genentech's cloning of insulin. After 1980, when universities were permitted to patent the engineered organisms and other products of publicly funded medical research by the Chakrabarty decision and the Bayh-Dole act, they had leverage for negotiating a part of such mutually profitable arrangements with industry. Furthermore, the timing of these legal changes coincided with a decline in federal support for basic life science research, which had grown rapidly from about 1950 to the early 1970s, giving molecular geneticists fewer non-industrial options for the funding of ambitious projects. Thus by the 1980s the conditions under which drug companies and molecular geneticists, with their universities, would mutually benefit from the sort of relations that were common between these same firms and endocrinologists in the 1920s and 1930s, had returned. [30]

Quite apart from the nature of collaborative arrangements between industrialists and biologists, the general ambitions to manipulate life on an industrial scale, frequently pursued through such collaborations, also extend back to the early twentieth century. Philip Pauly traces these ambitions to Jacques Loeb, avatar of the engineering ideal in biology in turn-of-the-century America. Loeb himself took the first steps toward human cloning with his sea urchin parthenogenesis work, or so contemporary newspapers perceived it at least half-seriously. [31] More significantly, many specific projects of today's biotechnology actually date in conception to the interwar period; that is, schemes to control life incompletely achieved in the 1930s have been refined or revived, now using genetic engineering. I have discussed the construction of artificial tissues with biomolecules assembled in the test tube, the quickening of crop growth, and the deliberate alteration of plant morphology. Bonner in a sense fulfilled the biotechnological ambitions of his early career at its end in 1980, by founding the successful plant genetic engineering firm Phytogen. Kraus never saw the day of enhanced plants, but genetic engineers today are trying to accelerate plant growth by improving photosynthetic enzymes, and Monsanto's herbicide resistant soybean is clearly an extension of Kraus's own 1940s-style weed control technique. The most successful recombinant DNA drugs to date, Lilly's human insulin, Amgen's erythropoietin, human (and bovine) growth hormones developed by Genentech and others, and various interferon products are all, somewhat loosely speaking, hormones. The first of these, insulin, is obviously just an old drug produced by the new process of fermentation using genetically engineered bacteria. Major pharmaceutical firms were already considering interferon, erythropoietin, and human growth hormone for development as products by the early 1960s, and were only deterred by the impracticality of producing these protein hormones in cultured human cell lines. [32] Genetic engineering entered an already structured marketplace. Thus the new biotechnology continues the technological trajectory of the old, while the old patterns of industrialist-biologist interaction making possible such developments have become more frequent, at least until industry assimilates the new science and technology unleashed by recombinant DNA.

One historiographic implication of the view I am advancing, that molecular genetics is not the first life science to become biotechnology, is that we need to expand our vision of the history of "molecular biology" – if this problematic term refers to the sort of science dealing with the molecules governing life processes. Extending Robert Kohler's view, we could regard Warren Weaver, the Rockefeller Foundation's famous promoter of mechanistic life science and coiner of the term "molecular biology", as simply encouraging an already growing linkage between biology and industry in the 1930s. Weaver was not only, or even especially, interested in fostering molecular genetics above other experimental life sciences. [33] Thus Weaver funded not just biologists studying the gene, but also large numbers of endocrinologists, physiologists, biochemists, microbiologists, and even plant physi-

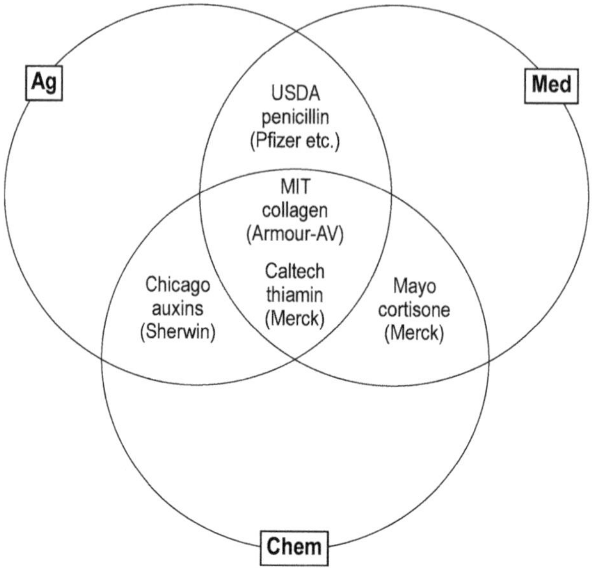

Figure 11.2 Industrial life science projects of the late 1930s and 1940s discussed in this essay, mapped in the manner of Figure 11.1. for comparison.

ologists, though he tended to avoid the areas attracting the most industrial sponsorship because here his influence would be diluted. As I (and others) have shown, in the interwar period many American scientists from these other fields were engaged in collaborative work with industry, availing themselves of the techniques of the day to bring about an overlap in medical, agricultural, and chemical sectors strikingly similar to that attributed by pundits to today's genetic engineering (Figure 11.2). Indeed, in the 1930s and 1940s, molecular geneticists were mostly, with rare exceptions like Beadle, poor cousins of colleagues studying life's master molecules in ways more useful to industry. Before the middle 1970s few molecular geneticists found their way into the rich territory near the central zone of the "Venn" diagram. When they began to arrive there in large numbers in the 1980s, this was an exotic land for many of them, but it was hardly unoccupied terrain.

So, in the end, what's so new about today's "life science industry"? Little more, I would contend, than that with recombinant DNA technique, molecular geneticists joined the club of life scientists able to manipulate the biomolecules they studied, and attracted much attention in the process of doing so. With the already noted legal changes promoting the commercialization of this new knowledge, *existing* industries engaged in biotechnology were thus stimulated, and induced to take up the newly useful knowledge of genetic engineering. Quite possibly, the populating of the center of our "Venn" has accelerated since 1980, but there has not been any qualitative change in the map. This undramatic view of continuity and constrained, merely incremental change raises the question of why a brand-new age of biotechnology has been heralded with so much fanfare. True, the notions that living things could now be altered at will in a genetically permanent way, and that life forms could now be intellectual property, do seem to have struck a cultural nerve in a different way than hormones and wonder drugs. This symbolically loaded technological novelty may well have promoted expectations of concomitant sociological and economic novelty. Furthermore, industrial life science *was* actually new to molecular geneticists in the late 1970s; to stimulate investment in their genetic engineering projects from big firms or capitalists they emphasized this and otherwise did what they could to create dramatic press. [34] Indeed, everyone seems to benefit from the myth that today's biotechnology is different and unprecedented in every way: to business concerns invested in genetic engineering, the claim of great novelty heightens the promise and glamour of their products and thus helps increase chances of public acceptance; to activists opposing the industrial exploitation of life, the claim of novelty helps justify demands for special attention and dramatic action; and for nations involved, talk of joining in a new industrial revolution may stimulate the economy.

This reference to the politics of the historiography of molecular genetics and biotechnology raises one final issue which needs to be addressed here. What are the politics of my own argument? The de-mythifying case for continuity and against novelty in biotechnology advanced here does certainly undermine a line of attack that activists have taken against biotechnology, the one playing on fears of unprecedented corruption in academia. It nevertheless has no intrinsically laissez-faire political valence. My goal is simply to raise the intellectual tone of the debate about the new biotechnologies. Thus, acceptance of biotechnology's lack of great novelty in its ambitions and in its promotion of business-academia linkages does not entail that stricter regulations are not required. Proponents of current biotechnologies might argue along such lines (and indeed, have done), [35] suggesting that the foreseen negative impacts on agriculture or on academia should already have materialized if they are continuous with the industrial initiatives of the past half century. But opponents of biotechnology can equally take up the empirical challenge of continuity as well, looking for negative impacts on agriculture and on life science fields brought under strong industrial influence during advances in agricultural technology half a century ago. Probably, few molecular geneticists would relish comparison with entomologist champions of chemical insecticide technology in the 1950s. [36] Still more certainly, continuity with the style of agricultural technology actually transferred from Fort Detrick to the farm in the 1940s serves as no unambiguous endorsement of current directions. I believe that critics and proponents alike could better serve the interest of the public in reaching wise decisions about current biotechnology if they refrained from basing their arguments on dramatic myths, and I am optimistic enough to hope that the best arguments need no embellishment with falsehoods in order to win popular assent.

Acknowledgments

This article owes much to help from archivists at the California Institute of Technology, the Massachusetts Institute of Technology, Merck, the National Archives and Records Administration, Princeton University, the University of Chicago, and the University of Pennsylvania. The author is grateful to them for their assistance, to the Dibner Institute for the History of Science for a visiting fellowship in autumn 1998, to the National Science Foundation for grant SBR-9729131, which supported much of the research. The author would also like to thank colleagues at the School of Science and Technology Studies of the University of New South Wales (where some elements of the research were also done), the Dibner Institute, the University of Minnesota, the Yale History of Medicine group, and especially at the Munich IUHPS/CHMC meeting in May 1999, for comments on spoken presentations. Very special thanks are due as well to Soraya DeChadarevian and Robert Olby for their detailed critiques.

References and Notes

1 M. Kenney, *Biotechnology: The University-Industrial Complex* (New Haven, 1986); id., "Biotechnology and the creation of a new economic space," in A. Thackray (ed.), *Private Science* (Philadelphia, 1998), 131–143; J. Enriquez, "Genomics and the world's economy," *Science* 281 (1998): 925–926. For similar accounts of dramatic change in life scientist-industrialist relations, see E. Yoxen, *The Gene Business* (New York, 1983), and S. Krimsky, *Biotechnics and Society: The Rise of Industrial Genetics* (New York, 1991). For a broader range of analyses of current technology transfer in various sciences, see N. E. Bowie (ed.), *University-Business Partnerships: An Assessment* (Lanham, Mo., 1994).

2 On the history of the term "biotechnology" and its close cognates, see R. Bud, "Biotechnology in the twentieth century," *Social Studies of Science* 21 (1991): 414–457.

3 J. Liebenau, *Medical Science and Medical Industry* (Baltimore, 1987); J. P. Swann, *Academic Scientists and the Pharmaceutical Industry* (Baltimore, 1988), chap. 2; T. Mahoney, *The Merchants of Life* (New York, 1959).

4 E. F. Keller, "The force of the pacemaker concept in theories of aggregation in cellular slime mold," *Perspectives in Biology and Medicine* 26 (4) (1983): 515–521.

5 See Mahoney, *Merchants*, in ref. 3, 73–74; E.C. Kendall, *Cortisone: Memoirs of a Hormone Hunter* (New York, 1971), chap. 7; J. J. Abel, "On epinephrine and its compounds," *American Journal of Pharmacy* 75 (1903): 301–325.

6 See Kendall, *Cortisone*, in ref. 5, chaps. 3–4. On Kendall's advice to Macleod, see M. Bliss, *The Discovery of Insulin* (Chicago, 1982), 132–133 (quote); also see R. Apple, "Patenting university research: Harry Steenbock and the Wisconsin Alumni Research Foundation," *Isis* 80 (1989): 375–394. On the history of the Mayo Clinic, see L. Wilder, *The Mayo Clinic* (Rochester MN, 1936). On thyroid hormone, see E.C. Kendall, "The isolation in crystalline form of the compound containing iodine, which occurs in the thyroid, its chemical nature and physiologic activity," *Journal of the American Medical Association* 64: 2042–3 (1915); id., "Isolation of the iodine compound which occurs in the thyroid," *Journal of Biological Chemistry* (1919) 39: 125–147; id., *Thryroxine* (New York, 1929). See anon., 1919 (otherwise n.d.), copy of licensing agreement between Squibb and University of Minnesota regarding US patent application 102,302, dated 7 June 1916, in E. C. Kendall papers, Princeton University Archives, collection c0669 (hereafter, Kendall Papers), carton 8, folder "Patents – Thyroxin."

7 R. Tattersall, "Pancreatic organotherapy for diabetes, 1889–1921," *Medical History* 39 (1995): 288–316; Swann, *Academic Scientists*, chap 5; Bliss, *Discovery*.

8 On the role of industry in providing materials for sex hormone research, see N. Oudshoorn, "United we stand: The pharmaceutical industry, laboratory, and clinic in the development of sex hormones into scientific drugs, 1920–1940," *Science, Technology, and Human Values* 18 (1993): 5–24, and A. Clarke, "Research materials and reproductive science in the United States, 1910–1940," in G. Geison (ed.), *Physiology in the American Context, 1850–1940* (Baltimore, 1987), 323–369. On Laqueur and Organon, see N. Oudshoorn, *Beyond the Natu-*

ral Body (London, 1994), chap 5 (quote is Robert J. Frank,1929, cited on 88); on Doisy, see A. E Doisy, "An autobiography," *Annual Review of Biochemistry* 45 (1976): 1–9; also anon., *Nobel Lectures in Physiology or Medicine 1942–1962* (Amsterdam, 1964), 3–7, 27–28. On Parke-Davis's use of Doisy's patents (US Pats. nos. 1,967,350 and 1,967,351, of 1934) and name in the production and marketing of Theelin for 'female disorders' both physical and mental see, e.g., Parke-Davis advertisements "Theelin-Theelol" and "What's come over Sally?" in the advertising sections of *The New England Journal of Medicine* vol. 221, nos. 6 and 11 (1939) respectively. A more detailed analysis of Parke-Davis and Doisy's interaction in product development will be forthcoming from the author.

9 For details and archival sources on Bonner's 1930s thiamin research described in this following paragraphs, see N. Rasmussen, "The forgotten promise of thiamin: Merck, Caltech biologists, and plant hormones in a 1930s biotechnology project," *Journal of the History of Biology* 32 (1999): 245–261.

10 A. E. Hitchcock, "The practical use of root hormones," *Horticulture* 15 (1937):239–240. See advertising campaign for Hormodin A in *Horticulture* from 1937–1939, e.g., advertisements in *Horticulture* 15 (1937): 224, and *Horticulture* 16 (1938): 242. R. R. Williams to A. N. Richards, 16 April 1936, University of Pennsylvania Archives collection UPT50/R514 (hereafter, Richards Papers), box 16, folder 19.

11 J. Bonner and J. Greene, "Vitamin B1 and the growth of green plants," *Botanical Gazette* 100 (1938): 226–237 (quote, 226); J. Bonner and P. Devirian, "Growth factor requirements of four species of isolated roots," *American Journal of Botany* 26 (1939): 661–665.

12 H. B. Logan, "News of an exciting discovery," *Better Homes and Gardens* (October 1939): 13, 94, 104–105; quote p. 94. On the role of the Research Corporation in mediating between academia and industry in the first half of the twentieth century, see C. Weiner, "Patenting and academic research: Historical case studies," *Science, Technology, and Human Value* 12 (1987): 50–62. In this case the Research Corporation funds channeled to Bonner were almost certainly royalties paid by Merck to Caltech accounts from their licensing of vitamin B related patents such as those belonging to Caltech chemist Edwin Buchman.

13 H. Marks, *The Progress of Experiment* (Cambridge, UK, 1998).

14 J. Bonner, "Effects of application of thiamin to cosmos," *Botanical Gazette* 104 (1943): 475–479. "Amaze your friends" advertisement, *Horticulture* 18 (1940): 59.

15 M. Fishbein, "Medical patents," *Journal of the American Medical Association* 109 (1937): 1539–1543, on 1542.

16 Swann, *Academic Scientists*, chap. 3. On the OSRD see H. Dupree, "The great instauration of 1940: The organization of scientific research for war," in G. Holton (ed.), *Twentieth Century Sciences: Studies in the Biography of Ideas* (New York, 1972). For an assessment of the overall impact of wartime government contracts on life science in the United States, see N. Rasmussen, "Of 'small men', big science, and big business: The second world war and biomedical research in America," submitted to *Journal of the History of Medicine and Allied Sciences.*

17 E. C. Kendall, *Cortisone* (New York, 1971); also see N. Rasmussen, "Steroids in arms: Hormones of the adrenal cortex and US military research, 1940–45," submitted to *Medical History.*

18 See Rasmussen, "Steroids in arms", in ref. 17; G. G. Hazen, "Cortisone: The quest for oxygen at C-11," *Journal of Chemical Education* 57 (1980): 291–293; G. Hetenyi, Jr., and J. Karsh. "Cortisone therapy: A challenge to academic medicine in 1949–1952," *Perspectives in Biology and Medicine* 40 (1997): 426–439; E. C. Kendall, *Cortisone*; H. Marks, "Cortisone, 1949: A year in the political life of a drug," *Bulletin of History of Medicine* 66 (1992): 419–439. D. Cantor, "Cortisone and the politics of drama, 1949–1955," in J. V. Pickstone (ed.), *Medical Innovations in Historical Perspective* (New York, 1992), 165–184.

19 Information for this and the following paragraph on penicillin are based on the following works. J. P. Baxter, *Scientists Against Time* (Boston, 1946), chap. 22; P. Neushul, "Science, government, and the mass production of penicillin," *Journal of the History of Medicine and Allied Sciences*, 48 (1993):

371–395; J. Sheehan, *The Enchanted Ring: The Untold Story of Penicillin* (Cambridge, MA, 1982); J. P. Swann, "The search for synthetic penicillin during World War II," *British Journal for the History of Science* 16 (1983): 154–190; Y. Aharonowitz and G. Cohen, "The microbial production of pharmaceuticals," *Scientific American* (September 1981): 141–152.

20 F. O. Schmitt, *The Never-Ceasing Search,* chap. 9; C. E. Hall, "Recollections from the early years: Canada – USA," *Advances in Electronics and Electron Physics, Supplemental* 16 (1985): 275–296, and N. Rasmussen "Making a machine instrumental: RCA and the wartime beginnings of biological electron microscopy," *Studies in the History and Philosophy of Science* 27 (1996): 311–349. Also see F. O. Schmitt, "Interim Progress Report," 25 October 1943, United States National Archives Record Group 227, Entry 29, box 95, folder OEMcmr-286.

21 Ibid. Also see CMR Meeting Minutes, 17 December 1942; Richards Papers, carton 12, folder 36, and CMR meeting minutes 23 September 1943, Richards Papers, carton 12, folder 42.

22 The nerve sheaths were first requested and tried to especially good effect by Paul Weiss, then of University of Chicago. See F. O. Schmitt, "Final Report, Contract OEMcmr-286," 7 January 1942, United States National Archives Record Group 227, Entry 29, box 95, folder OEMcmr-286; also E. L. Howes, "Experimental wound healing," in E. C. Andrus (ed.), *Advances in Military Medicine* (Boston, 1948), 111–122.

23 For details and archival sources concerning the herbicide-related research discussed in this and following paragraphs, see N. Rasmussen, "Plant hormones in war and peace: Science, industry, and intellectual property in the development of herbicides in 1940s America," forthcoming in *Isis.*

24 P. Marth and J. W. Mitchell, "2,4 Dichlorophenoxyacetic acid as a differential herbicide," *Botanical Gazette* 106 (1944): 224–232; C. L. Hamner and H. B. Tukey, "Selective herbicidal action of midsummer and fall applications of 2,4 – Dichlorophenoxyacetic acid," *Botanical Gazette* 106 (1944): 232–245.

25 *Proceedings of the Second Annual North Central States Weed Control Conference,* 1945, 75, on 78. On the extent of 2,4-D use in wheat production, see L. M. Stahler, "The national picture in aerial spraying for weed control in 1949," *Proceedings of the Sixth Annual North Central States Weed Control Conference* (1949), 37–41 (these acreage figures for the Plains region include Canada, while the production figures did not). On the various agricultural use of herbicides, see e. g., R. J. Aldrich and C. J. Willard, "The effect of post-emergence 2,4-D on Corn," Ibid., 55–56; R. F. Carlson, "Control of weeds in strawberry plantings by the use of 2, 4 dichlorophenoxyacetic acid," *Proceedings of the American Society for Horticulture Science* 49 (1947), 221–223; C. J. Willard, "Plant regulators for weed control," in H. B. Tukey (ed.), *Plant Regulators in Agriculture* (New York, 1954), 184–201.

26 Kraus to Emery Filbey, 15 August 1947, and Emery Filbey to Robert Hutchins, 21 August 1945; President's Papers, University of Chicago Archives, 1945–1950, box 7, folder 13.

27 The concept of the "habitus", which can be defined as the set of cultural dispositions informing and underlying the replication of patterns of work, economic exchange, and social interaction, of course derives from Pierre Bourdieu. See P. Bourdieu, *Homo Academicus,* trans. P. Collier, (Palo Alto, 1988), and id., *The Logic of Practice,* trans R. Nice (Palo Alto, 1990).

28 On OSRD management of wartime life science, see Rasmussen, " 'Small men,' big science" in ref. 16. According to one historian this form of research contract traces its origins to World War I chemical warfare research brokered by the NRC. See D. P. Jones, "Chemical warfare research during World War I: A model of cooperative research," in J. Parascandola and J. Whorton (eds), *Chemistry and Modern Society* (Washington, 1983), 165–185. The NRC acted as the main clearinghouse for industrial and philanthropic funding to academic science in the interwar period; see, e. g., G. Bugos, "Managing cooperative research and borderland science in the National Research Council, 1922–1942," *Historical Studies in the Physical Sciences* 20 (1989): 1–32.

29 The concept belongs to historian E. Hawley, *The Great War and the Search for a Modern Order* (New York, 1979). For history

of science work illustrating this form of cooperation, see Bugos, "Managing cooperative research"; R. Kargon, "Temple to science: cooperative research and the birth of the California Institute of Technology," *Historical Studies in the Physical Sciences* 8 (1977): 3–31; L. Kay, "Cooperative individualism and the growth of molecular biology at Caltech 1928–1953," Ph. D. diss., Johns Hopkins University, 1987.

30 Much in this paragraph's argument has been said before by others. See S. Wright, "Recombinant DNA technology and its social transformation, 1972–1982," *Osiris* 2 (1986): 303–360; Swann, *Academic Scientists*, Epilogue. On Genentech see S. Hall, *Invisible Frontiers: The Race to Synthesize a Human Gene* (New York, 1985).

31 P. Pauly, *Controlling Life: Jacques Loeb and the Engineering Ideal in Biology* (New York, 1987).

32 For instance, on the early consideration of these protein drugs see William Castle to A. N. Richards, 14 May 1958, Richards Papers, box 16, folder 24; Anon., 8 June 1961, "Report of Merck Institute Scientific Advisors Meeting," Ibid., box 21, folder 11; Anon., 23 December 1963, "Report of Merck Institute Scientific Advisors, Meeting of October 27, 28 and 29," Ibid., box 16, folder 50. On erythropoietin, interferon, human insulin, and the nature of recent biologist-industrialist interactions see, e.g., A. Kornberg, *The Golden Helix: Inside Biotech Ventures* (Sausalito, 1995).

33 R. Kohler, *Partners in Science: Foundations and Natural Scientists, 1900–1945* (Chicago, 1991). For critiques of too-narrow historiographic constructions of "molecular biology," see D. Zallen "Redrawing the boundaries of molecular biology: The case of photosynthesis," *Journal of the History of Biology 26* (1993): 65–87, and N. Rasmussen, "The midcentury biophysics bubble: Hiroshima and the biological revolution in America, revisited," *History of Science* 35 (1997): 245–293. For standard historiography deriving the biotechnological turn primarily from molecular genetics, see E. Yoxen, "Giving life new meaning: The rise of the molecular biology establishment," in N. Elias, H. Martins, and R. Whitley (eds), *Scientific Establishments and Hierarchies* (Dordrecht, 1982), 123–143; L. Kay, *The Molecular Vision of Life* (Oxford, 1993), M. Morange, *A History of Molecular Biology*, trans. M. Cobb (Cambridge, MA, 1998).

34 See Wright, "Recombinant DNA technology" in ref. 30.

35 For instance, in the mid-1980s Hoechst made such an argument in its unsuccessful court battle to operate a fermentation factory for recombinant human insulin under older environmental regulations; see H. Gottweiss, *Governing Molecules* (Cambridge, MA, 1999), chap. 3.

36 On changes in entomology, see P. Palladino, *Entomology, Ecology and Agriculture: The Making of Scientific Careers in North America, 1885–1985* (Amsterdam, 1998); R. Van den Bosch, *The Pesticide Conspiracy* (New York, 1978).

12.
Polymer Science: From Organic Chemistry to an Interdisciplinary Science

Yasu Furukawa

Polymer science, or macromolecular science, grew to be an important research area in the twentieth century, thanks to its scientific importance in the study of matter and life as well as its far-reaching industrial applications. While its central core has remained in the domain of chemical science, the field grew by incorporating and interacting with other sciences, such as physics and biology.

Polymer science has been divided roughly into two major branches: polymer chemistry and polymer physics. This division is illustrated by the *Journal of Polymer Science*, a primary journal for the field, which is published in the United States. According to its editorial policy, the "Polymer Chemistry Section" deals with "the synthesis of polymers, reaction mechanisms, kinetics, and other areas of the organic and physical organic chemistry of macromolecules", while the "Polymer Physics Section" contains "those papers dealing with macromolecules in solid state and solution", specifically studies of "such topics as crystallization and crystallinity; polymer morphology; crystal physics; mechanical, optical, and other physical properties; molecular conformation; thermodynamics, and statistical thermodynamics."[1] Thus, polymer chemistry consists of organic chemistry and some physical chemistry related to polymerization reactions; polymer physics includes not only physics but also a good portion of physical chemistry. Generally, the polymer science community has adopted this demarcation which, though conventionally, is not rigorously applied.[2] "Polymer biology" has yet become dominant. Still, polymer science has increasingly embraced the subject of biopolymers, such as nucleic acids and proteins.

This essay seeks to examine how the science of polymers came to acquire its interdisciplinary nature. It will do so by illuminating the historical vicissitudes of such key elements as concepts, methodology, terminology, pedagogy, textbooks, professional journals, and scientific debates. The focus will be upon the German and American scientific communities, the cradle of polymer science.

12.1
Macromolecular Chemistry as a New Branch of Organic Chemistry

In his lecture at the 1982 Goodyear Innovation Conference, Paul J. Flory, an American polymer scientist and Nobel laureate, described what he believed to be fundamental to polymer science:

> Polymer science dates from the recognition that polymers – such as rubber, cellulose, polystyrene, to name a few – consist of the very large molecules that we call macromolecules. More specifically, they consist of long chains of atoms linked by chemical bonds. The number of atoms in these chains usually runs into the thousands. This rudimentary, but fundamental, conception of the molecular constitution of polymeric substances is the cornerstone of modern polymer science. Without it, a science of polymers could not have been founded and elaborated. [3]

Flory's perception gets straight to the intellectual origin of a discipline; it stresses how this science owed its birth to a new concept. Given that polymer science is inseparable from the conception of the macromolecule, it is an irony that the founder of this concept struggled to dismiss the word "polymer" throughout his scientific career. The German organic chemist Hermann Staudinger first proposed the long-chain molecular structure for such polymers as rubber and polyoxymethylene in 1917. [4] Five years later he introduced the term "macromolecule" (Makromolekül) to designate the long-chain molecules. [5] In so doing, he began a lengthy struggle against the prevailing notion of polymers.

The study of polymers was not new in Staudinger's time. Ever since the nineteenth century, a number of chemists, with both practical and scholarly concerns, had been working on rubber, cellulose, resins, and proteins. These materials were known to be "polymers" (from the Greek word for "many parts", as introduced by Jöns J. Berzelius in 1832) – that is, bodies having molecules in which the same atomic groups were arranged repeatedly, without regard to the size of the molecules. As they possessed a colloidal nature, they were also the major subject for colloid chemistry, a field which grew under the influence of physical chemistry. Leading colloid chemists like Wolfgang Ostwald suggested that a colloid was a physical state of matter into which any substance could be brought under appropriate conditions, and that colloid particles in solution were too small to be seen, yet too large to be identified as molecules. In parallel with the colloid concept, organic chemists (including Emil Abderhalden, Carl D. Harries, Max Bergmann, Paul Karrer, Rudolf Pummerer, Hans Pringsheim, and Kurt Hess) elaborated a theory that explained the sub-particulate structure of colloid polymers, a task which colloid chemists had largely left aside. According to this theory, polymers were the aggregates of relatively small cyclic molecules held together by certain physical forces – "secondary" or "partial" valences – that seemed to cause such peculiar properties of polymers as colloidality and elasticity. On the basis of the data from X-ray diffraction measurements of rubber and fibers, a number of crystallographers supported this view. [6]

The conservative organic chemist Staudinger was an anti-physicalist to the extent that he distrusted almost any method or concept of physical chemistry and physics. A staunch believer in Kekuléan organic structural chemistry, he maintained that all physical and chemical properties of organic matter must stem from the inner molecular structure and not from the physical forces outside the molecules. Polymers, constituting carbon-rich organic substances, were by no means an exception. He suggested that colloid particles were themselves macromolecules, composed of a considerable number of atoms linked together by the Kekulé valence bonds. The colloidal nature of polymers was due to the structure of macromolecules. Colloid phenomena should therefore not be interpreted on principles of colloid chemistry, but rather on those of organic chemistry.

By using this approach, Staudinger attacked the orthodoxy of colloid chemistry and the physicalist perception of polymers. The German Colloid Society's symposium, "Organische Chemie und Kolloid Chemie", held in Frankfurt in 1930, illustrates the ensuing territorial clash between Staudinger, the defender of organic chemistry, and the colloid chemical community.[7] In addition, Staudinger's 1940 textbook, *Organische Kolloidchemie*, documented his manifesto in which he rewrote colloid chemistry in terms of organic chemistry.[8]

Together with a large following, Staudinger struggled to demonstrate the macromolecular structure of various polymers (such as rubber, cellulose, polysaccharides, polyoxymethylene, polystyrene, and polyindene) by mobilizing organic analysis and synthesis (using such reactions as hydrogenation, bromination, and saponification). By the mid-1920s, he was convinced that the field of polymers was not simply a part of classical organic chemistry but a promising new direction in organic chemistry. Unlike ordinary organic substances, macromolecular substances were composed of molecules, not of identical but various sizes owing to their enormity. They therefore evaded representation by a single molecular formula; their molecular weights could only be expressed by average values rather than by precise numbers. The shape of macromolecules also affected the physical and chemical properties of substances far more strongly than was the case with ordinary compounds. All these aspects, which sprang from the macromolecular concept, necessitated a departure from the conventional concepts of organic chemistry. Thus, by the time the Society of German Natural Scientists and Physicians met in Düsseldorf in 1926, Staudinger was able to declare that the concept of macromolecules was opening up a new horizon for organic chemistry: "Despite the large number of organic substances which we know today, we are only standing at the beginning of the chemistry of true organic compounds and have not reached anywhere near a conclusion."[9]

Despite his isolation from the rest of the chemical community during the late 1910s and the 1920s, Staudinger launched a coherent research program along the lines of macromolecular organic chemistry at the Eidgenössische Technische Hochschule in Zurich and the University of Freiburg (training 25 doctoral students by 1930 and 32 more by 1940) and began acting as a discipline-builder of this new field.

On the other side of the Atlantic, the American organic chemist Wallace H. Carothers began applying basic reactions of organic chemistry, such as condensations, to the synthesis of macromolecules. His fundamental research at the Du Pont Company, which began in 1928, strongly supported Staudinger's position and eventually resulted in the invention of the synthetic fiber nylon. [10]

By the early 1930s, after numerous stormy debates, the concept of macromolecules had been accepted by the majority of scientists. The Faraday Society general discussion on polymerization – the first international conference devoted exclusively to general studies on polymers, and held at Cambridge in 1935 – convinced the audience of the birth of a new field which, as one participant put it, "had grown into a full-scale science with unexpected new vistas for intensification of understanding and expansion of application." [11]

12.2
From Macromolecular Chemistry to Polymer Science: Staudinger, Mark, and the Naming of a Discipline

Staudinger preferred to call his science "macromolecular chemistry (makromolekulare Chemie)" rather than "polymer chemistry." Although during the 1920s he sporadically called the field "the chemistry of high-molecular [weight] organic substances (Die Chemie der hochmolekularen organischen Stoffe)", Staudinger dismissed that name in favor of "macromolecular chemistry" after the mid-1930s. As editor of several journals, he applied the latter appellation to their titles: *Journal für praktische Chemie: Unter Berücksichtigung der makromolekularen Chemie* (from 1940), *Journal für makromolekulare Chemie* (run from 1943 to 1944), and *Die makromolekulare Chemie* (established in 1947). [12]

Staudinger's dislike of the term "polymer chemistry" and dismissal of "the chemistry of high-molecular organic substances" is understandable in the light of their historical context. As we have seen, before Staudinger chemists had widely used the term "polymers"; however, during the 1920s many of them considered polymers to be made up of small molecules. The term "hochmolekulare organische Stoffe" had also been used in the early decades of the twentieth century, but again, it was not intended to mean the substances of large molecules. Many chemists did not regard the reported high values of molecular weights of polymers as evidence for the big "chemical molecules." Instead, they considered them to be apparent values of physical aggregations of small molecules. In short, at the time both "polymers" and "high-molecular substances" meant compounds of small molecules.

Staudinger's coinage of "macromolecule", which made a very clear distinction from previous usage, was remarkably effective in avoiding unnecessarily anachronistic terminology. The meaning behind the jargon term "polymer" changed after the acceptance of the macromolecular theory. In other words, in scientists' minds the chemistry of polymers became synonymous with the chemistry of macromolecules only after they adopted the reinterpretation of the polymer structure. Thus, to Staudinger, "polymer chemistry" was but a relic of the past. Under his influence, "makromolekulare Chemie" came into wide use in German scientific circles.

Despite Staudinger's advertisement, "macromolecular chemistry" did not prevail as the official terminology in the United States. A fairly wide acceptance of "polymer chemistry" by English speakers is, again, related to the history of its usage. Throughout his scientific career in the 1920s and 1930s, the American counterpart Carothers occasionally called his field "the chemistry of macromolecular materials." [13] We find no mention of "macromolecular chemistry" or "polymer chemistry" – even in the obituaries on Carothers written by his colleagues in 1939 and 1940, in which they described his study of polymers and polymerization at some length. [14]

Undoubtedly, it was Herman F. Mark who was responsible for the widespread usage of "polymer chemistry" and "polymer science." Earlier, the Austrian-born chemist Mark favored the aggregate theory, but then around 1928 adopted aspects of Staudinger's macromolecular theory. Together with Kurt H. Meyer at I. G. Farbenindustrie, Mark developed the so-called new micelle theory, an alternative long-chain molecular view which was similar to Staudinger's theory, but which embraced a compromise between the macromolecular theory and the aggregate theory. This choice caused a bitter controversy with Staudinger, which lasted from the late 1920s until well into the early 1940s. After serving as professor at the University of Vienna, Mark fled from Nazi-occupied Austria to Canada and eventually settled in the United States. As professor at the Polytechnic Institute of Brooklyn, New York since 1940, he became America's foremost leader and an influential educator in this new field. [15]

Given this background, Mark was loath to call the field "macromolecular chemistry," a name associated with Staudinger. So, during the late 1930s, he coined an alternative name, "high polymer chemistry (hochpolymere Chemie)", as employed in his 1937 textbook and other papers. [16] In 1940, he launched a monograph series in this field, *High Polymers,* published by Interscience Publishers. This series was to become widely-circulated, and to date has amounted over forty volumes. A decade later, Mark played a central role in establishing the American Chemical Society's Division of High Polymer Chemistry which, omitting the word "high", was soon renamed "the Division of Polymer Chemistry" (today, the second largest division in the American Chemical Society). [17] American chemists followed this convention, as exemplified by the titles of such dominant textbooks as Paul Flory's *Principles of Polymer Chemistry* (1953) and Fred Billmeyer's *Textbook of Polymer Chemistry* (1957). [18]

When he received a belated Nobel Prize in 1953, Staudinger proudly confirmed that, "Macromolecular chemistry is the youngest branch of organic chemistry." [19] Despite his confidence, by this time Staudinger's assertion looked somewhat outdated to many of his contemporaries. Whereas to Staudinger "macromolecular chemistry" meant essentially a branch of organic chemistry, to Mark and his American followers "polymer chemistry" meant more than that: a combination of the organic chemistry and physical chemistry of polymers. With this broader definition in mind, Mark went further by applying the term "science" to the field. When in 1946 he established the first English periodical devoted to the field, published by Interscience Publishers, he entitled it the *Journal of Polymer Science* –

perhaps his earliest public use of this general term. He also launched the *Journal of Applied Polymer Science* in 1959 and edited the monumental *Encyclopedia of Polymer Science and Technology*, which appeared between 1965 and 1977.[20]

Mark's naming of "polymer science," in which no rigorous distinction was made from his usage of "polymer chemistry," stemmed in part from his scientific personality as a generalist as well as his own interests in both chemistry and physics. Here again, we see a reflection of marked contrast in scientific style between the two leaders, which affected both the naming of the discipline and the nature of its pedagogy. As a purely organic chemist, Staudinger on the one hand explored the field of macromolecules from the standpoint of organic chemistry, with a strong distaste for physics and physical chemistry. Mark on the other hand, while adept at organic chemistry, approached polymers primarily as a physical chemist, utilizing such physical methods as X-ray crystallography. As Mark recalled:

> We [Staudinger and Mark] both favored the concept of long-chained molecules. He [Staudinger] did, on the basis of organic chemistry; and I did, on the basis of X-ray diffraction. He only trusted organic chemistry. I said trust both (techniques); we have two methods which do not contradict. My God, they could have contradicted![21]

While Staudinger had trained exclusively organic macromolecular chemists at Zurich and Freiburg, Mark was quick to make his polymer education interdisciplinary. His core curriculum at Vienna in the late 1930s consisted of the organic chemistry, physical chemistry, and physics of polymers. In the United States, by the late 1940s, Mark had made Brooklyn Polytechnic a center for a graduate program that covered all areas of polymer science: organic chemistry, physics, physical chemistry, biochemistry, and industrial applications. A number of newly educated chemists worked and helped to teach in this new field at the Institute, under Mark's directorship. In addition to Mark's introductory course, specialized courses offered then included "Polymerization Kinetics" by Turner Alfrey (Mark's 1943 Ph. D. student), "Solution Properties of High Polymers" by Frederick R. Eirich (Mark's former assistant at Vienna), "Organic Polymer Chemistry" by Charles G. Overberger (Carl S. Marvel's 1944 Ph. D. and an organic chemist whom Mark hired in 1946), and "Chemistry of Proteins," by Douglas McLaren (a specialist in the physical chemistry of proteins). The Brooklyn school was eventually to train a large number of versatile polymer scientists.[22]

12.3
The Rise of Polymer Physics

Mark's use of the word "science" not only arose from his own personal taste but also reflected the emergence of polymer physics as a new subdiscipline. By the late 1930s, the limitations of Staudinger's organic-chemical approach were recognized by physical chemists and physicists alike. Organic chemistry alone could not solve the whole problem of macromolecules. Clearly, the organic chemists' static views of molecules had failed to take into account the dynamics of macromolecules. Indeed, it were the physical chemists and physicists who most welcomed and favored the name "polymer science."

Then, after adopting the macromolecular concept, physicists and physical chemists objected to Staudinger's notion that macromolecules were as rigid as thin rods – a notion deriving from his viscosity formula, the single elementary physical device he allowed to be employed for measuring the molecular weight of macromolecular compounds. On the basis of physical considerations (such as spectroscopic data), these scientists advanced the view that macromolecules were flexible, and it was here that they found a golden opportunity to exploit their mathematical skills and physical methods. The study of the distribution of molecular weights, sizes, and shapes, and of the behavior of macromolecules seemed amenable to probability, statistics, kinetics, thermodynamics, and hydrodynamics, an aspect which "normal" small-molecular substances did not possess.

Physical studies of polymeric substances had a long history. For example, as early as the 1800s, the British scientist John Gough studied the elasticity of rubber, in terms of the principles of heat and mechanics of his day. In 1906, the physicist Albert Einstein calculated the viscosity of a suspension of colloid spheres in a hydrodynamic continuum. Colloid chemists carried out investigations on the morphology and mechanical properties of colloid systems, diffusion, dialysis, filtration, precipitation, electrophoresis, surface energies, and viscosity. In the mid-1920s the physical chemist The Svedberg developed the ultracentrifuge, a powerful physical apparatus for the molecular-weight measurement of proteins. A number of scientists, including physicists, physical chemists, and supporters of the aggregate theory, applied X-ray analysis to their studies of fibers and rubber. All these physical studies had been advanced even before Staudinger began formulating macromolecular chemistry.

Given this context, it would be tempting to say that polymer chemistry and polymer physics co-existed from the very beginning of polymer science; in other words, this field was an "interdisciplinary science" from the outset. However, the existence of physical studies of polymers and the establishment of polymer physics as a discipline are not necessarily one and the same. As I have shown earlier, historically, the concept of the macromolecule originated from organic chemistry, not from physical chemistry or physics. This concept then became an essential starting point from which scientists would formulate macromolecular chemistry as a new organic chemistry. To some who had employed the physical approach, the ultimate constitution was not a primary consideration at all. To others, polymers were composed of quite different entities from what polymer scientists later perceived. Only after polymers were recognized as macromolecular substances, did scientists give new meanings to these precursory physical studies, which led to further studies. It is true that soon after the establishment of macromolecular chemistry, polymer physics emerged. And that remarkably short time lag may well be attributed to the fact that, by then, physical studies had already attained a certain maturity in their own right.

Mark may reasonably be called the initiator of polymer physics. With his long-chain molecular concept, he carried out physical work on polymers even in the late 1920s. Still, it would be difficult to claim that polymer physics was established with the realization of Mark's study at that time. Mark's co-worker, Eugene Guth, considered a theory of rubber elasticity – based on statistical mechanics of flexible macromolecules – to be the definitive step toward the formation of what we now call polymer physics. As Guth claimed: "This was inaugurated in 1934 [in their joint paper] by Guth and Mark, who, therefore, are the founders of the physics and physical chemistry of polymers." [23] Whatever the justification of this claim of priority, it is clearly agreed upon that, in the eyes of polymer physicists, a solid theoretical and methodological ground or paradigm of polymer physics was established after the mid-1930s, and not before.

Critical pioneering studies that would shape polymer physics appeared from the mid-1930s to the late 1940s. [24] The physical properties of rubber formed one of the most vital subjects of this time. By the early 1940s the cause of rubber elasticity had been fully explained, in terms of thermodynamics and statistics. This was then a virgin field, where almost everything physicists and physical chemists attempted could result in a new and original study. Thus, the period between the late 1930s and 1940s marked an exciting formative stage "when polymer science looked easy", as the American physical chemists Walter Stockmayer and Bruno Zimm recalled. [25]

The first generation of physical chemists and physicists, who were to emerge in the 1930s, benefited much from what organic chemists had already accomplished in the field. The former exploited the data which the latter had compiled and also utilized a wide variety of polymers that had been synthesized by the latter. Indeed, physical chemists and physicists typically began their studies by collaborating with macromolecular organic chemists, even before establishing themselves as independent polymer scientists. For example, the physicist Werner Kuhn worked with the cellulose chemist Karl Freudenberg at Heidelberg, and the physical chemist Paul J. Flory with Carothers at Du Pont. However uncomfortable the physical approach may have made him, even Staudinger could not ignore its importance. To catch up with the new trend, he had to employ Günther V. Schulz who remained his only physical chemist – a marriage of convenience which turned out to be fruitful. Even Mark, a physical chemist, invited the able physicist Eugene Guth at Vienna to better his physical studies.

Perhaps industry was more inclined to overstepping disciplinary boundaries than academia. For example, we know that in the 1930s the research management of the Du Pont Company encouraged interdisciplinarity in its study of polymers – an environment in which organic chemists, colloid chemists, physical chemists, and physicists frequently exchanged opinions and worked closely together. [26] Such industrial undertakings required a teamwork atmosphere, whereas disciplinary boundaries at the university were kept more rigid, and the holder of a specialty chair could be expected to dominate the department. At Du Pont, it was common practice for physical chemists to review organic chemists' draft manuscripts before publication and vice versa. While the internal review system served company's intention to protect technological know-how, the cross-disciplinary check – which would not have been customary at the university – no doubt also benefited the industrial research on polymers. In large measure, polymer science owed its growth to the rise of industrial research.

The political situation of the 1940s facilitated cooperation between organic chemists, physical chemists, and physicists and between industry and academy. The U.S. government's wartime synthetic rubber research program, launched immediately after Japan's occupation of the Pacific area in 1942, provided chemists and physicists unique opportunities to work closely together on polymers. The large rubber companies, such as U.S. Rubber, Goodrich, and Goodyear, joined the program, and turned out to mobilize a mix of their chemists and physicists in the absence of trained polymer scientists. On the academic side, the organic chemist Carl S. Marvel at Illinois, the Dutch émigré physicist Peter J. W. Debye, and Flory, both at Cornell were, to name a few, among those involved in this program. In cooperation with industrial chemists Debye, in particular, developed the pathbreaking light scattering method that allowed a quick and accurate molecular weight determination of polymers. [27]

Given this background, there appeared a growing number of physics-related studies of macromolecules during the 1940s. In the United States, these papers were published in various physics-related journals, including the *American Journal of Physics*, the *Journal of Applied Physics*, and the *Journal of Chemical Physics*, as well as the *Journal of the American Chemical Society*. For example, Hubert M. James and Guth's improved theory of the elastic properties of rubber appeared in the *Journal of Chemical Physics* in 1943; and Debye's landmark paper on light scattering in the *Journal of Applied Physics* in 1944. [28] It was in this context that Mark founded an independent journal in 1946, titling it the *Journal of Polymer Science* rather than the *Journal of Polymer Chemistry*, to carry papers dealing with all aspects of polymers.

By the 1960s, polymer physics had increasingly grown to be a vital part of polymer research. From its inauguration in 1960, the British journal *Polymer* had a subtitle, namely, *The Chemistry, Physics and Technology of High Polymers*. In 1966, Mark's journal was split into three sections: "Polymer Chemistry", "Polymer Physics", and "Polymer Letters", a further indication of the independence of polymer physics. The 1966 editorial explained:

> Certainly one of the notable developments of the past few years, as far as the *Journal of Polymer Science* is concerned, has been the great success we have achieved in attracting an increasing number of papers in the area of polymer physics. In the April, 1964, issue of the Journal we indicated editorially our plan to give stronger identification to the growing field of polymer physics ... Since then the influx of manuscripts in polymer physics has gained such momentum that the Editors and publisher of the Journal decided that the physical separation of such papers from the bulk of the material was both practical and desirable. [29]

Each section began publishing a separate volume, and to promote the division of editorial labor, a separate editor and advisory board were appointed for the physics section. By the 1960s, physical chemistry and physics had thus proven not to be foes of organic chemistry (as Staudinger had once feared), but complementary partners in perfecting the science of polymers. At the same time, the polymer community was able to expand the size and scope of its professional activity by winning over to its side physical chemists and physicists who now could share a disciplinary identity as polymer scientists.

12.4 The Biological Nexus

As we have seen, the term "science" was applied to the area of polymer research as it came to involve not only chemistry, but also physics. Biology remained peripheral to the discipline of polymer science. The biological significance to the field, however, had been well recognized from an early stage. After all, polymer scientists were dealing with such biological macromolecules as cellulose, lignins, starch, and proteins. It is not surprising that Staudinger, Meyer, Mark, and Carothers all discussed at some length the biological implications of polymers in their publications.[30] As mentioned before, the generalist Mark, in particular, introduced a biochemical course into his polymer science curriculum at Brooklyn Polytechnic as early as the 1940s.

During the 1930s, a gulf existed between polymer researchers on the one hand and biochemists and biologists on the other, as indicated by the literature of the latter group, which did not even mention the work of the former. This was not due simply to the differentiation in approach and concerns, but also to the lack of communication between the scientific communities. Biochemists and biologists were, by and large, indifferent to the heated macromolecular debate in the polymer science community, just as polymer scientists were unfamiliar with the developments in the biological community.[31] Yet, whether directly or indirectly, polymer science did influence the biological fields, notably molecular biology, a subject searching for a molecular basis to biological processes. The macromolecule formed a conceptual ground that sustained the intellectual framework of molecular biology.

In his autobiography, Staudinger wittily recalled that as a university student he had entered the world of chemistry in order to be a botanist because the best botanist must have a strong chemical background; therefore, his life-long chemical studies were, after all, only a prelude to his true goal.[32] His second marriage, to the plant physiologist Magda Woit, renewed an interest in biology in relation to the macromolecular theory. Falling behind in physico-chemical issues, Staudinger's research interest in the postwar years turned more and more to the biological aspects of macromolecules. In 1947 he published a monograph, *Makromolekulare Chemie und Biologie*, in which he argued that macromolecular chemistry was the key to understanding biological phenomena.[33] This slim volume did not have as much impact on his fellow chemists as the quantum physicist Erwin Schrödinger's 1944 book *What is Life?*, which was to transform physicists into molecular biologists.[34] In tackling the enigma of life, Staudinger aimed to challenge biochemists and biologists to scrutinize the structure of proteins and nucleic acids more intensively. A propaganda piece rather than a rigorous technical work, the book advocated a new vision of biology in which macromolecular chemistry could play a fundamental role in the life sciences.

Makromolekulare Chemie und Biologie, however, incorporated some recent findings by Staudinger's eminent student, Rudolf Signer. In his study, the University of Bern professor Signer had successfully demonstrated the macromolecularity of deoxyribonucleic acid (DNA) by means of flow birefringence. The enormous size of DNA molecules (having a molecular weight between 500 000 and 1 000 000), (as revealed by Signer) would convince biologists that DNA could store genetic information – a conviction that helped dismiss the dominant assumption of proteins as genes. [35]

During this time, we find some fruitful interactions between polymer scientists and future molecular biologists, which resulted in remarkable consequences. For example, it was Mark who inspired the young Linus C. Pauling and Max F. Perutz in their respective works on the precise three-dimensional structures of proteins, the pioneering works that won them each a Nobel prize. [36] When molecular biologists began their search for the precise structure of the DNA macromolecule to clarify the mechanism of replication and transmission of genetic information, Signer helped them by offering an improved DNA preparation. It was Signer's specimen, evaluated as "the best DNA preparation" that enabled Rosalind Franklin to produce a fine X-ray photograph of the paracrystalline B-DNA. [37] Inspired by this picture, James Watson and Francis Crick went on to build the double-helix model of DNA in 1953, a discovery that was followed by the unprecedented growth of research in molecular biology in the 1960s.

Just as molecular biologists benefited from polymer scientists, the rise of molecular biology in return stimulated a number of polymer scientists to work on biopolymers. For example, the polymer chemist Charles Price at the University of Pennsylvania, who began seriously considering the synthesis of life and evolution in the 1960s, studied the alkylation of DNA and the ribonucleic acid (RNA) synthesis. His book, *The Synthesis of Life*, was an outcome of this work. [38] In 1963, Mark's *Journal of Polymer Science* added a section on "Biopolymers," which was superseded by a separate new journal, Biopolymers, with Mark, Melvin Calvin, Marshall Nirenberg, and Pauling on the editorial board. It would become a common practice that a textbook on polymer science included, even to date, a chapter on biological macromolecules (such as polypeptides, proteins, nucleic acids, polysaccharides, and synthetic biopolymers), discussing their structural details, conformations, and functions.

12.5 The Problem of Interdisciplinary Science

As polymer science enlarged its scope, strengthening more and more its interdisciplinary character, research activities within the discipline became increasingly specialized. The *Journal of Macromolecular Science*, an American periodical established in 1967, aimed to "facilitate a complete overview of the science of macromolecules." Thus, the publisher Marcel Dekker stated:

> A broad approach is intended, ranging through the chemical, physical, biological, and engineering sciences. It is the aim of the *Journal of Macromolecular Science* to create a new dialogue and broader awareness of inter- and intradisciplinary problems and progress. [39]

The journal began serving as a means to unify the diverse knowledge in polymer science. By the 1960s, there had been strong appeals from both academic and industrial scientists for the expansion of polymer science education in America. The 1962 *National Register of Scientific and Technical Personnel*, prepared by the National Science Foundation, listed 25% of all organic chemists in the United States as engaged in the chemistry of polymers, such as plastics, resins, and rubber, and 15% of all physical chemists in polymer chemistry. [40] However, most of them were not university-trained polymer chemists but those who had to learn through their own efforts by means of a rather painful process of many years' duration. University education lagged behind the rapid growth of the polymer industry. Even when polymer science began flourishing in graduate programs at American universities (including Brooklyn Polytechnic, Illinois, Case-Western Reserve, and Akron), the undergraduate curricula in chemistry remained far from meeting urgent needs of the industrial sector. As a consequence, the American Chemical Society organized several symposia during its national meetings to discuss the matter. At one such symposium, the inclusion of the subject of polymers in existing courses, either in organic chemistry or in physical chemistry, was recommended as a realistic first step toward the adoption of this field in the curricula. [41] At another symposium, held in 1968 in Chicago, Robert W. Cairns, Vice President of Hercules Company and President of the American Chemical Society, made a more demanding proposal: the creation of a well-designed, independent undergraduate course in polymer science in every university:

> Any student seriously interested in chemistry as a career should be expected to include this polymer science course in his curriculum. This course should be treated as a special opportunity to counteract the increasing fragmentation of knowledge. Background information and contributions should be included on physics, chemistry, biology, and basic engineering principles as they apply to macromolecules. The interplay and interrelationships of these sciences should be shown by specific examples, thus reinforcing the students' belief in the integrity of the real world and the unity of man's knowledge of it. Interdisciplinary science has become a goal of the modern university; polymer science is an example par excellence. [42]

The undergraduate curriculum in chemistry should provide a basis for understanding specialized knowledge in many fields of research and development, for example, organic synthesis, chemical kinetics, thermodynamics, biochemistry, biophysical chemistry, genetics, chemical processing, and solid state physics. Polymer science would be best suited to this end. Thus, Cairns intended to spread polymer science in university education by appealing to its interdisciplinary nature, just as physical chemistry as a "general chemistry" (an *allgemeine Chemie* that would serve as a framework for all chemical knowledge) had been successfully installed into undergraduate curricula in American universities a half century ago.[43] Cairns' claim, however, failed to attract solid support, as the very nature of interdisciplinary polymer science increasingly posed serious problems to specialized university teachers in their field. The paucity of polymer science in chemistry curricula would continue to be an issue well into the 1970s.[44]

12.6
Polymer Science versus Macromolecular Science: Continuing Strife

Finally, we shall return to the problem of the naming of the discipline. As we have seen, Staudinger named his field "macromolecular chemistry" to designate it as a new organic chemistry. Throughout his career, he did not venture to use the name "macromolecular science." But, as the field acquired an interdisciplinary character, the latter name was adopted in some quarters as a natural extension of "macromolecular chemistry."[45] "Polymer science," which had the connotation of interdisciplinarity from the outset of its usage, had been employed by Mark and his followers decades before "macromolecular science" came into use.

The predominant usage of the name "polymer science" over "macromolecular science" persists to the present. While "*Polymerwissenschaft*" has been increasingly adopted in Germany, the German equivalent to "macromolecular science", "*Makromolekulare Wissenschaft*", does not exist even to this day. Magda Staudinger-Woit, Hermann's widow and loyal collaborator, pleaded for her husband's term "macromolecule," which she feared was falling into disuse. After Staudinger's death in 1961, she remained an influential figure in German chemical circles as well as on the editorial board of *Macromolecular Chemistry and Physics*, an English-language German journal which was renamed *Die makromolekulare Chemie* in 1994. Uncomfortable with the official wide usage of the word "polymer" by the Nomenclature Commission of the International Union of Pure and Applied Chemistry (IUPAC), the authoritative organ for chemical nomenclature, she published her critical comments in *Terminology* in 1996:

> Time has come to give careful consideration to the historical development of both terms: "polymer" and "macromolecule". The decision should definitely be in favor of the well defined term macromolecule. This term should be applied to the titles of publications, including journals and books, and particularly to the names of institutions of scientific research. They should be named "for macromolecular science" and no longer include the term "polymer."[46]

She called on the Commission to adopt officially "macromolecular science" as the central science of all related areas of chemistry, physics, and biology:

> Originating from organic chemistry, macromolecular science developed several branches: macromolecular chemistry and physics, materials sciences, molecular biology, supramolecular structures, etc. Therefore, the central term comprising

> the whole domain derived from macromolecular chemistry should be placed under the heading "Macromolecular Science." [47]

"It is a pity," she explained in terms of history, "that due to World War II a separation of scientists took place, thereby interrupting a normal development of nomenclature, for this reason the old term "polymer" continued to exist." [48] She was right in identifying the war as one reason for contributing to the break-up of the consensus of the scientific community. But the old rivalry between Staudinger and Mark, which determined the original usage, had failed to be recognized by the scientific community.

In reply to Staudinger-Woit, Aubrey D. Jenkins, on behalf of the IUPAC Commission, wrote a "Letter to the Editor" which appeared in the subsequent issue of the same journal. Evading the matter of naming the discipline, he instead discussed the definitions of the two terms, polymer and macromolecule. He pointed out that both terms had separate definitions that identified them unequivocally: "macromolecule" referred essentially to a molecule while "polymer" referred to a substance. [49] The letter concluded:

> The commission does not share Frau Dr. Staudinger-Woit's view that the term polymer is redundant, but it does believe that macromolecule is the best word to denote what may less elegantly be called a polymer molecule, and the use of Staudinger's word will be perpetuated if the polymer-science community follows the recommendations of the commission. [50]

Magda Staudinger-Woit did not live to see this letter, as it was published shortly before her death in April 1997 at the age of 94. Whereas her claim appealed to an historical justification, the committee gave priority to the present-day conventional usage in the scientific community. After all, the Commission provided a definition that a polymer was "a substance composed of macromolecules," a pragmatic solution which, however, blanks the history of strife surrounding the two words. [51]

Acknowledgments
This essay is a revision of my paper read at the 1999 Munich CHMC conference, "Between Physics and Biology: Chemical Sciences in the Twentieth Century." I have benefited from the insightful comments of Peter J. T. Morris and Teiji Tsuruta. Thanks also to Elizabeth Sandager for her kind assistance.

References and Notes

1 "Editorial," *Journal of Polymer Science*, Part A-1, 4 (1966): 1.
2 Given this context, it is not surprising that polymer physics is today usually taught in chemistry departments and not physics departments at universities, except at places where independent department of polymer science exist.
3 P. J. Flory, "Innovation and polymer science." Paper read at the Goodyear Innovation Conference, held in Akron, Ohio, April 20, 1982, Paul J. Flory Papers, Chem-

ical Heritage Foundation. Additions Box 6. Flory's underscoring.

4 H. Staudinger, "Über Isopren und Kautschuk: Kautschuk-Synthese." Paper read at the 36th general meeting of the Schweizerische Gesellschaft für chemische Industrie, held October 7, 1917; reprinted in H. Staudinger, *Das wissenschaftliche Werk von Hermann Staudinger: Gesammelte Arbeiten nach Sachgebieten geordnet*, vol.1: *Arbeiten über Isopren, Kautschuk und Balata* (eds), M. Staudinger, H. Hopff, and W. Kern (Basel and Heidelberg: Hüthig & Wepf Verlag, 1969), 22–39. Cf., id., "Über Polymerisation," *Berichte der deutschen chemischen Gesellschaft* 53 (1920): 1073–1085.

5 H. Staudinger and J. Fritschi, "Über Isopren und Kautschuk: Über die Hydrierung des Kautschuks und über seine Konstitution," *Helvetica Chimica Acta* 5 (1922): 785–806, especially p. 788.

6 See Y. Furukawa, *Inventing Polymer Science: Staudinger, Carothers, and the Emergence of Macromolecular Chemistry* (Philadelphia: University of Pennsylvania Press, 1998), Chapter 1.

7 Papers for this symposium were published in *Kolloid-Zeitschrift* 53 (1930).

8 H. Staudinger, *Organische Kolloidchemie* (Braunschweig: Verlag Vieweg & Sohn, 1940).

9 H. Staudinger, "Die Chemie der hochmolekularen organischen Stoffe im Sinne der Kekuléschen Strukturlehre," *Berichte der deutschen chemischen Gesellschaft* 59 (1926): 3019–3043, on 3043.

10 On Carothers' work, see Furukawa, *Inventing Polymer Science*, in ref. 6, Chapter 3.

11 H. Mark, "Coming to an age of polymers in science and technology," *History of Polymer Science and Technology*, ed. by R. B. Seymour (New York and Basel: Marcel Dekker, 1982), 1–9, on 5.

12 Cf., H. Staudinger, "Die Chemie der hochmolekularen organischen Stoffe im Sinne der Kekuléschen Strukturlehre, I." *Zeitschrift für angewandte Chemie* 42 (1929): 37–40, 67–73; id., "Zur Entwicklung der makromolekularen Chemie: Zugleich Antwort auf die Entgegnung von K. H. Meyer und A. van der Wyk," *Berichte der deutschen chemischen Gesellschaft* 69 (1936): 1168–1185; id., "Über die makromolekulare Chemie," *Zeitschrift für angewandte Chemie* 49 (1936): 801–813.

13 E. g., W. H. Carothers, "Review of 'Die hochmolekularen organischen Verbindungen: Kautschuk und Cellulose' by Hermann Staudinger," *Journal of the American Chemical Society* 54 (1932): 4469–4471, on 4471.

14 Roger Adams, "Biographical Memoir of Wallace Hume Carothers, 1896–1937," *Biographical Memoirs of the National Academy of Sciences, U. S. A.* 20 (1939): 293–309; J. R. Johnson, "Wallace Hume Carothers, 1896–1937," *Journal of the Chemical Society* 143 (1940): 100–102.

15 On Mark, see H. F. Mark, "Polymer chemistry in Europe and America – how it all began," *Journal for Chemical Education* 58 (1981): 527–534; id., *From Small Organic Molecules to Large: A Century of Progress, Profiles, Pathways, and Dreams: Autobiographies of Eminent Chemists* (Washington, D. C.: American Chemical Society, 1993); H. Morawetz, "Herman Francis Mark, May 3, 1895-April 6, 1992," *Biographical Memoirs of the National Academy of Sciences* 68 (1995): 3–16; Furukawa, *Inventing Polymer Science*, in ref. 6, 76–82 and 197–200. For the controversy between Staudinger, Mark, and Meyer see C. Priesner, *H. Staudinger, H. Mark und K. H. Meyer. Thesen zur Struktur und Größe der Makromoleküle. Ursachen und Hintergründe eines akademischen Disputes* (Weinheim: Verlag Chemie 1980).

16 E. g., K. H. Meyer and H. F. Mark, *Hochpolymere Chemie* (Leipzig: Akademische Verlagsgesellschaft, 1937); H. F. Mark, "Aspects of high polymeric chemistry," *Nature* 142 (1938): 937–939; H. F. Mark and G. Stafford Whitby, "Introduction to 'Collected Papers of Wallace Hume Carothers'," in H. F. Mark and G. S. Whitby (eds), *Collected Papers of Wallace Hume Carothers on High Polymeric Substances, High Polymers* (New York: Interscience Publishers, 1940), VIII-X, on VIII.

17 On the history of this division, see R. D. Ulrich, "The history of ACS division of polymer chemistry," in R. D. Ulrich (ed.), *Contemporary Topics in Polymer Science*, vol. 1: *Macromolecular Science: Retrospect and Prospect* (New York and London: Plenum Press, 1978), 1–30.

18 P. J. Flory, *Principles of Polymer Chemistry*

(Ithaca, New York: Cornell University Press, 1953); F.W. Billmeyer, Jr., *Textbook of Polymer Chemistry* (New York: Interscience Publishers, 1957). The later edition of Billmeyer's book was entitled *Textbook of Polymer Science* (1962).

19 H. Staudinger, "Macromolecular chemistry: Nobel lecture, December 11, 1953," in *Nobel Lectures: Chemistry, 1942–1962* (Amsterdam, London, and New York: Elsevier, 1964), 397–419.

20 H.F. Mark (ed.), *Encyclopedia of Polymer Science and Technology: Plastics, Resins, Rubbers, Fibers* 16 vols. (New York: Interscience Publishers, 1964–1977).

21 G.A. Stahl, "Interview with Herman F. Mark." *Journal of Chemical Education* 56 (1979): 83–86, on 84.

22 See C.E. Carraher, Jr., "Polymer education and the Mark connection." in G.A. Stahl (ed.), *Polymer Science Overview: A Tribute to Herman F. Mark* (Washington, D.C.: American Chemical Society, 1981), 123–142; id., "History of polymer education – USA." in R.B. Seymour (ed.), *History of Polymer Science and Technology* (New York and Basel: Marcel Dekker, 1982), 173–197.

23 E. Guth, "Birth and rise of polymer Science – myth and truth," *Journal of Applied Polymer Science* 35 (1979): 1–12, on 2. Cf., E. Guth and H.F. Mark, "Zur innermolekularen Statistik, insbesondere bei Kettenmolekülen I." *Monatshefte für Chemie und verwandte Teile anderer Wissenschaften* 65 (1934): 93–121.

24 E.g., W. Kuhn, "Über die Gestalt fadenförmiger Moleküle in Lösungen," *Kolloid-Zeitschrift* 68 (1934): 2–15; id., "Beziehungen zwischen Molekülgröße, statistischer Molekülgestalt und elastischen Eigenschaften hochpolymerer Stoffe," ibid. 76 (1936): 258–271; Guth and Mark, "Zur innermolekularen Statistik, insbesondere bei Kettenmolekülen, I,"; E. Guth and H.M. James, "Elastic and thermoelastic properties of Rubberlike Materials," *Industrial and Engineering Chemistry* 33 (1941): 624–629; P.J. Flory, "Molecular size distribution in linear condensation polymers," *Journal of the American Chemical Society* 58 (1936): 1877–1895; id., "The mechanism of vinyl polymerizations," ibid. 59 (1937): 241–270; id., "Kinetics of polyesterification: A study of the effects of molecular weight and viscosity on reaction rate," ibid. 61 (1939): 3334–3340; id., "Molecular size distribution in three dimensional polymers, I. Gelation," ibid. 63 (1941): 3083–3090; id., "Molecular size distribution in three dimensional polymers. II. Trifunctional branching units," ibid. 63 (1941): 3091–3096; id., "Molecular size distribution in three dimensional polymers, III. Tetrafunctional branching units," ibid. 63 (1941): 3096–3100; id., "Constitution of three-dimensional polymers and the theory of gelation," *Journal of Physical Chemistry* 46 (1942): 132–140; id., "Thermodynamics of high polymer solutions." *Journal of Chemical Physics* 10 (1942): 51–61; id., "The configuration of real polymer chains," ibid. 17 (1949): 303–310; Günther V. Schulz, "Über die Beziehung zwischen Reaktionsgeschwindigkeit und Zusammensetzung das Reaktionsproduktes bei Makropolymerisationsvorgängen," *Zeitschrift für physikalische Chemie* B 30 (1935): 379–3978; id., "Über die Verteilung der Molekulargewichte in hochpolymeren Gemischen und die Bestimmung des mittleren Molekulargewichtes," ibid. B 32 (1936): 27–45; id., "Osmotische Molekulargewichts-Bestimmungen in polymerhomologen Reihen hochmolekularer Stoffe," ibid. A 176 (1936): 317–337; M.L. Huggins, "The viscosity of dilute solutions of long-chain molecules, I," *Journal of Physical Chemistry* 42 (1938): 911–920.; P. Debye, "Light scattering in solutions," *Journal of Applied Physics* 15 (1944): 338–342; R. Houwink, "Zusammenhang zwischen viscosimetrisch und osmotisch bestimmten Polymerisationsgraden bei Hochpolymeren," *Journal für praktische Chemie* 157 (1940): 15–18.

25 W.H. Stockmayer and B.H. Zimm, "When polymer science looked easy," *Annual Review of Physical Chemistry* 35 (1984): 1–21.

26 M. Chayut, "New site for scientific change: Paul Flory's initiation into polymer chemistry," *Historical Studies in the Physical and Biological Sciences* 23 (1993): 193–218. D.A. Hounshell, J.K. Smith, *Science and Corporate Strategy. Du Pont R&D, 1902–1980* (Cambridge: Cambridge University Press 1988).

27 For more on the U.S. government's wartime synthetic rubber research program, see P.J.T. Morris, *The American Synthetic*

Rubber Research Program (Philadelphia: University of Pennsylvania Press, 1989).

28 P. Debye, "Light scattering in solutions," *Journal of Applied Physics* 15 (1944): 338–342.

29 "Editorial," *Journal of Polymer Science*, Part A-1, 4 (1966): 1.

30 E. g., H. Staudinger, "Die Chemie der hochmolekularen organischen Stoffe im Sinne der Kekuléschen Strukturlehre," *Berichte der deutschen chemischen Gesellschaft* 59 (1926): 3019–3043; W. H. Carothers, "Polymers and polyfunctionality," *Transactions of the Faraday Society* 32 Pt. 1 (1936): 39–49; K. H. Meyer and H. F. Mark, *Der Aufbau der hochpolymeren organischen Naturstoffe* (Leipzig: Akademische Verlagsgesellschaft, 1930).

31 Cf., C. Transford and J. Reynolds, "Protein chemists bypass the colloid/macromolecule debate," *Ambix* 46 (1999): 33–51. It is true that some protein chemists had conceived a concept of large molecules for proteins while not involved in the macromolecular debate. However, it is misleading to identify their views as the macromolecular theory which polymer scientists established as a coherent system of study. On the early concept of large molecules and its difference from Staudinger's, see Furukawa, *Inventing Polymer Science*, in ref. 6, 18–23 and 87–91. On the stance of some protein researchers of the 1920s, see ibid., 117–123.

32 H. Staudinger, *From Organic Chemistry to Macromolecules: A Scientific Autobiography Based on My Original Papers*. Translated from the German by J. Fock and M. Fried (New York, London, Sydney, and Toronto: Wiley-Interscience Publishers, 1970), 1.

33 H. Staudinger, *Makromolekulare Chemie und Biologie* (Basel: Verlag Wepf & Co., 1947).

34 E. Schrödinger, *What is Life? The Physical Aspect of the Living Cell* (Cambridge: Cambridge University Press, 1944). The role of the Schrödinger book in the emergence of molecular biology has been controversial among scholars. See R. Olby, "Schrödinger's problem: What is life?" *Journal of the History of Biology* 4 (1971): 119–148; E. J. Yoxen, "Where does Schrödinger's 'What is life?' belong in the history of molecular biology?" *History of Science* 17(1979): 17–52; E. F. Keller, "Physics and the emergence of molecular biology: A history of cognitive and political synergy," *Journal of the History of Biology* 23 (1990): 389–409.

35 R. Signer, T. Caspersson, and E. Hammarsten, "Molecular shape and size of thymonucleic acid," *Nature* 141 (1938): 122. See also Y. Furukawa "Macromolecules, their structures, and functions," in M. J. Nye (ed.), *The Cambridge History of Science*, vol. V: *Modern Physical and Mathematical Sciences* (Cambridge: Cambridge University Press, forthcoming).

36 Mark inspired Pauling to study biological macromolecules when the latter visited his laboratory in Germany in 1930. Pauling learned from Mark the X-ray analysis of polymers (such as crystallized rubber and fibrous proteins) as well as recent developments in the chemistry of macromolecules. The Cambridge physicist John D. Bernal was among the first to obtain clear images of X-ray diffractions by protein crystals. His picture of the enzyme pepsin impressed Mark (then professor at the University of Vienna) who met Bernal at the 1935 Faraday Society meeting in Cambridge. Mark persuaded his chemistry student Perutz to join Bernal's laboratory. The Viennese went to Cambridge the next year, and stayed there during and after the war, eventually working with the physicist Lawrence Bragg at the Cavendish Laboratory. See Furukawa, "Macromolecules, their structures, and functions" in ref. 35.

37 H. R. Wilson, "The double helix and all that," *Trends in Biochemical Sciences* 13 (1988): 275–278. Interview with R. Signer by T. Koeppel, 30 September 1986, Chemical Heritage Foundation, transcript, 17–18.

38 Interview with C. Price by L. Gorter, 26 April 1979, Chemical Heritage Foundation, transcript, 42–45. C. Price, *The Synthesis of Life* (Stroudsburg, Pennsylvania: Dowden, Hutchinson, and Ross, 1974).

39 M. Dekker, "From the publisher," *Journal of Macromolecular Science*, Part A: *Chemistry*, Vol. A1, no.1 (1967): front-page. More recently, the American Chemical Society decided to publish a new journal *Biomacromolecule*, which was inaugurated in 2000.

40 M. Morton, "Polymer chemistry: Introduction," *Journal of Chemical Education* 42 (1965): 2.

41 J. L. Kice, "Polymer chemistry instruction

in small colleges and universities," *Journal of Chemical Education* 36 (1958): 168–170. Kice reported that at the University of South Carolina "We concentrate almost all our undergraduate polymer work in the organic chemistry course." Ibid., 168.

42 R. W. Cairns, " . . . Over one-half of industry's concern . . .," *Journal of Chemical Education* 45 (1968): 503–504, on 504.

43 See J. Servos, *Physical Chemistry from Ostwald to Pauling: The Making of a Science in America* (Princeton: Princeton University Press, 1990), Chapter 2.

44 See Furukawa, *Inventing Polymer Science*, 2.

45 The current use of macromolecular science or macromolecular chemistry is illustrated by the titles of such journals as the *Journal of Macromolecular Science* (U. S. A.) and *Macromolecular Chemistry and Physics* (Germany). The American Chemical Society's official journal for the field, established in 1968, is entitled *Macromolecules.*

46 M. Staudinger-Woit, "Macromolecule and polymer," *Terminology* 3 (1996): 343–347, on 345. The coveted new Max-Planck institute, founded in 1984 in Mainz, was named Institut für Polymerforschung.

47 Ibid., 346.

48 Ibid.

49 A. D. Jenkins, "Letter to the editor," *Terminology* 4 (1997): 157–159. See also id., "Problems in composing definitions of terms for polymer chemistry," ibid. 2 (1995): 351–364.

50 Ibid., 158. Cf., Commission on Macromolecular Nomenclature, Macromolecular Division of the International Union of Pure and Applied Chemistry, "Minutes of the meeting in Sydney, Australia, 19–22 July 1998," 4 (Courtesy of Teiji Tsuruta); id., "Glossary of Basic Terms in Polymer Science (IUPAC Recommendations 1996)," *Pure and Applied Chemistry* 68 (1996): 2287–2311.

51 "Glossary of Basic Terms," 2288.

13.
At the Boundaries: Michael Polanyi's Work on Surfaces and the Solid State

Mary Jo Nye

The professional career and the personal achievements of Michael Polanyi (1891–1976), from the 1920s to the 1960s, lay at the boundaries of the sciences and the humanities, just as his scientific work lay at the boundaries of chemistry and physics. Polanyi worked mainly within the disciplinary domain of physical chemistry until the 1940s. After that, he focused mostly on the disciplinary domain of economics, social studies, and philosophy of science from the 1940s through the 1960s. All through his life he crossed back and forth across intellectual disciplinary boundaries, with wide-ranging interests and wide-ranging reading, just as he crossed back and forth between the national boundaries of Hungary, Germany, Great Britain, and the United States.

Polanyi's scientific work lay most squarely within a physical chemistry that encompassed thermodynamics, X-ray crystallography, the study of reaction rates, and the application of quantum mechanics to the study of molecular forces and transition states. In two particular areas, the investigation of solid-surface adsorption phenomena and X-ray diffraction studies of the properties of solids, Polanyi helped establish new scientific specialities, at the boundaries of physics and chemistry, for studying the solid state. He also turned his research experiences in these fields into a basis for the formulation of a new philosophy of science centered on scientific practice, rather than scientific ideas.[1] It is these themes that I would like to explore, with remarks in my conclusion on Polanyi's influence in solid-state science.

13.1
Polanyi on Scientific Ideals and Scientific Practice

At a conference in Oxford in 1961, Thomas S. Kuhn presented a paper on "The Function of Scientific Dogmas in Scientific Research" in which he summarized his thesis on paradigms, normal science, and scientific revolutions which was about to be published by the University of Chicago Press.[2] At the beginning of the Oxford paper, Kuhn noted the similarity of his and Michael Polanyi's views on the "importance of quasi-dogmatic commitments as a requisite for productive scientific research." [3] On that occasion, sitting in Oxford, Polanyi remarked to Kuhn that he had tried in vain for many years to call attention to the steadfastness of scientists' commitment to established beliefs, the view that Kuhn now was arguing. [4]

Indeed, in essays from the 1940s through the 1960s, Polanyi, writing from Manchester and then from Oxford, cautioned against what he called the false presumption of "scientific detachment." The presumption, or prescription, of the scientist's freedom from authority, he argued, is a false ideal since the very basis of scientific knowledge lies in the constraints of authority which are developed and shared among members of the scientific community. Science is a community of belief, in which the community imposes beliefs about the relative interest and value of scientific investigations on its members, as well as judgments about plausibility or truthfulness of ideas. [5]

In talking about science in Oxford in 1961, Polanyi spoke with authenticity because he was first of all a scientist who had directed laboratory work for more than two decades. [6] Born in Budapest in 1891, Polanyi had studied medicine during the period 1908–1913, first embarking on experimental work in biochemistry. He entered the Technische Hochschule in Karlsruhe in 1913 in order to study physical chemistry after completing his medical degree in Budapest. In August 1914 Polanyi entered the Austrian army as a military surgeon, but he spent much of the war period on leave or on light duty, largely for reasons of ill-health. In 1917 he completed a doctoral thesis in physical chemistry, which he defended at the University of Budapest. [7]

In September of 1920, Polanyi took a position in Berlin at the Kaiser Wilhelm Institute for Fiber Chemistry, which was housed in the buildings of the Kaiser Wilhelm Institute for Physical Chemistry and Electrochemistry directed by Fritz Haber. [8] By this time Polanyi had published his doctoral thesis and papers in several areas of thermodynamics, including papers on Nernst's heat theorem and Einstein's quantum theory for specific heats. [9] In the next 13 years, before he was forced to leave Germany in 1933, Polanyi worked in several areas of physical chemistry in Berlin, afterwards heading Manchester's physical chemistry laboratory for fifteen years. In 1948 Polanyi exchanged his professorship in chemistry at Manchester for a chair in social studies, thus formally becoming a philosopher.

In developing his later views on the social practice of science, Polanyi explicitly drew upon his earlier career experiences in the 1920s and 1930s. Two series of investigations at the boundaries of chemistry and physics were prominent among his examples of scientific practice and the distribution of merit within the scientific community. I turn now to these two cases: the potential theory of adsorption on solid surfaces and X-ray studies of the solid state.

13.2
The Potential Theory of Adsorption, 1914–1932

Heterogeneous adsorption is a process in which gases are attracted and held to the surface of a solid. Polanyi's doctoral thesis of 1917, like some of his early papers, focused on the adsorption of gases, using data in published literature on adsorption of CO_2 by charcoal.[10]

Polanyi conceived of the forces of adsorption as working through a potential gradient. In order to derive an adsorption isotherm, i.e., a representation of adsorption results at a fixed temperature, he plotted the volume of gas adsorbed (adsorbate) per gram of adsorbing material (adsorbent) against the equilibrium pressure at constant temperature.[11] Polanyi's notion of a "potential gradient" assumed the existence of long-range, inter-molecular attractive forces between the adsorbing solid and the atoms or molecules of gas that locate themselves in layers at the surface of the solid.[12] In this, he was working within the framework of nineteenth-century classical thermodynamics, but at a time when there was no good explanation for the non-electrostatic Van der Waals-type attractions that he was invoking.

In 1916, Irving Langmuir, at the General Electric Laboratory in Schenectady, New York, began publishing experimental results on the adsorption of gases on mica surfaces and on water, arguing that the surface layer is monomolecular with a structure determined entirely by electrostatic forces. Langmuir, who had taken his doctoral degree with Walther Nernst in 1906, framed his work within the new electron-pair theory of chemical valence, first clearly articulated in print by Gilbert N. Lewis that same year, 1916. Langmuir argued that the force which retains the adsorbed gas particles on the solid surface results from the electrical valence forces of the atoms or molecules in the outermost layer of the solid surface. Chemically adsorbed gas particles fit into a single layer at the solid's surface, as if on a chessboard where each square can only be occupied by a single gas particle. Thus, adsorption ceases when the surface is fully occupied.[13]

In contrast to Polanyi's theory, Langmuir's theory constituted a break with classical thermodynamic theory, as well as with the prevalent assumption that a gas adsorbate is relatively thick, becoming less dense as the distance from the solid adsorbent surface increases. Langmuir was saying that the layer is very thin – just one molecule thick. Langmuir's approach was familiar to physical chemists in its use of the language of kinetic theory, and it was innovative in its use of the language of the electrical, or electron-pair, chemical bond, a theory to which Langmuir contributed by introducing the terms "covalent" and "electrovalent" in 1919, leading to the identification of the electron valence bond theory as the "Lewis-Langmuir" theory.[14]

In autumn 1921, shortly after Polanyi joined the staff of the Kaiser Wilhelm Institute, Haber invited Polanyi to give a full account of adsorption theory at Haber's colloquium.[15] The result was considerable criticism from both Haber and Albert Einstein, who faulted Polanyi for disregarding in his lecture the new electrical theories of the structure of matter. Polanyi later said, "professionally, I survived the occasion only by the skin of my teeth."[16] Although Herbert Freundlich, who headed the colloid department at the Kaiser Wilhelm Institute,[17] gave an account of Polanyi's adsorption theory in the 1922 edition of his textbook *Kapillarchemie*, Polanyi later recalled that Freundlich expressed some ambivalence, saying "I am heavily committed now to your theory myself; I hope it is correct."[18] From 1914 to 1922 Polanyi published twelve papers on adsorption,[19] but following the Haber colloquium, Polanyi did little work in the field for six years, focusing instead on chemical reaction rates, as well as on X-ray diffraction studies of fibers, crystals, and metals, to which I will return below.

Polanyi's interest in adsorption revived around 1928, about the time Fritz London came to Berlin.[20] Polanyi began to think that a theoretical justification for his potential function might be found in the new Heitler-London approach to binding energy.[21] London was interested in studying the Van der Waals forces between atoms and molecules, forces, which, unlike valence forces between atoms, are additive and relatively unaffected when a third molecule is brought in the vicinity of the two molecules. These were precisely the kinds of forces Polanyi earlier had proposed as acting across his adsorption potential gradient. In 1930, London gave the name "dispersion forces" to the long-range forces which exist between non-polar molecules.[22]

By then, Polanyi had renewed laboratory experiments on adsorption with several co-workers,[23] substituting for his original adsorption theory a picture that covered both unimolecular and multimolecular adsorption.[24] Polanyi began to argue that Langmuir's formula represented too simple an idealization, which was not obeyed in all cases, and that his own revised theory could be used to derive Langmuir's isotherm as a special case.[20, 25]

This was a theme that Polanyi repeated in a colloquium that he organized at Berlin in the spring of 1929 and in a paper in June 1929,[26] shortly before co-authoring a paper with London in 1930, demonstrating that the adsorption potential of an adsorbent decreases with the distance from the adsorbent wall just as Polanyi had first argued in 1914. The old-fashioned potential gradient now had a firm theoretical basis in the new quantum mechanics.[27]

A 1932 conference in Oxford, sponsored by the Faraday Society, promised to decide the outcome of the Langmuir-Polanyi dispute. The three keynote speakers were supposed to be Eric Rideal from Cambridge, and Freundlich and Polanyi from Berlin but, as it turned out, neither Freundlich nor Polanyi was able to attend the meeting.[28]

Hugh S. Taylor, who was on leave from Princeton at Manchester University, and who had participated in Polanyi's Berlin colloquium in 1929, introduced the agenda for the Oxford symposium. He noted the importance of Polanyi's recent theoretical work with London, but Taylor offered no hint that Langmuir's approach might be abandoned.[29]

> As to adsorption, [said Taylor] one can summarize the situation by saying that the thick compressed film has during the last decade become progressively thinner until now the tendency is to reinterpret the ideas of the compressed film in terms of the unimolecular layer.[30]

Freundlich's precirculated paper suggested, as in his remarks at the 1929 Berlin colloquium, [31] that current experimental results did not permit a clear decision between the rival theories, [32] and Polanyi's paper argued the usefulness of both Langmuir's and his approaches. [33]

If the Oxford symposium of January 1932 was inconclusive, deliberations of the Swedish Academy of Sciences were not. Langmuir's work, unlike Polanyi's, fit squarely into recent preoccupations in physical chemistry, including the Swedish physical chemist Theodor Svedberg's concern with establishing molecular dimensions and what came to be called "molecular reality", i. e., a simple, visual kinetic-mechanical picture in physics and chemistry. [34] In the fall of 1932 the Swedish Academy announced that Langmuir would receive the 1932 Nobel Prize in Chemistry. The prize was awarded "for his discoveries and investigations in surface chemistry", a new scientific specialty and, literally, a boundary science. [35]

13.3
Diffraction and the Solid State

During the period when Polanyi was working on surface adsorption, he pursued investigations in other areas as well, including, from the very beginning of his tenure in Berlin, X-ray diffraction studies of solid materials. On first arriving in Berlin, when Polanyi had wanted to work in the field of reaction velocities, Haber admonished him: "Reaction velocity is a world problem. You should cook a piece of meat." [36]

Reginald Herzog had just the problem for Polanyi. Cotton fiber is almost pure cellulose. The chemist Hermann Ost had shown a few years earlier that cellulose undergoes conversion into dextro-glucose upon acid hydrolysis. In order to study the structure of cellulose, Herzog and his collaborator Willi Jancke made X-ray diffraction photographs of cellulose fiber, using a new technique that Jancke learned from Paul Scherrer in Göttingen. [37] Polanyi's assignment was to study the new cellulose X-ray diffraction photographs and interpret them.

Polanyi had no experience with X-ray crystallography and its interpretation, although his good friend Alfred Reis was working in this field in Karlsruhe. [38] However, Polanyi turned his attention to learning how to interpret the photographic results obtained with monochromatic radiation from a Coolidge X-ray tube over a good number of hours. [39] The photograph of a bundle of ramie fibers showed a pattern of spots symmetrical to two mirror-planes, one plane passing through the primary beam and the axis of the fibers, and the other oriented to the normal. [40] The photograph was primitive and its interpretation was hardly straightforward. Polanyi concluded that the four-point diagram was due to a group of parallel crystals arranged around the axis of the fiber and that the fiber as a whole thus had rotational symmetry around its axis. His results appeared in a joint publication with Herzog and Jancke. [41]

Herzog was sufficiently impressed that he put Polanyi in charge of a research group that was to develop X-ray diffraction studies of both natural fibers and metals.

Polanyi soon began to see the possibility of exploiting the analogy between the diffraction patterns for extended fibers and for metals under stress, a program of investigation that he carried out over the next decade, thereby helping to establish the field of solid-state and materials science. His group, first housed in the Fiber Institute and, from 1923, in the Physical Chemistry Institute, was one of three teams in Berlin-Dahlem in the X-ray field, included Hermann Mark, Erich Schmid, Ernst von Gomperz, E. Schiebold, Karl Weissenberg, and briefly Eugene Wigner. [42, 43]

Using both the powder and rotating crystal methods, with improvements made by Weissenberg, Polanyi and his co-workers undertook a program of research in solid-state analysis. They constructed a variety of machines for treating fibers, particularly cellulose and silk, by swelling, stretching, relaxing, and drying them. Their studies of rigidity, extensibility, elasticity, melting, softening, and swelling were applied to metal wires, as well as to fibers. They drew out molten metals into single-crystal wires and studied stress-strain properties using a stretching apparatus ("*Dehnungs-apparat*") devised in 1925. [43]

Polanyi and Mark discovered the slip properties of single-crystal tin, while Schmid worked out a law for the shear stress component along the slip direction in a slip plane. [44]

These studies had clear industrial, as well as theoretical, interest and Polanyi continued them throughout the 1920s and 1930s, developing in 1932 the concept of dislocation ("*Versetzung*") for describing the strength of crystals. [43, 45] He found that every process that destroyed the ideal structure of crystals increased material strength: in dislocation "ten atoms on one side are opposed by eleven atoms on the other side of the line." [46]

Polanyi presented his theory of dislocation in April 1932 to members of Abraham Joffe's research institute in Leningrad. Polanyi's theory noted the fact that diffraction lines from powdered samples of cold-worked metals are diffuse and that diffraction spots in single-crystal photographs are elongated. While a material like diamond is a reasonably perfect crystal, metallic crystals were now proposed to consist of small units that are slightly out of alignment with one another. Thus, many properties of metals depend on imperfection of the crystal structures, for example reducing the tensile strength of rocksalt 200-fold from its theoretical values. [47]

On returning to Berlin from Leningrad, Polanyi learned that a similar idea on dislocation was to appear in a thesis by the Hungarian engineer Egon Orowan, under the direction of Paul Becker. Orowan was aware of Polanyi's work on the subject and suggested that they write a joint paper, but Polanyi instead published separately, in the *Zeitschrift für Physik*. In 1934 Geoffrey I. Taylor, Royal Society Professor at Cambridge, published another version of the theory of dislocation. [36, 48] Gradually, dislocation, including the concept of the point defect, consisting of missing, misplaced or foreign atoms, ions or molecules came to be well understood, especially through publications of Robert W. Pohl's research group in Göttingen during the period 1920–1940. [49]

Polanyi's solid-state work, carried out in an institute with significant ties to commerce and industry, had clear industrial applications. Yet perhaps partly on account of that very fact, Polanyi found the reactions of most of his colleagues to this work disappointing. Among chemists, organic chemists seemed largely uninterested in the application of X-ray diffraction to molecular structure, not only Polanyi's early proposal of a large-molecule structure for cellulose, but also his investigations of the mechanical behavior of organic solids. [50] As Hermann Mark wrote:

> We, in our fiber research, were interested in ... strength, elasticity, water absorption, and abrasion resistance. Until this point, for solid organic compounds, interest was focused on melting point, solubility, color, surface activity, ... but never on mechanical behavior. We soon recognized that the solution viscosity of polymeric systems is not their most important property, but the enormous influence of polymer chain length on all physical and mechanical characteristics is ... [important, as can be seen in] ... the hardness of ivory and ebony, the elasticity of kangaroo tails, the toughness of alligator skins, and the softness of cashmere and vicuna wool. [51]

Even more disturbing to Polanyi than the views of his colleagues in organic chemistry was the reaction by his colleagues in physics and physical chemistry to his work in X-ray diffraction studies. Again, relying on Mark:

> Truthfully, the results of our studies failed to impress the leading members of the scientific community in the Kaiser Wilhelm Institute, including Max von Laue, Fritz Haber, O. Hahn, Lise Meitner, James Franck, K. F. Bonhoeffer, and others who were preoccupied with radioactivity, atomic and molecular quantum phenomena, and catalysis. [52]

Accordingly, in correspondence, Polanyi wrote despondently that the scientific problems of the strength of materials ("*Festigkeit*"), magnetism, electrical conduction, viscosity, etc., were areas in which no physicist wanted to study unless they had some relation to atomic physics. [53] "Dirt physics" was the term of opprobrium used by Wolfgang Pauli for the study of processes in real solids. [54] Orowan, by original training an electrical engineer, later recalled of his turn from electrical engineering to materials science, "plasticity was a prosaic and even humiliating proposition in the age of De Broglie, Heisenberg, and Schrödinger, but it was better than computing my 60th transformer." [55]

Polanyi's innovative work on solid materials, at the boundary of chemistry and physics, like the work on surface adsorption, simply did not seem to fit into the principal preoccupations considered to be the leading edge of innovation and originality among physical chemists and physicists of the 1920s and 1930s.

13.4
Rewards and Recognition in the Scientific Community

In reflecting in later years, in Manchester and Oxford, on the nature of scientific discovery, the norms of scientific practice, and the awarding of priority and recognition within the scientific community, Polanyi concluded that the idealization, and the idolization, of the scientist hero was misleading. "The example of great scientists is the light which guides all workers in science," wrote Polanyi in 1962, "...but we must guard against being blinded by it."[56]

Drawing upon his own experiences, Polanyi came to characterize his work on solid materials, as well as other of his work, as *typical*, rather than atypical of the scientific process. He came to think that his investigations had won him neither the prizes nor the accolades he most coveted because the work he was doing failed to fit within the current predominantly accepted scientific view of the nature of things and the current view of what counted as pathbreaking research. What matters in the attribution of scientific discovery and scientific originality, is not simply plausibility, but intrinsic interest at the time within the scientific community.[57] Polanyi's approach to the study of reactivity at solid surfaces and his studies of the physical characteristics of the solid state were not held to be of fundamental theoretical interest in the 1920s and 1930s.

Polanyi came to feel himself at the boundaries, or at the margins, of the leading edge of theoretical work in physical chemistry and physics by the late 1930s. He turned increasingly to writing about science rather than doing science, although, finding himself now speaking at the margins of philosophy to philosophers whose interests lay in the great heroes and the great ideas of science. In contrast to their preoccupations, Polanyi's firm counsel, speaking as an everyday scientist, was that philosophers and historians must study the everyday practices of typical science, rather than the extraordinary practices of heroic science.

In the matter of solid-state studies, a heroic age was to come after World War II, most dramatically in the chemistry and physics of semi-conductors, transistors, and integrated circuits on silicon chips in a new information age.[58] The defining text of modern solid-state science is widely recognized to be Frederick Seitz's *Modern Theory of Solids* (1940), written during 1936 to 1940 in direct descent from Eugene Wigner's lectures on the subject at Princeton University. Wigner had worked with Polanyi's X-ray research group in Berlin in 1925 before switching to studies on chemical reaction rates, and in the early 1930s, Wigner, still a close friend of Polanyi's, spent half his time in Princeton and half in Berlin.[59]

Fifty years later, by the 1980s, mainstream physics had become largely solid-state physics. In 1970 the *Physical Review* split into two separate journals, with *Physical Review B* devoted to solids and containing more articles than in nuclear and high-energy physics.[60] Thus, the "dirt physics" that Polanyi pursued in the 1920s and 1930s had become a leading edge of physics and chemistry in the 1970s and 1980s, with solid-state science and materials science becoming well-established fields of both theoretical and practical import.

It was Wigner, Seitz, Pohl, and others who pioneered the solid-state field rather than Polanyi, largely because of their mastery of the new quantum mechanics, in which he never was at ease, and their theoretical application of quantum mechanics to solids.[61] Polanyi's pioneering work on surfaces and solids became part of the background, rather than the foreground of textbook science. Yet his legacy lay, too, not only in the scientific practice about which he wrote from a philosophical and social point of view, but as well in the scientific practice into which he introduced Wigner and other students and colleagues.

Acknowledgments

This paper makes considerable use of the Michael Polanyi Papers (MPP in the notes), which are held at the Special Collections of the Regenstein Library at the University of Chicago. I am grateful for permission to have consulted these Papers. Research for this project was supported by National Science Foundation grant no. SBR-9321305 and by the Thomas Hart and Mary Jones Horning Endowment in the Humanities.

Parts of this paper have been presented and discussed with colleagues at the Dibner Institute for the History of Science and Technology, the University of California at Berkeley, the University of Chicago, and Yale University. See my essays "Laboratory Practice and the Physical Chemistry of Michael Polanyi," in F. L. Holmes and Trevor Levere, eds, *Instruments and Experimentation in the History of Chemistry* (Cambridge, MA.: MIT Press, 2000), 367–400 and "Michael Polanyi's Theory of Adsorption: How Premature?" in Ernest Hook, ed., *Prematurity in Scientific Discovery* (Berkeley: University of California Press, in press).

References and Notes

1 See especially, M. Polanyi, "The potential theory of adsorption," *Science* 141 (1963): 1010–1013, reprinted in M. Grene (ed.), *Knowing and Being* (London: Routledge and Kegan Paul, 1969); and M. Polanyi, "My time with x-rays and crystals," in P. P. Ewald (ed.), *Fifty Years of X-Ray Diffraction* (Utrecht: Oosthoek, 1962), 629–636, also reprinted in *Knowing and Being*, 97–104. And M. Polanyi, *Science, Faith and Society* (Chicago: University of Chicago Press, 1946); id., *Personal Knowledge: Towards a Post-Critical Philosophy* (Chicago: University of Chicago Press, 1958); and id., *The Tacit Dimension* (New York: Doubleday, 1966).

2 T. S. Kuhn, *The Structure of Scientific Revolutions* (Chicago: University of Chicago Press, 1962).

3 T. S. Kuhn, "The function of dogma in scientific research," in A. C. Crombie (ed.), *Scientific Change* (New York: Basic Books, 1963), 347–369, note 1 on 347.

4 M. Polanyi, "Commentary by Michael Polanyi," in Crombie, *Scientific Change*, 375–380, on 375.

5 Polanyi, *Personal Knowledge.*

6 See R. A. Hogdkin and E. P. Wigner, "Michael Polanyi, 1891–1976," *Biographical Memoirs of Fellows of the Royal Society* 23 (1977): 421–448, and W. T. Scott, "Michael Polanyi's creativity in chemistry," in R. Aris et al. (eds.), *Springs of Scientific Creativity* (Minneapolis: University of Minnesota Press, 1983), 279–307.

7 On the early years in Budapest, see G. Pallo, "Michael Polanyi's early years in science," *Bulletin for the History of Chemistry* 21 (1998): 39–43.

8 M. Polanyi, "Curriculum Vitae," composed June 1933, in M. Polanyi Papers (MPP), University of Chicago Regenstein Library Special Collections, Box 2, Folder 12 (2:12). Of great utility is J. M. Cash, *Guide to the Papers of Michael Polanyi* (Chicago: The Joseph Regenstein Library, September 1977). Also, Wigner and Hodgkin, "Michael Polanyi, 1891–1976," 413–415.

9 M. Polanyi, "Eine neue thermodynamische Folgerung aus der Quantenhypothese," *Verhandlungen der deutschen physikalischen Gesellschaft* 15 (1913): 156–161, esp. 157; and "Neue thermodynamische Folgerungen aus der Quantenhypothese," *Zeitschrift für physikalische Chemie* 83 (1913): 339–369,

discussed in W. T. Scott, "Michael Polanyi's Creativity in Chemistry," in R. Aris et al. (eds.), *Springs of Scientific Creativity* (Minneapolis: University of Minnesota Press, 1983), 279–307, on 282–283. See also Wigner and Hodgkin, "Michael Polanyi, 1891–1976," 416.

10 M. Polányi, "Gázok absorptiója szilárd, nem illanó adszorbensen" (Adsorption of Gases by a Solid Non-Volatile Adsorbent), University of Budapest Ph. D. thesis, 1917.

11 Polanyi, "The potential theory of adsorption," 1013, n. 2. Polanyi noted that Arnold T. Eucken had introduced the term "Adsorptionspotential" in 1914, a few months before Polanyi's first paper on the subject.

12 M. Polanyi, "Ueber die Adsorption vom Standpunkt des dritten Wärmesatzes," *Verhandlungen der deutschen physikalischen Gesellschaft* 16 (1914): 1012–1016; and "Adsorption von Gasen (Dämpfen) durch ein festes nicht flüchtiges Adsorbens," ibid. 18 (1916): 55–80. See Scott, "Michael Polanyi's creativity in chemistry," 283; Hodgkin and Wigner, "Michael Polanyi, 1891–1976," 417; but especially S. Brunauer, *The Adsorption of Gases and Vapours*, Volume I, *Physical Adsorption* (Oxford: Oxford University Press, 1944), 95–99.

13 I. Langmuir, "The constitution and fundamental properties of solids and liquids. I. Solids," *Journal of the American Chemical Society* 38 (1916): 2221–2295, and "II. Liquids," ibid., 39 (1917): 1848–1906. See G. Gaines, "Irving Langmuir (1881–1957)," in L. K. James (ed.), *Nobel Laureates in Chemistry, 1901–1993*, (Washington, D.C.: American Chemical Society, 1993), 205–210.

14 I. Langmuir, "The arrangement of electrons in atoms and molecules," *Journal of the American Chemical Society* 41 (1919): 868–934; "Isomorphism, isoterism and covalence," ibid., 1543–1559; "The Structure of Molecules," *BAAS Reports, 1921, Edinburgh* (London, 1922), 468–469.

15 The talk is cited by Polanyi as *Festschrift der Kaiser-Wilhelm-Gesellschaft*, 1921, 171, in M. Polanyi, "Grundlagen der Potentialtheorie der Adsorption," *Zeitschrift für Elektrochemie* 35 (1929): 431–432, on 431.

16 Polanyi, "The potential theory of adsorption," 1011.

17 On Freundlich and other researchers at the institutes of the Kaiser Wilhelm Gesellschaft in Berlin, see E. Henning and M. Kazemi, *Dahlem--Domain of Science* (Munich: Max Planck Society, 1998).

18 Polanyi, "The potential theory of adsorption," 1010. The chemist Hermann F. Mark, Polanyi's colleague in the Institute for Fiber Chemistry, interested himself in the work on adsorption and its implications for catalysis, later recalling that most organic chemists found Polanyi's theory perfectly satisfactory, but that they were not much interested in electrons or the new physics. Scott, "Michael Polanyi's creativity in chemistry," 284; also see Hermann Mark, "Recollections of Dahlem and Ludwigshafen," in *Fifty Years of X-Ray Diffraction*, 603–607, on 603.

19 Perhaps the most important was M. Polanyi, "Ueber Adsorptionskatalyse," *Zeitschrift für Electrochemie* 27 (1921): 142–150, noted in G.-M. Schwab, *Catalysis from the Standpoint of Chemical Catalysis*. Translated from 1st German ed. (1931) by H. S. Taylor and R. Spence (New York: Van Nostrand, 1937), 241–243.

20 See Hodgkin and Wigner, "Michael Polanyi, 1891–1976," 418.

21 E. g., see K. Laidler, *The World of Physical Chemistry* (Oxford: Oxford University Press, 1993), 347.

22 See K. Gavroglu, *Fritz London: A Scientific Biography* (Cambridge: Cambridge University Press, 1995), 67–68; and F. London, R. Eisenschitz, "Über das Verhältnis der Van der Waalsschen Kräfte zu den homöopolaren Bindungskräften," *Zeitschrift für Physik* 60 (1930): 491–527; followed by other papers.

23 F. Goldmann and M. Polanyi, "Adsorption von Dämpfen an Kohle und die Wärmeausdehnung der Benetzungsschicht," *Zeitschrift für physikalische Chemie* 132 (1928): 321–370; M. Polanyi and K. Welke, "Adsorption, Adsorptionswärme und Bindungscharakter von Schwefeldioxyd an Kohle bei geringen Belegungen," ibid., 371–383; W. Heyne and M. Polanyi, "Adsorption aus Lösungen," ibid., 384–398.

24 This new approach employed two-dimensional equipotential lines, instead of three-dimensional equipotential surfaces, using an interpretation of "islands" of compressed gas or liquid that grow in places of lower potential on the adsorbent surface.

As pressure increases, new islands form, and some of them flow together until the whole surface becomes covered with adsorbed liquid when the vapor pressure is reached. Using this model, Polanyi and Goldmann concluded that adsorption of vapors on charcoal was not unimolecular, although the data obeys the Langmuir equation. Goldmann and Polanyi (1928), in ref. 23. See Brunauer, *The Adsorption of Gases and Vapours*, 116–119.

25 Hodgkin and Wigner, "Michael Polanyi, 1891–1976," 417. See M. Polanyi, "Betrachtungen über den Aktivierungsvorgang an Grenzflächen," *Zeitschrift für Elektrochemie* 35 (1929): 561–567.

26 Polanyi, "Grundlagen der Potentialtheorie der Adsorption," *Zeitschrift für Elektrochemie* 35 (1929): 431–432. This is in part a response to H. Zeise, "Die Adsorption von Gasen und Dämpfen und die Langmuirsche Theorie," ibid., 426–431. See Brunauer, *The Adsorption of Gases and Vapours*, 75–76, including Figure 38, on 75, reproduced in the text.

27 F. London and M. Polanyi, "Ueber die atomtheoretische Deutung der Adsorptionskräfte," *Die Naturwissenschaften* 18 (1930): 1099–1100.

28 "The adsorption of gases: A general discussion (12–13 January 1932)," *Transactions of the Faraday Society* 28 (1932): 129–447. Introductory remarks by Dr. Robert Mond, 130.

29 Taylor in "The adsorption of gases: A general discussion," 138, citing H. Eyring and M. Polanyi, "Über einfache Gasreaktionen," *Zeitschrift für Physikikalische Chemie* 12B (1931): 279–311; and H. Eyring, "The energy of activation for bimolecular reactions involving hydrogen and the halogens, according to the quantum mechanics," *Journal of the American Chemical Society* 53 (1931): 2537–2549.

30 Taylor, "The adsorption of gases: A general discussion," 132. In his footnote to the published text, Taylor cited Polanyi's 1928 articles with Goldman and Welke.

31 H. Freundlich, in "Diskussion," *Zeitschrift für Elektrochemie*, 35 (1929): 585.

32 Freundlich,"Introductory paper to section II," *Transactions of the Faraday Society* (1932): 195–201, on 198

33 Polanyi,"Introductory paper to section III," *Transactions of the Faraday Society* (1932): 316–333, on 321–322

34 On the nominations, see E. Crawford et al. *The Nobel Population. 1901–1937* (Berkeley: Office for History of Science and Technology, 1987).

35 I. Langmuir, "Surface chemistry," in *Nobel Lectures. Chemistry 1922–1941*, 287–325; and *Journal of the American Chemical Society* (1932): 2798, cited in Brunauer, *The Adsorption of Gases and Vapours*, 62.

36 Polanyi, "My time with x-rays and crystals," in *Knowing and Being*, 97–104, on 97.

37 On Jancke, see E. E. Hellner and P. P. Ewald, "Schools and regional development: Germany" in P. P. Ewald (ed.), *Fifty Years of X-Ray Diffraction* (Utrecht: Oosthoek, 1962), 456–468, on 461–462. And, R. O. Herzog and W. Jancke, "Roentgenspektrographische Beobachtungen an Zellulose," *Zeitschrift für Physik* 3 (1920): 196–198.

38 See A. Reis, "Zur Kenntnis der Kristallgitter," *Zeitschrift für Physik* 1 (1920): 204–220; see comment about Reis by H. Mark, *From Small Organic Molecules to Large: A Century of Progress* (Washington, D. C.: American Chemical Society, 1993), 80.

39 P. P. Ewald, "The immediate sequels to Laue's discovery" in *Fifty Years of X-Ray Diffraction*, 57–80, on 80.

40 R. O. Herzog and W. Jancke, "Röntgenspektrographische Beobachtungen an Zellulose," *Zeitschrift für Physik* 3 (1921): 196–198.

41 R. O. Herzog, W. Jancke, and M. Polanyi, "Roentgenspektrographische Beobachtungen an Zellulose II," *Zeitschrift für Physik* 3 (1920): 343–348.

42 Scott, "Michael Polanyi's creativity in chemistry," 288; H. Mark, "Recollections of Dahlem and Ludwigshafen" in *Fifty Years of X-Ray Diffraction*, 603–607, on. 603. E. Wigner, *The Recollections of Eugene P. Wigner as told to Andrew Szanton* (New York: Plenum, 1992), 78–80.

43 See Scott, "Michael Polanyi's creativity in chemistry," 291.

44 Hellner and Ewald, "Schools and regional development," in *Fifty Years of X-Ray Diffraction*, 462.

45 Scott, "Michael Polanyi's creativity in chemistry," 292. M. Polanyi, "Ueber eine Art von Gitterstörung, die einen Kristall plastisch machen könnte," *Zeitschrift für Physik* 89 (1934): 660–664.

46 Polanyi, "Ueber eine Art von Gitterstörung." See E. Braun, "Mechanical properties of solids" in Lillian Hoddeson et al., *Out of the Crystal Maze: Chapters from the History of Solid-State Physics* (New York: Oxford University Press, 1992), 317–358, on 327–331.

47 See W. Hume-Rothery, "Applications of x-ray diffraction to metallurgical science," in *Fifty Years of X-Ray Diffraction*, 190–211, on 198–199 and Kathleen Lonsdale, "X-ray diffraction and its impact on physics," in *Fifty Years of X-Ray Diffraction*, 221–247, on 232–233.

48 Polanyi, "My time with x-rays and crystals," in *Fifty Years of X-Ray Diffraction*, 636. G. I. Taylor, "The mechanism of plastic deformation of crystals," *Proceedings of the Royal Society* A145 (1934): 362–415; E. Orowan, "Zur Kristallplastizität," *Zeitschrift für Physik* 89 (1934): 605–659. On recent studies of dislocation, see R. F. Service, "Materials scientists view hot wires and bends by the bay," *Science* 272 (1996): 484–485.

49 E. Braun, "Introduction," in D. L. Weaire and C. G. Windsor (eds.), *Solid State Science: Past, Present and Predicted* (Bristol: Adam Hilger, 1987), 1–9, on 3.

50 On macromolecules, including the work of Polanyi, Mark, and others in Berlin, see Y. Furukawa, *Inventing Polymer Science: Staudinger, Carothers, and the Emergence of Macromolecular Chemistry* (Philadelphia: University of Pennsylvania Press, 1998).

51 Mark, *From Small Organic Molecules to Large*, 25.

52 Mark, *From Small Organic Molecules to Large*, 29.

53 Carbon copy of letter to Wichard von Moellendorff [W], 27 March 1929 (MPP, 2:5).

54 M. Eckert and H. Schubert, *Crystals, Electrons, Transistors: From Scholar's Study to Industrial Research*, trans. Thomas Hughes (New York: American Institute of Physics, 1990), 184–185.

55 E. Orowan, "Dislocations in plasticity," in C. S. Smith (ed.), *The Sorby Centennial Symposium on the History of Metallurgy* (New York: Gordon and Breach, 1965), 359–376, quoted in Braun, in *Out of the Crystal Maze*, (1992), 331.

56 Polanyi, "My time with x-rays and crystalls," in *Knowing and Being*, 97.

57 Polanyi, "The potential theory of adsorption," 1012. Also see M. Polanyi, "The republic of science," *Minerva* 1 (1962): 54–73.

58 See M. Riordan and L. Hoddeson, *Crystal Fire: The Birth of the Information Age* (New York: Norton, 1997).

59 Interview of Eugene Wigner with T. S. Kuhn, 21 March, 4 December 1963, Session II, 3 December 1963, 5 (Sources for the History of Quantum Physics, American Institute of Physics Niels Bohr Library). Wigner, *Recollections*.

60 S. Weart, "The Solid Community," in Lillian Hoddeson et al. (eds.), *Out of the Crystal Maze*, 617–669, on 643. Also, Weaire and Windsor (eds.), *Solid State Science* (1987), xiv, on the U. K.

61 On this, and Polanyi's feeling that he could not keep up with Wigner in quantum mechanics, Polanyi's "Diary 1929," entry of 9 January 1929 (MPP 44:4).

14. The New Science of Materials: A Composite Field of Research

Bernadette Bensaude-Vincent

The title of this paper suggests a contrast between an old science of materials and a new one. This contrast is partly justified since we all know that Galileo's 1638 dialog about "Two New Sciences" laid the foundations of the mechanical approach to structural materials, a "new science" that we now regard as the "old science" of materials. Galileo's reflections on strength of materials began with several observations made during his visits to a Venetian arsenal and with considerations on the gap between the geometrical approach to matter and the practical knowledge about material structures and machines. From a geometrical point of view, the properties of a piece of wood mainly depend on its shape and should be independent from its size. In reality it is impossible to increase the size of structures to vast dimensions. Thus the elastic of mechanics bodies developed at the fringe of pure science, as a kind of mixed science. It later became integral part of the program of experimental philosophy developed by Robert Hooke at the Royal Society, by Edme C. Mariotte in France, and Jacob Bernoulli in Switzerland. However it is only in retrospect, through a whig perspective, that we identify these studies of wood, of iron or of steel as a proto-science of materials, because there was nothing like a notion of materials in general. [1]

It is only very recently that materials emerged as a collective field of scientific research. An annual review of materials science started in 1971, which indicated that it was a booming field of activity based on the general characteristics of materials. A Materials Research Society was founded in the USA in 1973, followed by a European Materials Research Society founded in 1983. [2] And there is now an International Union of Materials Research Societies. Thus materials went through the whole process of institutionalization with national learned societies and then internationalization that has become the standard way of disciplinary formation since the nineteenth century. Materials Institutes sprang up in many countries over the last decades. [3] At the same time, materials became a prime concern of national science policies. In the late 1970s, the federal government of the USA commissioned a report of the National Academy of Sciences which was followed by a series of political measures. In turn, the French Ministry of Research and Technology commissioned a report in 1982 and launched national programs through the *Centre*

National de la Recherche Scientifique. The OECD published a report on advanced materials in 1990, while NATO and the EC supported research programs such as the Brite-Euram Programme on Industrial and Material Technologies. So great is the public concern with materials that the phrase "age of materials" has often been used to characterize our epoch in journals and magazines.

What is exactly the impact of this plural "materials" in opposition to the single material which was said to mark an epoch in the "age of iron" or in "the plastics era"? Is it possible to set up a consistent research field including such diverse subjects as metals, wood, concrete, ceramics, polymers, and electronic materials? How does this field relate to traditional disciplines such as physics and chemistry?

This paper first examines the construction of this research field from a historical perspective. In such an attempt to point out the mainstreams that lead to materials science prior to the institutional existence of this field, I am forced to adopt a presentist perspective. I will not try to describe all the branches that finally merged into materials science. Rather, I will only retain those in which chemistry was involved, i. e., metallurgy and polymer chemistry. The main focus of this paper will be on the recent connection of materials research with biology. I try to analyze its status as an interdisciplinary field of research and will finally raise a prospective question "is there a future for chemists in materials science"?

14.1
From Metallurgy to Solid State Physics

Let me start with a simple comparison. Modern cosmology is said to be founded by Copernicus, modern physics by Galileo and Newton, chemistry by Lavoisier … Every branch of modern science, even every sub-discipline has its founding hero, duly celebrated by the disciplinary communities. Who is the founder of materials science? To my knowledge, there is no candidate for this prestigious position and this is not for want of genius or for lack of ambition among material scientists. Here is a founderless science, a domain that did not stem from any single identifiable root. In particular, materials science did not directly emerge from mechanical engineering. Despite tremendous advances in the mathematical treatment of the mechanical properties of material structures, neither Thomas Young, nor Henri Navier, nor Denis Poisson would be considered as the founders of materials science.

In his historical survey of metallurgy, Cyril Stanley Smith emphasized that chemistry did not help understanding the structure or predicting the properties of the different metals and alloys. [4] However, he mentioned a series of researches on metals throughout the eighteenth century, including "outstanding researches" by René-Antoine Ferchault de Réaumur on steel and iron in the early eighteenth-century, by Louis-Bernard Guyton de Morveau, and by Franz Karl Achard on the properties of metallic alloys. [5] The latter was a rich collection of experimental data comparable to Torbern Bergman's affinity tables as a tool for predictions. These outstanding contributions, Smith argued, remained unnoticed because the chemists' interest was focused on Lavoisier's new chemistry and language. Symmetrically, nineteenth-century practical metallurgists ignored the atomic and molecular theories developed in chemistry and hardly applied their methods of analysis to the study of alloys. In contrast, civil and mechanical engineers developed useful methods of calculus of the mechanical properties. However, nineteenth-century physicists and mechanical engineers mostly concerned as they were with the theory of elasticity were not interested in the contemporary study of crystalline structures and lattices.

> The physicist's important characteristic of considering only problems that could be solved by simple idealized mathematical models forced him in the nineteenth century to ignore all properties of matter that were related to structure – precisely those with which the metallurgist was mostly concerned. Elasticity was almost the only "metallurgical" property of metals that found its way into respectable physics textbooks. Physicists were relatively uninterested in real structure, and the beautiful mathematics of crystal symmetry and crystalline elasticity had little effect on metallurgy. Physics was not yet basically atomic in nature – it was molecular, and it invoked vaguely defined molecular changes whenever properties that depend upon the metallurgist's micro-crystals had to be discussed. [6]

Although typically whiggish, Smith's historical survey is extremely rich and useful because it emphasizes the interplay between various academic disciplines and the longstanding practical tradition of metallurgists. The quotation above points out at the two fundamental notions – structure and properties – of modern metallurgy which are also the basis of the conceptual framework of modern materials science. The earliest notions on metallic structures – polycrystallinity, solid solutions, segregation, diffusion, deformation – were not elaborated by theoreticians but rather by practitioners. [7]

A major turning point occurred in the 1910s when William H. and William L. Bragg, father and son, engaged in their work on X-ray diffraction because it opened a window to identify the arrangement of atoms in crystalline structures. The study of crystals and the determination of crystalline structures became a concern of the physicists or rather "physical metallurgists" as they were named in the 1920s. And from this moment the future of metallurgy lied in their hands and no longer in the hands of skilled metallurgists, or chemists. Whatever the importance of composition and chemical bonding for the study of alloys, chemists were marginalized. "Physics thus replaced chemistry as the crucial source of scientific insight into the behavior of metals and alloys," wrote Robert W. Cahn. [8] Investigating the microstructure became a priority because the geometrical and crystallographic features allowed understanding the mutual disposition of phases and the properties of the alloys. According to Cahn, "microstructure is a scientific category which is the peculiar contribution of physical metallurgy to the study of solid state physics, and later extended to materials science." [9] The notions of crystal lattice, of dislocation, of defect, were introduced by physicists.

The quantum theory soon reinforced the domination of physicists while it also lead to the emergence of solid state as an object of investigation in itself. This is a second crucial step on the way to materials science. The emergence of solid state physics as a branch of applied physics is referred to two dates by Spencer Weart. In 1936, Ralph P. Johnson and Frederick Seitz published a series of review articles on "Modern Theory of Solids" in the *Journal of Applied Physics.* [10] The authors justified their use of the general notion of solid by the recent unification of the interpretation of various solids such as copper, diamond or rocksalt. In 1944 a special division of the American Physical Society was created. The initiative came from Roman Smoluchovski who proposed to create a "division of the physics of metals." This choice was challenged by Léon Brillouin who "wanted to include all problems of crystal physics, metal or non-metal. After all, he explained, "the distinction between metals and other solids has no scientific basis ... so let us speak of Physics of Solids in general'." [11] If Brillouin's proposal was finally adopted after many debates, it was not only for the intellectual reasons that he put forward, but also for practical reasons: "A division of the physics of solids would attract people interested in ceramics, pigments, glass, and so forth, groups that are very large in number, and are no less well organized than the metallurgists." [12] The attempt to bring together people from different communities which, in the 1940s, counterweighed the process of balkanization of physics at the American Physical Society, remained the driving force behind the emergence of Materials Science 30 years later.

More directly, solid state physics contributed to the emergence of materials science, because of one of its foci. Spencer Weart identified three pillars on which solid state physics was erected: First, X-ray diffraction techniques provided precise atomic picture of solids; second, quantum mechanics provided the theoretical foundations for the description of solids; and the third, more subtle pillar was the attempt to discriminate between properties depending on the idealized crystal pattern and properties dependent on "accidents" of either the inner arrangement or the surface of the solid. [13] This focus on "structure-sensitive-properties" can be seen as the main "investigative pathway," to resume Frederic L. Holmes's concept, which lead to materials science.

To sum up this first section, the notion of solid state was the first step towards the emergence of the general concept of materials. Instrumentation played a crucial role at this stage and proved essential again for going further in the understanding of the fine structure of atomic arrangements with electron microscopes. The access to microstructures opened avenues of research and innovations. It became possible to conceive new materials at the micro-scale. It should be possible to create a material atom by atom. In the field of solid state physics, a new scientific style developed. The role of science is no longer to provide the theoretical framework from which applications can be derived. No longer is there a division of labor between pure and applied science. Though highly theoretical, material science is even more practice-oriented than solid state physics: it provides a variety of possible structures or virtual arrangements that can be effectively achieved for definite purposes.

14.2 From Reinforced Plastics to Composite Materials

The relation between microstructure and properties at the macro-scale is only one face of materials science. The second step consists in working on the relation between functions and properties. Structure/properties/functions are like the three summits of the base triangle of material science. [14] Contemporary materials are tailored for specific purposes, they are adapted to a set of specific tasks. In contrast to conventional materials that have standard specifications and a world market, more advanced materials are developed according to the functional demands of the final product and the services expected from the manufactured product. In other words, in the language of economics, instead of supplying commodities that would be finalized by the customers, new materials are the end-products of a cooperation between customers and suppliers.

It is the predominance of function over structure which led to composite materials, i. e., materials made of two or more heterogeneous components. Whereas traditional materials – stone, wood, glass, metals, plastics or concrete – were originally classified according to their physical structure, composite materials are mixtures of different structures: plastics reinforced with glass fibers were the pioneers and are still the most common. The composites gradually emerged from polymer chemistry and to a certain extent they can be considered as part of the evolution of chemistry. Glass fiber-reinforced plastics were manufactured as early as 1938 and commercialized in 1940 by the Pittsburgh Plate Glass Company. In the 1940s, glass fiber-reinforced plastics were developed for military purposes, such as radomes for aircraft and boats for the US Navy. [15] Although these early composites (laminates of polyester resins) were moulded at low pressures, they had few applications because the main difficulty was the moulding of large pieces. Reinforced plastics however started to be mass-produced in the 1950s for civil applications such as electric insulators. The practice of reinforced plastics led chemists to turn their attention towards the interface between two phases. Because the mechanical properties of heterogeneous structures depend upon the quality of interface between the fiber and the polymer, it was crucial to develop additive substances favoring chemical bonds between the glass and the resin. Interfaces and surfaces thus became one of the prime concerns of materials research. An entire new field of research – a molecular surface science – thus emerged over the past decades which is aimed at understanding electrical, magnetic, and optical properties of surfaces on the molecular level. [16]

However, the technology of composite materials cannot be fully considered as a new chapter in the history of chemistry. Originally the term “composite” was used in conjunction with “reinforced plastics.” The US Society for Plastic Industries had a Reinforced Plastics Division which was renamed Reinforced Plastics and Composites Division, in 1967. [17] In France, a bi-monthly magazine entitled *Plastique renforcé/Verre textile* published by the professional organization bearing the same name, started in 1963 and was rechristened in 1983 *Composites* with *Plastique renforcé/verre textile* as a subtitle. The shift from reinforced plastics to composite materials was not a radical break. It was nothing like a revolution since the technology of composites did not overthrow the more traditional reinforced plastics. However, gradually this process is a significant technological change and composites became an autonomous field.

What makes composites different from reinforced plastics? First, though a minority, composites can also be made up of metals or ceramics. There are metal/ceramic composites, metal/metal, and ceramic/ceramic composites. No matter the nature of the constituents, provided the structure associates two phases. The notion of composites clearly indicates that materials are more adequately defined as bearing certain properties than as sharing certain structures: Steel or iron are used as supports for toughness, plastics are useful for weight saving, and ceramics for heat resistance and stiffness. Creating a composite material means combining various properties that are mutually exclusive in one single structure in order to achieve a specific set of functions.

Moreover, whereas the reinforced plastics are basically aimed at adding the properties of the glass fiber to the plasticity of the polymer matrix, a composite is more than the sum or the addition of the properties of its individual components. A synergy can be obtained between the reinforcing fiber and the resin matrix which may reveal new possibilities and generate innovations. Let us take a familiar example to illustrate this point. The old chrome-steel bumpers of the automobile cars of the 1950s have been replaced by composite bumpers. The main reason for this change was that plastics were weight-saving and could offer comparable mechanical characteristics when adequately strengthened. Similar substitutions occurred in many technological items (skis, tennis rackets, window-frames). In the case of the bumper, however, the material substitution acted as a driving force generating a complex dynamic of change. The introduction of plastics in place of chrome steel did not immediately entail the cost reduction that was expected because this change involved heavy financial investments for R&D, for tests and trials, and new equipment. Eventually, the innovation costs were largely paid off, because the plastic material had opened new avenues for change. Plastics, whether reinforced with fibers or not, are moulded.[18] Unlike metals, they can be shaped in the process of hardening the resin. Whereas with metals manufacturing and shaping the material are two successive operations, in the case of composites they became one and the same process.

Car designers were consequently free to redesign the bumpers in accordance with the current styling of cars. The bumpers were curved and moulded along the line of the shell. The protective element became integral part of the body of the car. Instead of a separate part that had to be manufactured independently and then welded to the car, the shield was like a protective second skin wrapped around the body. Once the function of protection was integrated into the body of the car, other functions could also be similarly integrated. Thus ventilators and radiator grilles were combined with the same unit at the front. Integration proved useful, because it reduced the number of parts and assembly steps. New concepts thus emerged which gradually integrated more and more functions into the same structural part. The local change in the material structure of one part allowed a reconception of the whole automotive structure. Although there are rather few examples of such synergies, they provided a new paradigm for composite technologies. Whereas in the 1970s, composites were defined by the association of a matrix and reinforcing fibers, in the 1980s, the synergy effect was integrated as an essential attribute in their definition. For instance Philippe Cognard, the author of a French manual intended for training materials engineers wrote:

> A composite is a material whose assembly of constituent elements generates an effect of synergy within the properties of these elements. This bi- or tri- dimensional assembly is constituted by two or more basic elements, that can have all possible kinds of forms: matrices, fibres, particles, plaques, sheets ... It allows to obtain a resilient material, whose all elements are strongly and durably attached together. [19]

The emergence of the composites technology out of reinforced plastics was also a consequence of the introduction of new fibers in the 1970s. Carbon fibers which offer a higher modulus of elasticity than glass fibers were introduced as reinforcing components in the late 1960s. Initial development was made by the Royal Aircraft Establishment at Farnborough and Courtaulds. Very soon Japanese companies took over and now they hold the monopoly of carbon fibers. Hence a shift of leadership from Western to Eastern countries occurred. Moreover, carbon fibers required a different treatment of composites. Unlike glass fibers, they are used as long fibers. They are not spread all over the resin but are carefully arranged with a definite orientation, according to the main efforts during the functioning of the structure. Thus, unlike the glass-reinforced plastics, carbon composite materials are anisotropic structures.

In addition to the white glass fiber and the black carbon fiber, a "yellow" fiber was manufactured in 1971. Kevlar®, the first of the family of synthetic aramide fibers, was invented by a woman scientist, Stephanie Kwolek, working at Du Pont's laboratories at Wilmington Delaware. The development of this low-density and high-strength fiber is of special interest because it illustrates the importance of chemistry in the emergence of the materials science. To sum up a long story briefly: The strategy of the Pioneering Research Laboratory headed by Hale Charch was diversification through research. [20] They wanted other nylons and they had no doubts about the success of the products of their creativity. The emphasis was on the search for new products instead of improving rentability of existing processes or products. Unexpectedly, the path to the discovery began about 1950 when low-temperature processes for the preparation of condensation polymers were developed in Paul Morgan's laboratory, where Kwolek was working. There was quick recognition that these room-temperature processes could be useful in preparing polymers which are unmeltable or thermally unstable. They provided a basis for the preparation and commercialization of Lycra spandex fiber, Nomex and Kevlar® aramid fibers. This case thus exemplifies the close connection between products and processes in the technology of materials. The interplay between structure, properties, and functions is only one face of the problem. In fact, the triangle is only the basis of a pyramid whose summit is the process. The case of Kevlar® also illustrates a very specific configuration of relationships between fundamental science and industrial research. Stephanie Kwolek entered Du Pont with a bachelor's degree and she was unaware of the theoretical predictions made by Paul Flory on condensation polymers. However, her discovery of liquid crystalline solutions (in which the molecules all line up pointing in the same direction) proved to be a tremendous advance in fundamental science as well as the starting point for the large scale production of new fibers.

A third interesting aspect of this story is that Du Pont's aramide fiber was not specifically the result of market-driven research. When this fiber was patented in 1971 there was no commercial application in view. However, within ten years, three varieties of Kevlar® fiber were commercialized by Du Pont for dozens of reinforced plastic applications: in radial passenger tires, belts, in protective clothing, such as gloves or ballistic and flak vests, in ropes and cables in racing kayaks and canoes, and in commercial aircraft. Thus the Kevlar® fiber by no means resulted from the functional, bottom-up, approach which is sometimes considered as a major characteristic of materials science. The aramide fiber resulted from the traditional style of industrial research which was successful in the plastic era and confirms the leadership of chemistry in materials technologies.

However, the shift from reinforced plastics to composites seriously altered the leadership of chemical industries in this sector. The demand for high-performance materials stimulated by the space and computer programs launched by the leading industrial countries in the 1960s and 1970s prompted a multi-disciplinary approach. Special properties (high temperature and high pressure resistances for instance) were needed for uses in extreme conditions. Intense R & D led to the design of advanced materials for rockets, aircraft, and computers, regardless of the costs of production. In the course of these programs, new techniques and new processes were developed. Learning by doing became a major concern. Architects, mechanical engineers, and chemical engineers had to learn how to work together and how to design complex structures adapted to highly specific functions.[21] They became aware that they had to acquire "a composite way of thinking" that no academic training could provide. For instance, in the 1980s the French Dassault-Aviation Group designed and built up a unique prototype aircraft, the Falcon 10, with a maximum of composite materials in order to explore the potentials and the constraints of such materials. The conclusions of this "school exercise," as they call it, are kept secret and became part of the corporate competitive advantages. Composite materials thus generated new processes of learning that are beyond disciplinary formations. In this domain no one can feel "at home," although in most of the French industrial companies that I have consulted, the heads of the divisions materials were chemical engineers.

14.3
From Composite to Complex Structures ... Through Biomimetics

Since 1990, the keywords in materials technologies have been collaboration, cooperation, partnership, multidisciplinarity. Cross- and multidisciplinarity are the motto of all kinds of reports on materials research all over the world. [22] To be sure, the cooperation of physicists, chemists, mechanical engineers, computer scientists is necessary for the elaboration, design, and testing of composite structures. Remarkably the composite structure of materials is or should be mirrored by a composite arrangement of human competences. In both cases, the aim is to obtain a synergy of the various components: the properties of the whole structure should be more than the addition of the properties of individual components. What kind of synergies are expected and who are the key partners?

Research programs on smart or intelligent materials launched in the early 1990s essentially emerged from alliances between computer scientists, electronic engineers, and physicists. [23] Basically, intelligent materials are structures whose properties can vary according to changes in the environment or in operating conditions. For example, materials whose chemical composition varies according to their surroundings are used in medicine as prostheses. Materials whose structure varies according to the degree of damage due to corrosion or radiation are thus able to repair themselves. The whole problem is to have built-in intelligence, adaptive capabilities to external stimuli. This means at least having sensors (for strain, temperature, or light) as well as actuators embedded in the structure.

While striving to design smart materials, scientists realized that they already existed in nature. Living organisms suddenly revealed immense resources for technological creativity, for two major reasons. Nature displays a composite strategy: natural products are never as pure as synthetic products. They are always mixed with impurities and defects, which can result in very interesting properties. Like man-made composite materials, biominerals like mollusc shells closely associate inorganic and organic constituents. For instance, nacre is a kind of sandwich material made of layers of calcium carbonate crystal alternating with organic layers of proteins. Biological macromolecules form an intimate mix or composite of proteins and mineral phases at all levels of composition, starting at the scale of the nanometer. Thus, biomaterials present a complex hierarchy of structures with structural features occurring on different size scales, from the Angström scale to the micron scale and the millimeter scale.

Thanks to this complex structure, biomaterials are able to integrate a variety of functions into one structure and to achieve an optimal combination of them. Through the eye of material engineers, wood is a composite material made of long, orientated fibers immersed in a light ligneous matrix. Mollusc shells like sea-urchin or abalone shell are made out of a common raw material: calcium carbonate. Nevertheless they offer unchallenged high strength-to-weight ratio. [24] Similarly, spider's silk is both thin and extremely robust. How could nature design such efficient structures?

Over the past decade, material scientists have paid a lot of attention to the process of biomineralization. The key strategy evolved by living organisms is self-assembly. Nature builds the mortar before the bricks and self-assembles the components through the use of templates with a close control of the process at each level. The biological macromolecules are involved in controlling nucleation, growth, shaping the crystal, and adapting its mechanical properties to the functions of the biomineral. [25]

Because they intend to mimic nature, material scientists launched joint research programs with structural and molecular biologists. Cooperation may consist of interactions between researchers belonging to various departments of a campus, as is the case at the University of California, Santa Barbara, which takes a leading part in the investigation of the process of biomineralization: most of their articles are co-authored by members of the Materials Research Laboratory, the Departments of Chemistry and of Physics, the Department of Marine Biology, and the Department of Molecular and Cellular Biology. Sometimes cooperation can be materialized in a building, as is the case of Princeton. For the Princeton Materials Institute, established in 1990, a new 44 000 square-foot building was erected, equipped with state-of-the art laboratory facilities. [26] Whatever the structure of cooperation, there is no attempt at stabilizing this new and highly competitive research field in the form of a university discipline. Although materials science became part of programs in a number of engineering schools, although new materials divisions were established in research agencies, it is still a matter of debate whether Materials Science and Engineering are and should ever be a single discipline.

Multidisciplinarity is certainly a fashion, but how can such an unstable structure of research possibly work? It works as long as money floods in from industrial companies, state agencies, and military sponsors. However, the question remains about the future for chemists in biomimetic programs.

14.4 A Future for Chemists?

Fascinated as they were over the past decade by the structures achieved by nature, chemists and chemical engineers were eager to learn from natural scientists all the tricks that nature invented to optimize its structures. They were trying to discover in biomaterials immediate solutions to the technical difficulties they encountered in synthetic processes. [27] However, the smart complex structures of biomaterials are still beyond the reach of the traditional ways of synthesis. The major challenge for chemists is to control the structure from the molecular scale to the nanoscale and finally to the millimeter or macroscopic scale. Therefore, great expectations are derived from attempts at genetic engineering of polymeric materials. The great hunt of materials scientists is the protein sequence that would enable them to use a machine to synthesize a segment of DNA. The sequence would provide the recipe to use templates and self-assemble their components at a scale which is out of the reach of chemical technologies. It could thus turn out that chemistry would be overthrown by biotechnology in materials research. Significantly, when Du Pont advertised a new spider-like silk, produced by a genetically modified bacteria, its famous slogan "Better things for better living" appeared without its traditional end "through chemistry."

There are nevertheless a number of recent indicators that the biotechnological challenge has fostered the chemists' creativity. Some of them are able to self-assemble components, thanks to a clever use of surfactants as supramolecular templates. Inorganic complex structures presenting a variety of interesting shapes such as discs or helix can be obtained at a mesoscale or in some cases at the macroscopic level. [28]

Moreover, new biomimetic materials generate new perspectives on nature and artefacts that could help change the public image of chemistry. The development of synthetic materials – synthetic dyestuffs in the late nineteenth century and synthetic textiles in the twentieth century – has been celebrated as the triumph of artificial over natural products. In contrast to synthetic materials of the plastics era, composite materials (though still artificial and even more artificial than the conventional plastics) have turned the chemist's attention back to nature.

Biomaterials exemplify a deep change in the relations between nature and artefacts. Since the eighteenth century, since the time of the first artificial soda, the main objective was to invent substitutes for natural products. The production of ersatz foods in wartime represented the culmination of technological achievements. From artificial coffee to artificial rubber, the function of artefacts was above all to dispense with natural products in order to alleviate the shortages that resulted from a reduced supply of natural products. Welfare was achieved through emancipation from natural constraints. Current trends in advanced materials technologies are different. The main requirements for designers are to make environment friendly materials. No longer it is necessary to emancipate from nature. Nature provides a model for artefacts not only because it is highly functional but because it is viewed as a cooperating partner for the engineer designing artefacts. In this respect, material scientists and engineers are reviving the view of medieval alchemists who claimed that by using natural agents and processes they could accelerate the natural growth of metals in their laboratory.

In conclusion, how are we to define the "science of materials." Can we consider it as an emerging discipline? This is a controversial subject. Three different answers can be suggested, dependent on the meaning given to the term "discipline." If by discipline, one means a subject that is taught at school, certainly materials science has become a discipline in itself since the 1970s. It is now an integral part of many university courses in engineering. Certainly the textbooks published helped to coalesce a simple aggregate of metallurgy, chemistry, and other ingredients into a coherent mix. The terminology is already standardized. However, material science is not a fundamental discipline, since it presupposes a robust background in physics and chemistry. Hence the repeated debates about the best way to train students in materials science. A number of material scientists advocate an early course in material science at the undergraduate level, whereas others recommend a strong disciplinary formation in fundamental science prior to a Ph. D. in material science.

If by discipline one means a research field, materials science is a discipline clearly identifiable: first by conceptual parameters (with the basic framework structure/properties/functions/process), second by institutional parameters (learned societies, magazines, annual conferences), and third by a large number of research programs. However, it seems unlikely that this consistent research field will soon stabilize into a new boundary-discipline, like physical chemistry did for instance in the late nineteenth-century. There is nothing like a research school (comparable to the ionists' school who promoted physical chemistry) although there are many efforts to define a specific way of thinking and of reasoning, in other words a new culture. Finally materials science is by no means a young maturing discipline, if this term conveys the notion of constraint and authority. This field of research is currently characterized by a strong anti-discipline feeling. Through contacts and interviews with materials scientists, I perceived a kind of rebellion against the authoritarian aspects of "normal science" which up to now has prevented them from looking for a founding father. Rather than as specialists, they prefer to portray themselves as believers in a promising technology, sometimes as prophets of a new age. Whether this unstable and composite field of research will be short-lived (like cybernetics, for instance) or will become part of an already existing discipline (like radioactivity for instance) the near future will tell us.

References and Notes

1 As example of this perspective, see S. L. Sass, *The Substance of Civilization. Materials and Human History from the Stone Age to the Age of Silicon* (New York: Aracade Publishing, 1998)

2 The American Materials Research Society publishes a monthly journal entitled *MRS Bulletin.*

3 An institutional history of materials science is still to be done, and should be approached from an international perspective, in order to allow comparisons. In the French case, the pioneer was the *Centre des matériaux* of the National School of Mines in Paris (1967) and one of the most recent is the *Institut des matériaux de Nantes* created in 1988.

4 C. S. Smith, "The development of ideas on the structure of metals," in M. Clagett (ed.), *Critical Problems in the History of Science* (Madison: University of Wisconsin Press, 1959), 467–498; C. S. Smith, "Four outstanding researches in metallurgical history," *American Society for Testing and Materials* (1963): 1–35, on 11–14.

5 F. K. Achard, *Recherches sur les propriétés des alliages métalliques* (Berlin, 1788); see Smith, "Four outstanding researches in metallurgical history," 11–14.

6 Smith, "Four outstanding researches in metallurgical history," 26

7 C. S. Smith, "The interaction of science and practice in the history of metallurgy," *Technology and Culture* 2 (1961): 357–367.

8 R. W. Cahn, "Solid state physics and metallurgy," in D. L. Weaire, C. G. Windsor (eds.), *Solid State Science. Past, Present, Predicted* (Bristol: Adam Hilger, 1987), 79–108, on 82.

9 R. W. Cahn, "Solid state physics and metallurgy," 85

10 S. R. Weart, "The solid community," in Hoddeson L, Braun E., Teichman J., Weart S. (eds.), *Out of the Crystal Maze. Chapters from the History of Solid State Physics* (Oxford: Oxford University Press, 1992), 617–666, on 628.

11 Léon Brillouin to R. Smoluchowki, 25 January 1944, APS, Division of Solid State Physics records, quoted by Weart, *Out of the Crystal Maze*, 637.

12 Seitz to Smoluchowski, 3 March, 1944, APS, Division of Solid State Physics records, quoted by Weart, "The solid community," in *Out of the Crystal Maze*, 637.

13 Weart, "The solid community," in *Out of the Crystal Maze*, 623.

14 More developments on the emergence of composite materials can be found in B. Bensaude-Vincent, *Eloge du mixte. Matériaux nouveaux et philosophie ancienne* (Paris: Hachette littératures, 1998).

15 B. Parkyn "Fibre reinforced composites," in S. T. I. Mossman, P. J.T Morris (eds.), *The Development of Plastics* (London: The Science Museum, 1994), 105–114.

16 For a review of this field in 1997, see A. S. Gabor, "From surface materials to surface technologies," *MRS Bulletin* 23 (May 1998): 11–29.

17 J. H. Dubois, *Plastics History USA* (Boston: Cahners Books, 1972)

18 In Europe, car manufacturers first used a half product, a sheet molding compound (SMC) made of sheets of fibers pre-impregnated with resin. In the USA they used a different process named RIM (Reaction Injection Molding) which allowed the bumper to be painted the same color as the rest of the structure.

19 P. Cognard, *Les applications industrielles des matériaux composites* (Paris: Editions du moniteur, 1989), vol. I, introduction.

20 D. A. Hounshell, J. K. Smith, *Science and Corporate Strategy: Du Pont R&D, 1902–1980* (Cambridge: Cambridge University Press, 1989), 425–439. See also the transcript of the Interviews of Stephanie Kwolek, conducted by the Program for Oral History of the Chemical Heritage Foundation, Philadelphia.

21 B. Bensaude-Vincent, *Eloge du mixte. Matériaux nouveaux et philosophie ancienne* (Paris: Hachette, littératures, 1998), 195–207.

22 For instance "Progress through Partnership" was the title of the report published in 1995 by the Technology Foresight Panel on Materials of the British Office for Science and Technology

23 T. Tagaki, "A concept of intelligent materials," in I. Ahman, A. Crowson (eds.), *US/Japan Workshop on Smart/Intelligent Materials and Systems* (Lancaster, Penn.: Proceedings series, 1990), 1–10. A. Mc Donach, P.T. Gardiner, R. S.McEwen, B. Culshow (eds.), *Smart Structures and Materials, Second European Conference*, (Society of Photo-optical Instrumentation Engineers (SPIE), Proceedings series, volume 2361, 1994).

24 H. A. Lowenstam, S. Weiner, *On Biomineralization* (Oxford: Cahners Books, 1989); S. Mann, J. Webb, R. P. J. Willimas, *Biomineralization* (Weinheim: VCH Publishers, 1989); D. Perl-Treves, "Les cristaux du monde vivant," *La recherche* 305 (1998): 34–38

25 S. Weiner, L. Addadi, "Design strategies in mineralized biological materials," *Journal of Material Chemistry* 7 (1997): 689–702.

26 Pr Ilian Aksay thus presented the Princeton materials Institute "Princeton does not have a materials-science department and in some sense it can be said not to have any materials scientists, either. At least, not in the conventional sense of scientists who have earned degrees in materials science. Princeton never seriously considered starting a department of materials science, preferring instead to promote interaction of scientists in existing departments. PMI is a hybrid between a department and a center" (Princeton, *Alumni Weekly*, January 28, 1998, 13).

27 This attitude can be exemplified in Paul Calvert, "Biomimetic ceramics and composites," *MRS Bulletin* 17 (1992): 37–40.

28 For instance, H.Yang, N. Coombs, G. A. Ozin, "Morphogenesis of shapes and surface patterns in mesoporous silica," *Nature* 386 (1997): 692–695; N. Bowden, A. Terfort, J. Carbeck, G. Whitesides, "Self-assembly of mesoscale objects into ordered two-dimensional arrays," *Science* 276 (1997): 233–235.

Index

t

www.ingramcontent.com/pod-product-compliance
Lightning Source LLC
Chambersburg PA
CBHW080933240926
51849CB00003BA/444
9783527302710